Testing and Evaluation
of Agricultural Machinery

2nd Revised and Enlarged Edition

The Authors

Madan Lal Mehta, Retired as Director , CFMT&TI, Budni (MP) and he has experience of more than 40 years' experience in various organizations i.e. Govt. of India (Ministry of Agriculture and Farmers Welfare), ICAR, SFCI (GoI Undertaking), Foreign Assignment in Africa etc. besides 5 years in Tractor & Combine Harvester Industry. He has rich experience in testing & evaluation of tractors and agricultural machinery, training & development of machinery, mechanization in agriculture & management etc. Presently working as a consultant with Govt. of India in formulation & implementation of projects including selection of Farm Machinery for mechanization in agriculture in order to increase production & productivity and to make agriculture a profitable enterprise & pleasant job for farmers & younger generation.

Sewa Ram Verma , FNAAS, FISAE, FIIE, he did B.E. from Allahabad Agricultural Institute, Allahabad and M.Tech and Ph.D from Indian Institute of Technology, Kharagpur. He has been a passionate researcher in Farm Machinery & Power and has more than 55 years of experience in the development and evaluation of agricultural equipment as Professor & Dean, COAET, PAU, Ludhiana, Short-Term Consultant for ILO and Chief Technical Advisor, Commonwealth, Nigeria. Bestowed with many prestigious awards like Rafi Ahmed Kidwai Award, Jawaharlal Nehru Award, NRDC Invention Award, Punjab State 'Award of Honour', and ISAE Gold Medal. Also, have experience as a Member, Board of Management, of RPCAU, and MGCGV, Chitrakoot. He has numerous publications to his credit in the form of books, research bulletins, research papers in National & International journals.

Pradeep Rajan, did B.Tech (Agri. Engg.) from Allahabad University, Allahabad and M.Tech (Farm Machinery & Power) from GBPUA&T, Pantnagar, followed by Ph.D (Farm Machinery & Power) from IARI, New Delhi. Presently working as Principal Scientist in CSIR – CMERI Centre of Excellence for Farm Machinery, Ludhiana. He also holds a concurrent position of Honorary Assoc. Professor at Academy of Scientific & Innovative Research –a meta-university under CSIR. He is the Principal Investigator of a multi-disciplinary research programme (SERB& CSIR): Centre for Precision & Conservation Farming Machinery. He is a recipient of ISAE Team Award (2016). He has published a number of articles in various national and international journals.

Shashi Kumar Singh, did B.Tech. (Agril. Engg.), M.Tech. (Farm Power & Machinery) and Ph.D. (Farm Power & Machinery) from Punjab Agricultural University, Ludhiana (Punjab). He joined PAU in 1995 and presently working as Senior Research Engineer and in charge of Farm Machinery Testing Centre. He has published book chapter, research bulletin, many research papers in journals and presented papers in seminars/conferences. He received 'Innovative Students Project Award' awarded by INAE (2006). He taught many under graduate and post graduate courses and guided post graduate students. He is a life member of ISAE and also remained member of editorial board of Agricultural Engineering Today for the period 2014-15.

Testing and Evaluation of Agricultural Machinery

2nd Revised and Enlarged Edition

M L Mehta

Ex. Director
Ministry of Agriculture and Farmers Welfare (Government of India)
Central Farm Machinery Training & Testing Institute, Budni (MP)

S R Verma

Retd. Professor & Dean
College of Agricultural Engineering and Technology
Punjab Agricultural University, Ludhiana (Punjab)

Pradeep Rajan

Principal Scientist
CSIR-Central Mechanical Engineering Research Institute
Centre of Excellence for Farm Machinery, Ludhiana (Punjab)

S K Singh

Senior Research Engineer
Dept. of Farm Machinery & Power Engineering, COAE&T
Punjab Agricultural University, Ludhiana (Punjab)

2019

Daya Publishing House®

A Division of

Astral International Pvt. Ltd.

New Delhi – 110 002

Published by	: **Daya Publishing House®**
	A Division of
	Astral International Pvt. Ltd.
	– ISO 9001:2015 Certified Company –
	4736/23, Ansari Road, Darya Ganj
	New Delhi-110 002
	Ph. 011-43549197, 23278134
	E-mail: info@astralint.com
	Website: www.astralint.com
Digitally Printed at	: Replika Press Pvt. Ltd.

Foreword

The request made by Prof. S.R.Verma, a colleague of long standing, to write Foreword to the revised edition of the book 'Testing and Evaluation of Agricultural Machinery' offered me an opportunity to travel back in time to the era of intense discussion in nineteen sixties and seventies on the desirability of allowing mechanization of Indian agriculture. Some professional groups thought that small size of holdings, high capital cost of prime movers and machinery, lack of operating skills and maintenance facilities and exacerbated unemployment ruled out mechanization of Indian agriculture. Others, including the agricultural engineers, felt that improvement in equipment and power sources was indispensable for achieving production targets and sustainable growth. Ultimately the farmer settled the dispute and decided that mechanization was good for him. As the founder Head of Department of Agricultural Engineering at PAU, Ludhiana, I was privileged to have a strong team of highly qualified and devoted professionals who, through their participation in discussions, publications, design and development of a wide range of equipment, assistance to industries and field demonstrations, made invaluable contribution to promotion of agricultural mechanization, particularly in the North. Engineering groups at other agricultural universities or working in State and Central Government organizations like MERADO (now Centre of Excellence for Farm Machinery) of CMERI played similar role to support the programmes for mechanization of Indian Agriculture.

Success of a new technology is greatly dependent on the quality of related hardware. Quality of tractors, both imported and indigenously manufactured, has been well regulated by government through institutions like the Central Farm Machinery Training and Testing Institute at Budni (MP). Unfortunately testing of farm machinery remained poorly organized for many years. The situation has now changed. Under the auspices of Bureau of Indian Standards, test standards for a wide range of agricultural tools and machinery have been prepared and 35 institutions, including the 4 Farm Machinery Training and Testing Institutes of Central Government, have been authorized for testing. Manufacturers and fabricators of agricultural equipment need to be encouraged to utilize the test reports for improving their products. Similarly, the farmers should be trained to use test reports in selection of equipment. Testing gives clear insight into the design, construction and performance of a machine. It also helps in identifying scope for improvement through research and development. Testing and evaluation of agricultural machinery should be made an integral part of the training of agricultural engineers, both at under and post-graduate levels. This will need a hand-book of testing and evaluation of agricultural equipment. I congratulate the authors for revising and updating their book Testing and Evaluation of Agricultural Machinery to meet the need.

Prof. B.S.Pathak
FNAAS, FISAE, FISI
Ex. Director, Sardar Patel Renewable Energy Research Institute, Gujarat
Ex. Director, School of Energy Studies in Agriculture, PAU and
Ex. Dean, College of Agriculture Engineering, PAU

Preface

Agriculture and food production plays the key role in economic development of undeveloped and under-developed countries in Asia, Africa and South Pacific. The farmers in the past relied upon animate sources of power which included human beings and draft animals. With the passage of time and to meet the food requirement of rising population, agricultural machines and efficient power sources have been introduced to mechanize farming operations to attain higher productivity. Farm mechanization also helps to make farming a pleasant and profitable enterprise especially for the younger generation.

One of the most remarkable development in the history of agriculture during the 20th century has been the ushering-in of the Green Revolution. The phenomenal rise in agricultural production and productivity as a consequence is attributed to the biological, chemical and mechanical innovations. In other words, the Green Revolution in the mid-sixties resulted from the evolution and adoption of high yielding varieties (HYV), adequate water supply through canals, tubewells and pumps, use of higher and balanced doses of chemical fertilizers and pesticides for weed and pest control as well as introduction of modern farm power sources (stationary engines, electric motors, tractors, power tillers *etc.*) with matching farm equipment and tools. Along with the aforementioned technological factors, no less has been the contribution of the right government policies to ensure the timely provision and production of required inputs, provision of infrastructures like grain markets, rural roads, minimum support price (MSP) as well as the unprecedented contribution of the framers and entrepreneurs. This indeed has go down in the history of agriculture as an unparalleled event, which has saved a vast population of human kind especially in the developing and under-developing countries from mass starvation, hunger and malnutrition. As a well deserving acknowledgement of the contribution of different institutions and individuals, Dr. Norman Borlaug, who played the lead role in the Green Revolution was decorated with the *Nobel Peace Prize* in 1970.

Productivity of farm depends largely on the availability and judicious utilization of various inputs and farm power by farmers. Agricultural implements and machines increase the productivity of land and labour through timeliness, desired quality of farm operations, economically cost effective and increased work output per unit time. Besides its contribution to increase cropping intensity and diversification in agriculture, farm mechanization also enables efficient utilization of inputs such as seed, fertilizer, irrigation water, insecticide and pesticide *etc.* along with alleviation of drudgery.

The first edition of the book has been revised and enlarged by adding some new chapters along with major changes in text and contents of other chapters.

The revised edition has been organized in 20 Chapters. The first and second chapters deal with the status of agricultural mechanization in India and testing and evaluation system in vogue. The next 3 chapters (3 to 5) explain the detail testing procedure for the three major farm power sources, tractor, power tiller and IC engines respectively. Chapters 6 to 8 deal with the farm equipment required for seed bed preparation and selected hand tools. Testing methods of seeding, planting machines along with fertilizer broadcasters are included in chapters 9, 10 and 11. Chapter 12 deals with the testing and evaluation of sprayers and dusters followed by irrigation pumps in chapter 13. Harvesting machines like combine harvester, straw reaper combine, root harvesters, power thresher have been elaborated in the next 4 chapters. Recent advances in instrumentation for testing of agricultural machinery have been included in Chapter 18. Chapter 19 describes the safety aspects and the associated test procedures and cost estimate of agricultural equipment is also included as the last chapter.

The authors wish to express their gratitude to different organizations and individuals who have contributed directly or indirectly in preparation of the manuscript of this book. Our sincere thanks are due to Mr. J.J.R Narware, Director, CFMT&TI, Budni, Mr. P.K. Pandey, Director, NRFMT&TI, Hissar, Haryana and Er. Mukesh Jain, Testing Head, CCSHAU, Hissar for their guidance, support and help provided. We are also highly thankful to Er. C.K. Tijare, Er. R.K. Nema for their support in preparation of this book. Assistance of BRAIN-JAPAN, Bureau of Indian Standards, RNAM, CIAE, FAO, ISO, ICAR, Ministry of Agriculture and Famers Welfare and PAU, Ludhiana is also sincerely acknowledged. Thanks are also due to a large number of tractor and agricultural equipment manufacturers for providing us with the illustrations. We wish to acknowledge Prof. Gajendra Singh, Prof. Surendra Singh, Prof. D.V. Singh and Mr. S.S. Kohli for their continuous motivation and inspiration for the revised edition.

Special thanks are also due to Dr. V.K. Sharma and Dr. S. K. Misra for their insightful and multifarious inputs. Both of them had contributed immensely in the preparation of the text & manuscript in the first edition of the book. Grateful thanks to Prof. Harish Hirani, Director, CSIR-CMERI, Durgapur for his encouragement and support.

Thanks are due to Mr. Anshul Shrivastava, CFMTTI, Budni (MP) and Mrs. Jaspreet Kaur, who helped in preparation of the manuscript. We are also thankful to Daya Publishing House, A division of Astral International Pvt Ltd, New Delhi, for timely printing of this book.

We hope, the readers will find this book useful and informative in their profession. However, the authors would be grateful for receiving suggestions and constructive criticism on the material included in the book.

Sept., 2018

M L Mehta
S R Verma
Pradeep Rajan
S K Singh

Preface to the First Edition

Agricultural Mechanization is a sine qua non to remove drudgery, improve working comfort, enhance timeliness, reduce losses and increase production and productivity. Accordingly, use of better power viz. tractors and different types of agricultural machines in Indian agriculture has risen sharply on Indian farms to boost food and fibre production. But to safe guard the user's interest to ensure better quality and reliability of machines and for sustained growth of farm machinery industry, there is a need for sound scientific testing and evaluation of farm machines by using instrumentation and accepted methodology. Thus, testing and evaluation holds the proper key to standardization and quality control of agricultural machinery for better acceptability and sustained farm production. To satisfy the genuine need of different sectors, this book has been prepared. It is expected to serve as a Text-Book for the students of Agricultural Engineering degree and Postgraduate degree programme. It may also serve the needs of professional engineers, scientists, testing institutions and research organizations dealing with testing and evaluation of agricultural machinery. This book will also cater to the needs of tractor and agricultural implement manufacturing industries, consultants, Agricultural Universities/Colleges as a valuable reference for quality improvement and standardization. It is hoped this book will be a valuable reference for different groups in developing countries of Asia and Africa and Latin America.

The book has been organised in twelve chapters. The first and second chapters deal with the status of agricultural mechanization in India and testing and evaluation system in India respectively. Chapter-III deals with testing and evaluation of tractors. Chapter-IV to X deal with testing and evaluation of different agricultural implement, irrigation pumps and plant protection equipment. Chapter-XI deals with data acquisition, processing and analysis by personal computers (PCS). Chapter-XII deals with testing of agricultural machinery for safety.

The authors wish to express their gratitude to different organizations and individuals who have contributed directly or indirectly in the preparation of the manuscript of this book. Our sincere thanks are due to Dr. C.P. Singh, Dr. S.K. Sondhi, Dr. I.K. Garg, Dr. S.S. Ahuja, Dr. O.S. Taneja, Prof. N.S. Sandher and Prof. P.K. Gupta. Assistance of BRAIN-JAPAN, Bureau of Indian Standards, NRFMT& TI, Hissar, NEFMT& TI, Vishwanath Cheralli, P.A.U.,' Ludhiana, RNAM, CIAE, CFMT& Tl, Budni, FAO and Ministry of Agriculture is also sincerely acknowledged. We also thank National Agricultural Technology Information Centre, the publisher for timely printing of this book.

We are conscious of the time constraint each one of us had in writing this book. It is a maiden effort of its kind. It is possible that the method of presentation of the information may be at variance from that of some other authors. But what we thought best, we adopted. We hope, the readers will find this book informative and useful. The authors will no doubt appreciate the constructive criticism and suggestions from the readers.

August, 1995

Authors

Contents

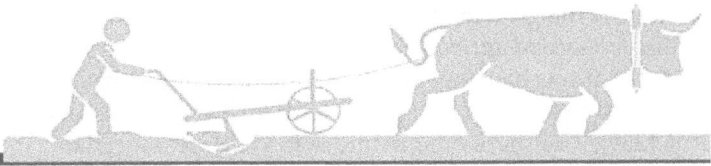

Chapter 1

Status of
Agricultural Mechanization

Agricultural mechanization is the process of improving productivity of farm inputs (seeds, water, fertilizer, chemical, labour) through the use of agricultural machinery, implements and tools. It involves the provision and use of different forms of power sources and mechanical assistance to agriculture, from simple hand tools, to animal draught power, and to mechanical power technologies. The choice of these power sources may depend on local circumstances and can complement each other in the same farm holding. In many developing countries up to 80% of farm power is provided by human beings [152, 174, 261]. In most developed countries human beings are used less as source of power and more for machine operation and control. The degree of automation and control is very high in countries practicing precision agriculture and the same level may not be suitable for most of the developing countries in Asia and Africa [208].

Development of agricultural mechanization in India is a fascinating story with many remarkable facets. The country has progressed over the past six decades from the nation having severe food shortages and famines to an exporter of food commodities and agricultural equipment including tractors. This is quite remarkable in spite of the burgeoning Indian population and a little increase of 7 percent in the arable land area. India is the second most populous country in the world with an estimated population of 1.21 billion in 2011 and an annual growth rate of 1.5 percent. About 70 percent of the population lives in rural areas with about 52 percent still depending on agriculture for their livelihood. The per capita income in 2010 was estimated at US $1200 and about 250 million people are still living below the poverty line. The literacy rate is 74 percent but there is higher illiteracy among women due to inadequate schools particularly in the rural areas. Other limitations include shortage of drinking water, poor sanitation, poor housing, inadequate hospitals and low per capita energy consumption.

The total land area of the country is 297 million hectares of which 142 million ha is classified as agricultural land. Whilst it has basically an agrarian economy with a Gross Domestic Product (GDP) growing at a rate of 7 to 9 percent annually from 1990 to 2010, the share of agriculture has now declined to 14 percent from a level of 56 percent in 1950. The agricultural sector is facing the biggest challenge of meeting the growing food demand to feed the ever increasing population of the country. Share of agriculture in Gross Domestic Product (GDP) is correlated with the level of mechanization and highly mechanized developed countries have less than 5% share in GDP compared with medium mechanized countries having 15 – 20% [259]. The share of agriculture in GDP with mechanization level is shown in Fig. 1.1 and country wise relative productivity is shown in Fig.1.2.

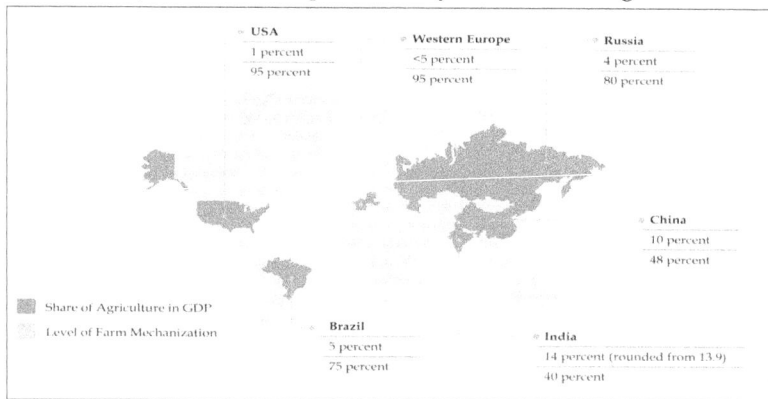

Fig. 1.1: Share of agriculture in GDP with mechanization level [260]

There has been more than a four-fold increase in grain production, since independence in 1947, due to the introduction of improved technologies and practices ushered in Indian agriculture through Green Revolution. However, the population has increased at a similar pace and there are still challenges for attaining full food and nutritional security. The country has a very diverse form of agriculture particularly due to varying soil and climatic conditions. India's climate is full of extremities; the temperature conditions vary from arctic cold to equatorial hot and rainfall from extreme aridity with less than 100 mm in the Thar Desert of Western India to the site of the world's maximum rainfall of 11,200 mm at Mowsinram in the northeast. The available rainfall has large spatial and temporal variations. Although there has been a significant increase in the area under irrigation, still more than 65 percent cent is devoid of assured irrigation and the agricultural productivity in the rain fed areas is low.

1.1 Benefits of Agricultural Mechanization

Farmers from all over the world benefited from agricultural mechanization which has provided a number of economic and social benefits to farmers. Primary among the economic benefits is the improved yield that comes as a

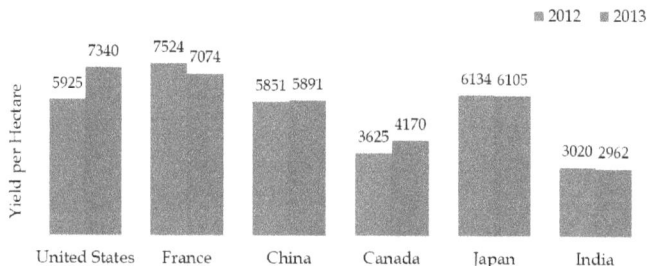

Fig. 1.2: Country wise relative productivity [260]

result of greater level of mechanization. With the burgeoning population along with looming water scarcity crisis and to ensure food security, the benefits of farm mechanization make it a crucial and critical component of shaping the future of agriculture.

There is direct relationship between the productivity level and farm mechanization. Countries with higher levels of farm mechanization are able to increase their productivity and therefore are better equipped to meet the food demand factors. Some of the direct benefits of adopting agricultural mechanization are:

a) Input savings

Studies have revealed a direct relationship between farm mechanization (farm power availability) and farm yield. Farm mechanization is said to provide the following input savings:

- Seeds (approximately 15-20 percent)
- Fertilizers (approximately 15-20 percent)
- Increased cropping intensity (approximately 5-20 percent)
- Reduction in labour (about 20 -30%)

b) Increase in efficiency

The use of appropriate farm machinery also helps in increasing the efficiency of farm labour, reducing drudgery and workloads. It is estimated that farm mechanization can help reduce time by approximately 15 - 20 percent. Additionally, it helps in improving the harvest and reducing the post-harvest losses and improving the quality of cultivation. These benefits and the savings in inputs help in the reduction of production costs and allow farmers to earn more income.

- Time saving (approximately 15 -20 percent)
- Productivity increase (approximately 10-15 percent)

c) Social benefits

There are various social benefits as well:

- Helps in conversion of uncultivable land to arable land through advanced soil tilling techniques and also in shifting grazing lands used for feed and fodder cultivation towards food production.
- Decrease in workload on women as a direct consequence of the improved efficiency of labour.
- Improvement in farm safety.
- Helps in encouraging the youth to join farming and attract more people to work and live in rural areas.

The adoption of farm mechanization is initiated by increasing the power availability in the form of mechanical, electrical or both. With the increased power, several crop production activities like seedbed preparation, planting, irrigation, spraying etc. can be mechanized. The main reasons for changing the power source for crop production from muscles (human or animal) to tractors and power tillers [174] are:

i)   Potential to expand the area under cultivation.

ii)   Ability to perform operations at the right time to maximize yield potential.

iii)   Multi-functionality of tractors, i.e. these can be used not only for crop production, but also for transportation, stationary power applications and infrastructure improvement (drainage and irrigation canals etc.).

iv)   Compensating seasonal labour shortages.

v)   Reduction of the drudgery associated with the use of human/power for tasks, such as primary tillage. This is especially important in tropical areas where high temperatures and humidity make manual work extremely arduous.

1.2 Size of Landholdings in India

The average size of landholdings in 2011 was 1.16 ha with only 0.7 percent (0.97 million) consisting of large farm holdings of more than 10 ha but constituting over 10.9 percent of the cultivated land. On the other hand, farms having less than 1 ha constitute about 67 percent of holdings with only 22.2 percent area in 2011 (Table 1.1). The three categories comprising semi-medium, medium and large farms (more than 2 ha) cultivate between them an area of about 55.7 percent of the cultivated land (15.1 percent of farm holdings). These farmers have been instrumental not only for the success of agricultural mechanization but for the overall success of green revolution and the unprecedented transformation in the food security situation.

Table 1.1 Size-group wise percentage number of operational holdings and area

S. No	Size –Group: Farm Holdings	Percentage number of holdings [percentage area under each category]		
		1991	**2001**	**2011**
1	Marginal (below 1 ha)	59.2 [15.0]	62.4 [18.7]	67.0 [22.2]
2	Small (1 – 2 ha)	18.7 [17.4]	19.1 [20.2]	17.9 [22.1]
3	Semi-medium (2 – 4 ha)	13.6 [23.2]	11.9 [23.9]	10.1 [23.6]
4	Medium (4 – 10 ha))	7.0 [27.1]	5.6 [24.0]	4.3 [21.2]
5	Large (10 ha & above)	1.5 [17.3]	1.0 [13.2]	0.7 [10.9]
	Average holding size (ha)	1.57	1.33	1.16
	All holdings (million)	106.6	119.9	137.8

Source: [5, 126]

Due to the laws of inheritance the number of holdings is increasing in many states, however, the situation in Punjab, with the highest level of mechanization and the highest productivity, a reverse trend has been witnessed with the marginal holdings declining from 38 percent in 1971 to 27 percent in 1991 and only 12 percent in 2001, cultivating less than 2 percent of the area. Many rural people migrated to cities are still counted as owners of the land holdings and their land is cultivated by other family members or rented by other farmers. Thus the actual number of operational holdings is much less than the number reported based on ownership of land. In most cases, land is rented by tractor owners making their operational holdings bigger to make the ownership of tractors more economically viable.

1.3 Mechanization, Power Sources and Productivity

The mechanization of Indian agriculture has preceded along the time tested two-pronged approach based on improved equipment and enhanced power supply. However, compared to the mechanization of western agriculture, which was motivated by the need to replace labour and draught animals with the mechanical prime movers, the guiding principle in mechanizing Indian agriculture has been to maintain a practical mix of labour, draught animal power, mechanical and electrical power. This practical mix has yielded good results in India and it has increased yields, cropping intensity and total production. It reduced family labour but increased overall employment. It also improved the quality of life of the farmers. With emphasis on timeliness, precision and general improvement in the quality of work, increase in cropping intensity and yields was expected. The 'power mix' approach eliminated the risk of labour displacement commonly associated with mechanization. On the other hand, mechanization changed the nature of farm tasks to make these more acceptable to the young generation.

The agriculture sector in India is witnessing a considerable decline in the use of animal and human power in agriculture related activities and this trend has paved the way for a new range of agricultural tools. A large number of these are driven by fossil fuel operated such as tractors, power tillers and diesel engines, resulted in a shift from the traditional agriculture process to a more mechanized process.

Farm mechanization in India stands at about 40-45 percent. This is still low when compared to countries such as the US, Brazil and even China. While the mechanization level lags behind other developed countries, the level of mechanization has seen strong growth. The farm power availability on Indian farms has grown from 1.47 kW/ha in 2005-06 to 2.02 kW/ha in 2013-14. However, power availability in the country is characterized by large variations, which in 2001 varied from 0.6 kW/ha of agricultural land in some states to 3.5 kW/ha in Punjab. There is a strong linear relationship between the farm power availability and agricultural output per ha (Fig. 1.3). This underscores the emphasis on the growth and development of power machinery systems in Indian agriculture.

Mechanization patterns in different regions are described based on the idea that private sector contractors are a driving force in the mechanization revolution. Power input in agriculture is forecasted to rise from 2 kW/ha now to 4.5 kW/ha by 2050 when there will be 7 million tractors in use in the country (compared with fewer than 8 thousand at independence). The agricultural engineering manufacturing industry is growing and maturing although traditional draught animal technology is expected to decline as draught animal numbers continue to fall. The importance of emphasizing and expanding agricultural engineering training and R&D is underlined.

In India, the level of mechanization varies greatly by region. Northern states (Punjab, Haryana and Uttar Pradesh) have high level of mechanization due to the highly productive land in the region as well as a declining labour force. The state governments have also provided timely support in promoting mechanization of farms. The western and southern states in the country have a lower level of mechanization due to the smaller land holdings prevalent in these regions as well as the land holdings being more scattered. As a result, in many cases, mechanization has been uneconomical leading to the lower development. Factors such as hilly topography, high transportation cost, lack of state financing and other financial constraints due to socio-economic conditions and dearth of agricultural machinery manufacturing industries have hindered the growth of farm equipment sector within north-eastern states resulting in low level of mechanization. Operation-wise, the level of mechanization varies from 42 percent for soil working and seed bed preparation, 29 percent for seeding and planting, 34 percent for plant protection and 37 percent for irrigation [1].

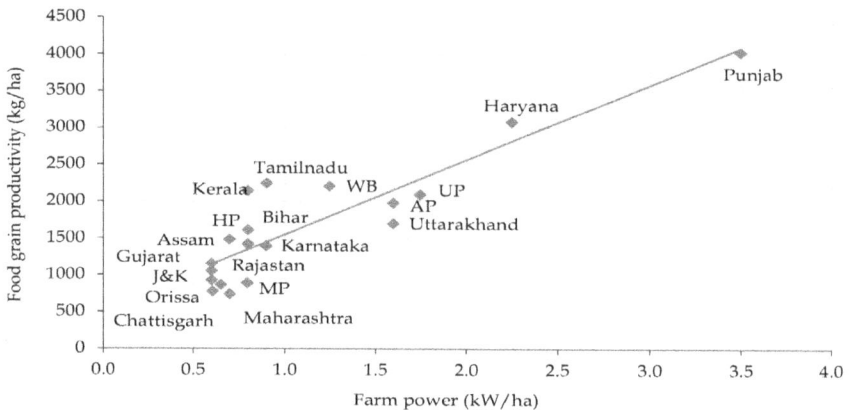

Fig. 1.3: Relationship between farm power and productivity in different Indian states [226]

The role of tractors in the Indian agriculture sector reflects the growing trend of tractorisation in the country. The economics of ownership of most tractors in India had been justified by custom hiring for on-farm work as well as for off-farm transport and construction activities. The use of tractors in transport activities accounted for about 60 percent of average annual use of 600 hours [212]. Custom hiring of farm equipment is a prevalent practice, especially among small land owners who find ownership of large farm machines expensive and uneconomical. The government is therefore promoting farm mechanization by subsidising purchase of equipment as well as supporting bulk buying through front-end agencies. The government also provides credit and financial assistance to support local manufacturing of farm mechanization equipment. Given the labour scarcity and the government's subsidy programs, adoption of farm mechanization is set to increase.

Level of mechanization also varies for different crops and its associated farm operations. For example, wheat cultivation is highly mechanized with more than 90% in seed bed preparation and above 80% in sowing, weeding, chemical spraying, harvesting and threshing [202]. However, for vegetables, the level is less than 10% for planting and almost zero in harvesting. Mechanization levels for various farm operations of selected crops along with mechanization levels of different Indian states are given in Fig. 1.4.

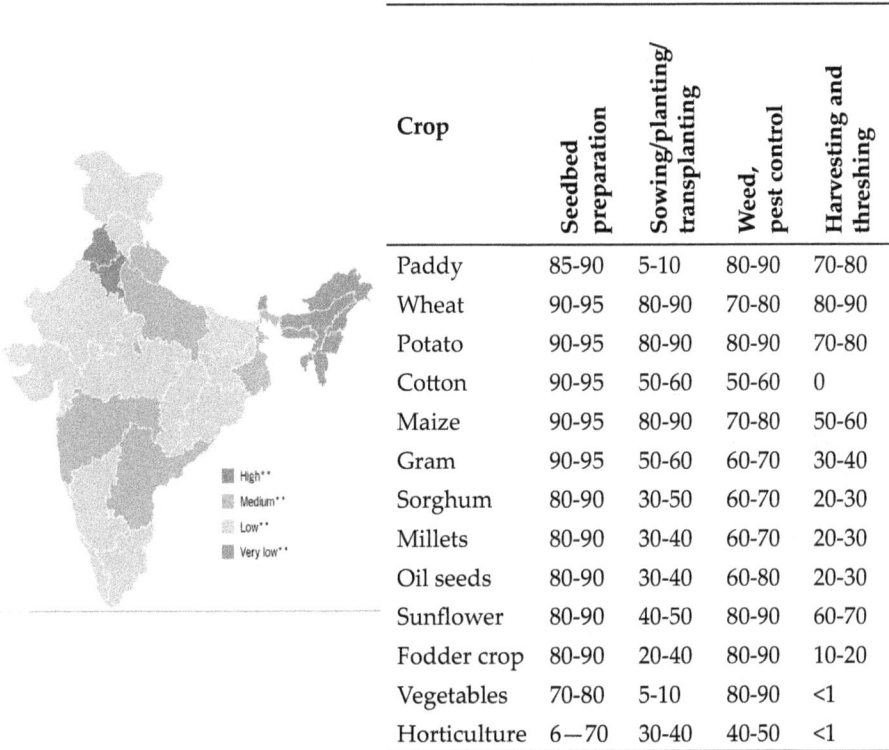

Crop	Seedbed preparation	Sowing/planting/ transplanting	Weed, pest control	Harvesting and threshing
Paddy	85-90	5-10	80-90	70-80
Wheat	90-95	80-90	70-80	80-90
Potato	90-95	80-90	80-90	70-80
Cotton	90-95	50-60	50-60	0
Maize	90-95	80-90	70-80	50-60
Gram	90-95	50-60	60-70	30-40
Sorghum	80-90	30-50	60-70	20-30
Millets	80-90	30-40	60-70	20-30
Oil seeds	80-90	30-40	60-80	20-30
Sunflower	80-90	40-50	80-90	60-70
Fodder crop	80-90	20-40	80-90	10-20
Vegetables	70-80	5-10	80-90	<1
Horticulture	6—70	30-40	40-50	<1

Fig. 1.4: Level of mechanization in percent, by crop and value-chain process [226]

Locally developed and improved agricultural hand tools and implements continue to play a critical role in agriculture despite the advances in agricultural machinery. This is due to various reasons: a) small and irregular farm sizes, b) lack of machinery available for smaller land holdings, c) lack of awareness and skills among farmers and d) inability of farmers to afford more advanced technologies. Hand tools have also been developed for all levels of the value chain. In 2010, when the size of the agricultural labour force was 269.74 million, the estimated number of hand tools in use was 809.22 million, which equates to about 3 hand tools per manual labour. However, the prevalence of these tools comes with the issue of safety [226]. About 34.2 percent of accidents in agriculture were due to hand tools, and 46 percent of farm injuries with sickles and spades were reported. Implications of injuries due to hand tools are severe as these injuries are very painful and disabling due to delayed treatment [153]. Also, 70 percent of agricultural hand tools injuries had a recovery time of more than seven days. Therefore, developing farm machinery more suited to the local conditions is essential so that injuries and problems that come with the use of hand tools can be abated while making agricultural practices more productive. Table 1.2 shows the increase in cropping intensity and food grain productivity with respect to increase in farm power availability in Indian farms. Agriculture progress in India pre and post Green Revolution is given in Table 1.3 and it also indicates the correlation between farm power availability with productivity increase.

Table 1.2 Power availability in Indian farms

Year	Cropping intensity, %	Food grain productivity, t/ha	Power available, kW/ha	Power per unit production, kW/t	Net sown area per tractor, ha
1965-66	114	0.636	0.32	0.5	2162
1975-76	120.3	0.944	0.48	0.51	487
1985-86	126.8	1.184	0.73	0.62	174
1995-96	130.8	1.499	1.05	0.7	82
2005-06	135.9	1.715	1.49	0.87	45
2010-11	140.5	1.93	1.78	0.92	34
2011-12	141.5	2.079	1.87	0.9	31
2012-13	140.9	2.129	1.94	0.91	29
2013-14	142	2.111	2.02	0.96	27

Source: [2, 114, 218]

Table 1.3 Agricultural growth in India – timeline

Pre-Green Revolution Era (before 1965)	Green Revolution Era (1965 – 1975)	Post Green Revolution Era (1975 onwards)
Farming by traditional methods	HYVs, fertilizer, irrigation, chemical inputs	Use of more scientific methods/ machinery/ implements
Farm power availability: 0.27 kW/ha	Farm power availability: 0.47 kW/ha	Farm power availability: 2.02 kW/ha [2014]
Share of animate power sources was 98%	Share of animate power sources decreased to 62%	Share of animate power sources decreased to 11.8%
Low productivity of food grain (0.58 t/ha)	Productivity of food grain increased (0.95 t/ha)	Present productivity of food grain is about 2.11 t/ha
Enhanced production through increase in cultivated area	Enhanced production/ productivity through adoption of HYVs, fertilizer, irrigation and chemical inputs	Enhanced production/ productivity through adoption of improved farm equipment and tools in addition to adoption of other agricultural inputs

Various farm power sources available in Indian farm for doing various mobile and stationary operations are animate and human power (men, women, children, and draught animals like bullocks, buffaloes, camels, horses, ponies, mules, donkeys), tractors, power tillers, self-propelled machines (combines, dozers, reapers, sprayers etc.), stationary power (diesel/oil engines) and electric motors. The time series population and power of agricultural power sources during the period 1960-61 to 2013-14 is given in Table 1.4. The values in the parenthesis show the total power availability (million kW) from the population of farm power sources.

Table 1.4 Farm power sources and power availability in India

Year	Population of farm power sources, million [power availability, million kW]					
	Agricultural Workers	Draft animal power	Tractors	Power tillers	Diesel engines	Electric motors
1960-61	131.10	80.40	0.037	0	0.23	0.20
	[5.80]	[30.60]	[1.00]	[0]	[1.298]	[0.74]
1970-71	125.70	82.60	0.168	0.0096	1.70	1.60
	[6.21]	[31.39]	[4.38]	[0.054]	[9.52]	[5.92]
1980-81	148.00	73.40	0.531	0.0162	2.88	3.35
	[7.46]	[27.89]	[13.86]	[0.091]	[16.13]	[12.39]
1990-91	185.30	70.90	1.192	0.0323	4.80	8.07
	[9.17]	[26.94]	[31.11]	[0.181]	[26.88]	[29.86]
2000-01	234.10	60.30	2.531	0.1147	5.90	13.25
	[10.70]	[22.90]	[66.06]	[0.642]	[34.86]	[49.03]
2010-11	263.00	53.50	4.207	0.3213	8.20	16.50
	[13.15]	[20.33]	[109.80]	[1.799]	[45.92]	[61.05]
2011-12	266.08*	53.00	4.553	0.3621	8.30	16.70
	[13.30]	[20.14]	[118.23]	[2.028]	[46.48]	[61.79]
2012-13	269.20*	52.80	4.858	0.4021	8.35	16.80
	[13.46]	[20.06]	[126.80]	[2.252]	[46.76]	[62.16]
2013-14	272.00*	52.00	5.237	0.4409	8.45	17.00
	[13.60]	[19.76]	[136.70]	[2.469]	[47.32]	[62.9]
CAGR of population of farm power sources, %						
1960-61 to 1990-91	1.16	-0.42	12.27	6.25	10.66	13.12
1991-92 to 2013-14	1.54	-1.33	6.65	12.03	2.5	3.29
1960-61 to 2013-14	1.38	-0.82	9.79	9.3	7.04	8.74

CAGR - Compound Annual Growth Rate, Source: [216]

1.4 Estimates of Investments by Farmers in Farm Machinery

Investment in machinery is on long term basis as compared to investment in other inputs such as seeds, fertilizer and chemicals. Investment in hand operated tools is growing at a slow pace with the increase in the population of agricultural workers. Investment in the animal operated implements is decreasing gradually due to the decrease in the number of draught animals. However, investment in power operated farm equipment is increasing rapidly.

The total investment in the farm machines in 2005 (Table 1.5) was estimated to be around Rs. 273 billion (US\$ 6 billion). This compares to an annual investment in 1997 of some Rs. 180 billion (US\$ 5 billion) [212]. Annual investment in 2005 in agro-processing and post- harvest equipment was estimated to be around Rs. 200 billion, bringing the total annual investment to Rs. 453 billion or US\$ 10 billion [130].

Production by top tractor manufacturers in India

Company	Domestic Volumes		
	FY16	FY17	FY18
M&M (M&M + Swaraj)	2,02,628	2,48,594	3,01,934
TAFE	1,13,500	1,19,200	1,35,800
Sonalika	58,900	69,000	86,700
Escorts	50,698	62,699	78,446
John Deere	29,500	43,300	62,300
New Holland	22,800	22,500	26,100
VST	7,801	9,635	11,369
Others	7,670	7,156	8,351
TOTAL Sector	**4,93,497**	**5,82,084**	**7,11,000**

Source: http://www.autopunditz.com, company data, IIFL Research

An average of over 5 lakh tractors per annum have been sold in India during the past three years. Power tillers are becoming popular in lowland flooded rice fields and hilly terrains and it is estimated that about 2,00,000 are currently in use. The production and sale of tractors and power tillers from 2004-05 to 2013-14 is given in Fig. 1.5.

The healthy growth in the tractor population has resulted in a corresponding growth in implement manufacturing particularly of rotavators, disc harrows, seed drills, potato diggers and trailers. Annual demand (estimate year - 2014) of major farm tools and equipment in India is given in Table 1.6. High capacity machines will be preferred in future, including rotary tillers, harrows, laser levellers, high clearance sprayers, planters, high capacity threshers and self-propelled and tractor drawn combines. The custom hiring of mechanical power will be preferred by those farmers who cannot afford, or prefer not to own machines. Animal operated implements will decrease due to the continued decrease in the number of draught animals. Sales data indicates a growing preference for 41 to 50 hp range tractors. Also, there is a significant increase in purchase of low hp tractors (around 20 hp) by small farmers and several leading tractor manufacturers are introducing new models in this range.

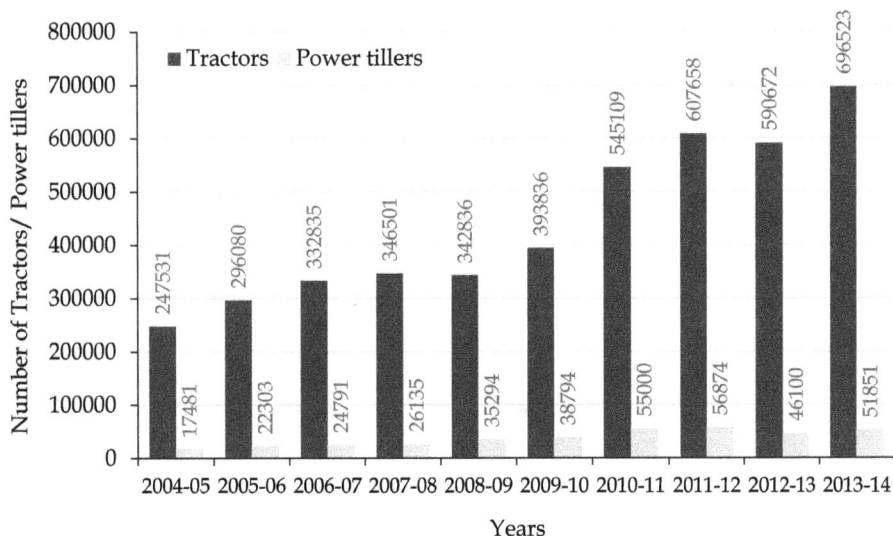

Source: [Tractor Manufacturers' Association (TMA);
Power Tiller Manufacturers' Association (PTMA)]

Fig. 1.5: Tractor and Power tiller sales

Table 1.5. Estimated Investment in Farm Machinery during 2005

S. No.	Type of Equipment	Annual Sales (Number)	Unit Cost (Rs.)	Total Cost (Million Rs.)
(1)	(2)	(3)	(4)	(5)
1.	Tractors with minimum equipment	250 000	5,00,000	1,25,000
2.	Bulldozers and earth moving machinery	500	20,00,000	1000
3.	Power tillers	15 000	1,00,000	1500
4.	Pump sets/ submersible pumps	1000 000	20,000	20,000
5.	Diesel engines	500 000	20,000	10,000
6.	Sprayers and duster (manual and powered)	NA	LS	20,000
7.	Power threshers	4000 000	30,000	12,000
8.	Combines	2500	8,00,000	2000

(1)	(2)	(3)	(4)	(5)
9.	Reapers	3000	50,000	150
10.	Straw combine	10 000	80,000	800
11.	Tractor drawn equipment (ploughs, harrows, cultivators, rotary tillers, seed drills/ planters etc.)	NA	LS	35,000
12.	Animal drawn equipment/ carts	NA	LS	25,000
13.	Hand tools and garden tools; manually operated equipment	NA	LS	10,000
14.	Sprinkler and drip irrigation equipment	NA	LS	2000
15.	Other agricultural equipment (stubble shavers, water tankers, land levelers, land places, forge harvesting equipment, manure spreaders etc.)	NA	LS	8550
	Total investment for 2005			2,73,000

NA-Not available, LS-Lump sum basis; Source: [130]

Table 1.6 Agricultural equipment annual market size

S. No.	Equipment	Annual market size (units)	S. No.	Equipment	Annual market size (units)
1.	Tractors	450000 - 500000	10.	Power tillers	50000 - 60000
2.	Mould board plough	45000 - 50000	11.	Rotavators	100000 - 120000
3.	Cultivators	150000 - 200000	12.	Harrows	120000 - 150000
4.	Seed-ferti drills	60000 - 75000	13.	Planters	15000 - 25000
5.	Rice transplanters	2000 - 3000	14.	Power weeders	35000 - 40000
6.	Reapers	10000 - 15000	15.	Threshers	60000 - 75000
7.	Combine harvesters	3500 - 4000	16.	Trailers	150000 - 175000
8.	Sprayers (TD)	10000 - 15000	17.	Laser land levellers	2500 - 3500
9.	Potato diggers	25000 - 30000	18.	Rotary hoes	20000 - 25000

Source: [216] Estimate year: 2014

1.5 Research and Development, Production and Extension System

1.5.1 Research and development

Department of Agricultural Research and Education (DARE) under the Ministry of Agriculture and Farmers' Welfare, Govt. of India coordinates and promotes agricultural research & education in the country. DARE provides the necessary government linkages for the Indian Council of Agricultural Research (ICAR), the premier research organization for coordinating, guiding and managing research and education in agriculture including horticulture, agricultural engineering, fisheries and animal sciences in the entire country and this is one of the largest national agricultural research systems in the world. It is also the nodal agency for International Cooperation in the area of agricultural research and education in India. ICAR institutes and state agricultural universities (SAUs) are located in several states. Most of the SAU's have an Agricultural Engineering College and Farm Machinery departments under these colleges conduct research and development in various farm implements and tools.

ICAR-Central Institute of Agricultural Engineering (CIAE), Bhopal [www.ciae.nic.in], and ICAR-Central Institute of Post-Harvest Engineering and Technology (CIPHET), Ludhiana [www.ciphet.in], are exclusively conducting research and development in the areas of farm machinery and post-harvest engineering respectively. Commodity institutes (for sugarcane, cotton, rice, fodder and horticulture) and several national institutes (for fish, dairy, dry-land agriculture and others) are also conducting research on mechanization on specific crops. The All India Coordinated Research Projects (AICRPs), with cooperating centres located in different states to cater mechanization needs of different agro-climatic zones, are implemented under the aegis of ICAR, which currently include:

- Farm Implements and Machinery
- Renewable Energy Sources
- Utilization of Animal Energy
- Ergonomics and Safety in Agriculture
- Post-Harvest Technology
- Application of Plastics in Agriculture

Recently, Council of Scientific and Industrial Research (CSIR) along with Science & Engineering Research Board (SERB) initiated a major programme in development of advanced agricultural equipment through its 'Centre for Precision & Conservation Farming Machinery (CPCFM)' at CSIR-Central Mechanical Engineering Research Institute - Centre of Excellence for Farm Machinery, Ludhiana [www.cmeri.res.in]. Major tractor manufacturers have set-up their own R&D facilities with well-equipped laboratory.

1.5.2 Efforts in improving extension system

Facilitation of the extension services concerning agricultural technologies in general and agricultural mechanization in particular have been focused on the following areas:

- Institutional arrangements to make the farmer driven extension system and far more accountable.
- Public Private Partnership (PPPs).

- Mass Media Support by providing location-specific broadcasts through FM and AM stations of All India Radio and the Doordarshan (DD), National TV Channel.

- Fee-based advisory services by graduates in agri-business development and through the establishment of agri-clinics.

- Kisan (Farmer) Call Centres through toll-free lines.

In addition, seventeen State Agro-Industries Corporations and Joint Venture Companies have been promoted by the Central Govt. and State Governments. The objectives of these corporations are to manufacture and distribute agricultural machinery together with other inputs to promote agro-based industries and to provide technical services and guidance to farmers and others.

The Ministry of Agriculture and Farmers Welfare carries out planning and activities at central level to promote mechanization in the country through various schemes and programmes. In the recent past, the government launched a major extension programme with financial aid from the World Bank in which mechanization was an important component. Similarly, the National Agricultural Innovative Project (NAIP) is being implemented in different parts of the country under the aegis of ICAR to improve agricultural productivity and rural livelihoods. Mechanization and value addition to agricultural produce have also been given major emphasis in this programme. In addition, promotion of mechanization is an important component of the National Horticulture Mission (NHM) being implemented by National Horticulture Board (NHB).

Considering the importance of agricultural mechanisation, the Ministry started the Sub-Mission on Agricultural Mechanisation in the year 2014-15 with an objective to promote agricultural mechanisation among small and marginal farmers and in the areas where the level of mechanisation is very low. In addition to the Sub-Mission on Agricultural Mechanisation (SMAM), farm mechanisation is also promoted through various other schemes and programmes of the Ministry such as RKVY, NFSM, NMOOP etc.

1.5.3 Agricultural machinery production system

Manufacturing of implements is reserved for small scale sector. The concentration of manufacturing units is in the States of Madhya Pradesh, Utter Pradesh, Punjab, Haryana, Bihar and Karnataka. There are about 250 medium to large scale, 2500 small scale, 15,000 tiny industries along with more than 100,000 village artisans engaged in manufacturing of agricultural equipment and implements [216]. The R&D institutions together with the farm machinery industry constitute two important pillars of agricultural mechanization and need to collaborate closely for their mutual benefit and that of the farmers. Manufacturers need training in terms of manufacturing processes, marketing and quality control. There is a sizeable network of training and testing institutes, this is insufficient to cater the needs.

Tractors are manufactured in the organized sector with about 21 manufacturing units. The production of tractors started in the year 1961-62 when 880 numbers were produced and sale of tractors reached a peak of 6.96 lakh units in 2013-14.

Power tillers are particularly useful in rice fields as puddling is a very important operation before paddy transplanting. The smaller size and light weight of the power tiller comes in very handy for marginal and small farm holding. There are about 5-8 medium manufacturing units and around 10 small scale units manufacturing power tiller in India. Tractor and power tiller sales for the last 10 years is given in Fig. 1.3.

1.6 Quality of Farm Machinery and Training

Standards and test codes formulated by the Bureau of Indian Standards (BIS) has been instrumental in the standardization and quality of implement manufacturing. However, small-scale manufacturers need to be trained for improving the quality of agricultural implements.

Four Farm Machinery Training & Testing Institutes (FMT&TIs) have been established in Central, North, South and Northeast India, respectively and these have now tested more than 2300 machines. Presently, 35 authorized testing centres, approved by Govt. of India for testing and certifying agricultural machineries and equipment, are conducting the testing of farm machinery.

In India, where agriculture is one of the dominating sectors of the economy, training on agricultural machines is arranged regularly. Various SAUs, ICAR Institutes, Krishi Vigyan Kendras (KVKs), Agricultural Engineering Colleges and Polytechnics, together with others have organized training in specific aspects of agricultural technologies for the benefit of thousands of artisans and professionals involved in the agricultural mechanization effort. Such training programmes have contributed significantly to expand farm mechanization to increase production and productivity to meet food requirements.

1.7 Adoption and Impact of Agricultural Mechanization

a) Tractor sales have significantly increased during last five years while the number of draught animals is declining rapidly. Statistics on various farm machinery and implement show similar trends. The states with higher farm power per hectare also have higher yields. Four wheel tractors and irrigation pumps have dominated the farm power sector in India with much less use of two wheel power tillers compared to other Asian countries.

b) Mechanization technologies were first adopted by farmers with large holdings (over 10 ha farm size) followed by medium scale farmers (with 4 to 10 ha farm size). The large numbers of such farmers in states like Punjab, Haryana and western Uttar Pradesh played a critical role in facilitating the creation of a viable agricultural machinery and implement distribution and services sector. Such farmers were also the ones who were able to provide mechanization and other services to the more numerous semi-medium (2 to 4 ha farm size), small holder farmers (1 to 2 ha) and marginal (<1 ha) farmers.

c) The availability of credit at subsidised rates has been catalytic to the rate at which farmers, especially the small and medium, were able to procure agricultural machinery and implements. In addition, assured support prices for the farmers' produce, as well as the availability of off and on farm custom hire possibilities where agricultural machinery could be used, further enhanced the profitability of acquiring agricultural mechanization inputs by farmers.

d) The high level of effective demand for agricultural machinery and equipment led to the creation of a competitive and viable manufacturing industry such that India became globally a leading player in this sector including becoming a net exporter.

e) The Govt. of India provided support services for research and development; testing and standards; as well as for human resources development in support of agricultural mechanization. The agricultural engineering programmes established in the numerous state agricultural universities and institutes were instrumental for the success of agricultural mechanization in India.

f) Business and enterprise friendly policies, laws, and regulations as well as physical and institutional infrastructures which encourage commercial activities and entrepreneurship in farming, input supply, produce handling, processing and marketing as well as in manufacturing were, and remain, the key factors to success of agricultural mechanization in the different states of India.

1.8 Focus of Policy Support for Food and Nutritional Security

a) Minimum support prices (MSP) for more crops and buffer stocking of food grains.

b) Fixing of ceilings and consolidation of land holdings through major agrarian reforms.

c) Investment in rural infrastructure such as rural roads, markets, major irrigation systems, rural electrification, water conservation and watershed development.

d) Agricultural research and education network coupled with an extensive extension system has to be strengthened.

e) Ensuring timely availability of quality inputs such as seeds, fertilizers, pesticides and farm machinery.

f) Ensuring credit availability and simplifying the subsidy process.

1.9 Future Strategies for Mechanization

Agricultural equipment and related technologies in the developed countries as well as in India has undergone a transformation in recent years. Farm implements and tools manufactured in India require a similar approach to provide more reliable machines in terms of reliability, versatility, pollution controlled, CMVR norms compliant, certification, economical in operation, comfort, safety, easy maintenance and higher efficiency.

India has emerged as a strong global hub in manufacturing of tractors and other agricultural machinery. The innovative machines e.g. tractor operated combine harvester (work like self-propelled combine harvester), maize special combine harvester, straw- reaper combine, happy seeder (for direct wheat sowing machine), straw management system (integrated with the combine harvester) have accelerated the pace of farm mechanization. Many multi-national companies have started production in India for export purpose and domestic use. Therefore, the tractor and other agricultural machinery manufactures are required to use latest production technology to bring out high quality machines at par with international standards to contribute towards India's progress as a developed country. The following points needs to be taken into account while planning future strategies for mechanization in India:

a) Location specific and crop specific technologies.

b) Mechanization of rice, cotton, sugarcane, plantation crops, horticulture, agro forestry and aqua-culture.

c) Green house and surface covered cultivation including application of electronics, computer science, automation and internet of things (IoT) in agriculture.

d) Multifunctional equipment for conservation of energy and turn around time.

e) Occupational Health Hazards and Safety including gender issues.

f) Multi-crop harvesters, to harvest many different grains and oilseeds with minimal adaptation, can capitalize on economies of scale in any given line of machinery production.

g) Adoption of large scale agricultural technologies for increase in level of mechanization which is not profitable or feasible for small farms.

h) Introduction of scale specific (based on farm holding size) agricultural machinery package in line with package of practices for crops.

i) New mechanical technologies to be coupled with new crop varieties (e.g. rice, sugarcane, cotton and vegetables). For example, commercialization of high-yielding sugarcane varieties was possible in Brazil and Australia because of the development and adoption of the mechanical sugarcane harvester that was capable of dealing with lodging associated with the new varieties.

j) Machines and equipment for agro-waste and straw management and introducing a total ban on burning of straw and other agro-waste.

k) Promotion of precision and conservation agriculture to check decline in the natural resources and soil fertility for long term sustainability.

l) Intensify research and infrastructure for post-harvest management, produce handling, storage, processing and value addition.

m) Intensification of R&D in horticultural mechanization.

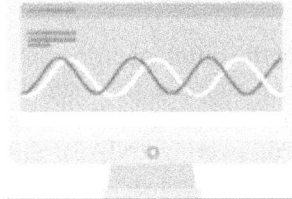

Chapter 2

Testing and Evaluation System of Agricultural Machinery

Agricultural inputs such as improved seeds, water, chemical fertilizers and pesticides are no doubt essential for increasing crop yields, but equally important are the implements, which ensure best possible use of these inputs and ensure their optimum application rates for higher productivity of land. Without implements, the efficiency of application of the inputs may be reduced and the overall productivity declines. For instance, howsoever effective a pesticide may be, if it is not uniformly and adequately sprayed in the field with the help of a suitable sprayer, its effectiveness is lost. Even for operations such as land preparation, sowing and harvesting, appropriate machinery can make all the difference although the cost of some of these machines may be higher. However, cost reduction at the expense of quality will always be undesirable. The manufacturing cost is primarily a commercial problem and is linked up with a number of parameters.

Testing and evaluation of agricultural equipment are undertaken to quantify the performance of the equipment for the desired operation. Testing is defined as 'the analysis of behaviour of machine when compared with standard codes/norms under ideal and repeatable conditions'. On the other hand, evaluation involves measurement of performance under actual field / working conditions. Evaluation also encompasses the economic and social aspect in addition to the functional performance of the machine.

Farmers must have made their own ad-hoc tests on agricultural hand tools and machines from the early days of mechanization, but formal testing is a more recent activity. New developments have sometimes been tried out publicly in competitive demonstrations, as was the case with the development of steam ploughing systems in England in the 1850's. More formal testing aimed at consumer protection was initiated in 1919 in the State of Nebraska for tractors, which had to pass the 'Nebraska Test' before they were allowed to be sold in the state. By 1960, most European countries had set up government funded machinery testing centres, The aim of such centres was assisting manufacturers and farmers to develop, test and select more productive agricultural equipment suited to national and local needs.

Formal testing of agricultural equipment in developing countries was started, during the development surge following the World War II of 1939/45, in recognition of the different crops, soils, climates and operating conditions in various countries. For example, the regional East African Testing Unit was set up at Nakuru, Kenya in 1956 with technical assistance from the National Institute of Agricultural Engineering (NIAE), Silsoe, UK. Such testing units generated information for government departments engaged in the development and extension activities related to mechanized agriculture and were thus indirectly, but not specifically, consumer oriented [195].

Machinery testing facilities have now been set up in many developing countries, most commonly as a result of government initiative and funding. Facilities may be incorporated within an appropriate unit of the Ministry of Agriculture, as in Thailand where testing is undertaken within the Agricultural Engineering Division of the Ministry, or jointly with a university, e.g. at the Agricultural Machinery Testing and Evaluation Centre (AMTEC) in the Philippines which operates in conjunction with the University of the Philippines at Los Bahos. Autonomous (usually quasi-government) organizations with the responsibility of machine testing have also been setup in some countries, e.g. the National Centre for Agricultural Mechanization (NCAM) in Nigeria. Organizations with the industrialized countries were able to arrange technical support as and when required. Example: Centre de Cooperation Internationale en Recherche Agronomique pour le Developpement: Department des Systems Agroalimentaires et Rureaux, (CIRAD-SAR) in France and development agencies such as Gesellschaft fur echnische Zusammenarbet (GTZ) in Germany. Overall, a considerable investment has been made in establishing and running machinery testing units in developing countries and a significant number of test reports have been prepared.

2.1 Standardization Efforts

The history of agricultural tractor testing in the world is only 75 years old. The first tractor with an internal combustion engine was introduced in the American agriculture in the year 1889. The tractors were judged for their performance mostly in demonstrations and fairs on the basis of comparative pulling power of the tractor for a certain number of bottoms of the mould board plough. The so called rating was termed as *"Plough Rating"*. It was for the first time in 1919 that an American farmer and senator of Nebraska State raised his voice for an Act for compulsory and official testing of tractors to enable a check on unjustified claims of the tractor manufacturers. The 'Tractor Testing Act' was passed in U.S.A. in 1919 to protect Nebraska farmers against such exploitations. The Agricultural Engineering Department in the Nebraska University was assigned the responsibility of establishing a Tractor Testing Laboratory and issuing the tractor test reports [154, 234]. Need for similar work in the interest of mechanized farming was felt simultaneously in other European countries. The National Institute of Agricultural Engineering or Testing Stations were established in important countries as under:

1. National Institute of Agricultural Engineering, U K

2. Swedish National Testing Institute of Agricultural Engineering, Sweden

3. Agricultural Machinery Testing Station, Germany

4. Centre National de Mechanique Agricole, Italy

5. University of Saskatchewan, Canada

6. Tractor Testing Station, School of Engineering, Victoria (Later shifted to Agricultural Engineering Department in the University of Melbourne), Australia

By 1959, many more countries felt the need to test tractors either with an export interest or for finding out the utility of imported tractors under their special agro-climatic conditions.

The introduction of tractors in Indian agriculture started with imports in 1940's in a small way. Imports gained a momentum in 1950's, when innumerable brands of tractors were imported. Till 1960, India had to depend upon the imported tractors and the evaluation of the performance of these imported tractors was known mostly from their field performance and from mechanical cultivation schemes of different States. In 1954, a need was recognized for having a National Tractor Testing Station in the country to test the suitability and performance of a tractor model before large scale import and also to test tractors which were proposed for indigenous production. After temporary location at Nagpur, the Tractor Testing Station was finally established at Budni, Madhya Pradesh in 1959. Effective work on testing started in 1961 when the first tractor test report was released.

The International Organization for Standardization (ISO) is the apex body in the area of standardization at international level and has its membership in National Standards Bodies of various countries. The Bureau of Indian Standard, BIS (formerly Indian Standard Institution) since its inception in 1947, realized the importance .of standardization at national and international level. The Bureau, therefore, from the beginning had been collaborating actively with ISO. In a developing economy like India, industrial development in general and agro/industrial in particular makes it imperative to make efficient use of all natural resources and organizational, technical and economic means. For this purpose, a coordinated, rational and efficient management in resource mobilization is of utmost importance. Experience has shown that standardization serves as an effective instrument in achieving this objective.

In the context of farm machinery, it has been observed that acceptance of farm machinery by the farmers largely depends on their quality. Hence, in order to reap the benefits of standardization including manufacture of high quality products, a need was felt for preparation of Indian standards for agricultural machinery. Organized efforts in this direction were made by the BIS in late 1950's by way of setting up a Technical Committee (TC) for formulation of standards for agricultural machinery. The committee is generally consisted of representatives of government departments, research, education and testing institutions and the manufacturing industries. So far, more than 400 standards have been published in this area. Some of the highlights of the work done by BIS is summarized below:

A. Standardization of specifications and interchangeability of components

When certain components of an implement or machine worn out, it remains under break-down for the period until the required component has been procured from the original manufacturer. This situation makes the repair, service and operation of the machine very difficult and uneconomical. The problem becomes more acute in case of imported items. Hence, the standardization of component was considered an important part of the work. Standards on components such as agricultural disc, reversible shovel, guards, knife sections, three-point linkage and power-take-off shaft of agricultural tractors, linch pin, ball and socket assembly, plough shares, sweeps, cultivator tynes, spools for harrow, seed and fertilizer metering mechanism, chaff cutter blades, roller and axle for sugarcane crusher, nozzles, lances and cut-off device for sprayer, etc. have been brought out from the angle of interchangeability.

B. Implementation of standards

All the efforts in formulation of standards would go waste if these are not implemented by the manufacturers, purchasers, leasing agencies and testing authorities. Implementation of standards would not only help in producing and procuring quality implements, but would also reveal difficulties, if any, in adopting them and thus provide a feedback to make the standard more realistic.

C. Certification Marks Scheme

To make quality implements and spares available to the farmers, BIS operates a Certification Marks Scheme under which licenses are issued to such manufacturers who apply for use of ISI mark on their goods, to indicate that the quality is in conformity with the relevant Indian Standard. The BIS Certification Marks Scheme is operated on voluntary basis. Availability of certified agricultural machinery in the country could only be promoted by developing a demand for quality certified equipment through organized purchasers insisting on ISI Marked products [26].

2.2 Advantages of Certification Scheme

a) For manufacturers:

- Streamlining of production processes and introduction of quality control system.
- Independent audit of quality control system by BIS.
- Reaping of production economies accruing from standardization.
- Better image of products in the market both domestic and overseas.
- Winning consumer confidence and goodwill for wholesalers, retailers and stockiest.
- Preference for standard-marked products by organized purchasers, agencies of Central and State government, local bodies, public and private sector undertaking etc. Some organized purchases offer even higher price for standard marked goods.
- Financial incentives offered by the nationalized banks.

b) For consumers:

- Conformity with standards by an independent, technical, national level organization.

- Helps in choosing a standard product.
- Free replacement of standard-marked products in case of substandard quality.
- Protection from exploitation and deception.
- Assurance of safety against hazards to life and property.

c) For organized purchasers:
- Convenient basis for concluding contracts.
- No need for inspection and testing of goods, saving time, labour and money.
- Free replacement of products with Standard Mark if found to be substandard.

d) For exporters:
- Exemption from pre-shipment inspection wherever admissible.
- Convenient basis for concluding export contracts.

e) For export inspection authorities:
- Elimination of the need for exhaustive inspection of consignments exported from the country, saving expenditure, time and labour.

2.3 Testing Standards

In order to obtain accurate and repeatable results, it is most important to conduct the testing under a set of standardized test conditions and to follow standardized test procedures and formats. Attempts in this direction were made to develop detailed test codes since unorganized testing of machinery was posing problems for rationalizing the performance of various equipment. Test codes and procedures for agricultural tractor, mouldboard plough, disc harrow, seed-cum-fertilizer drill, power thresher, maize sheller, sugarcane crusher, seed cleaner, chaff cutter, etc. have already been formulated and published by the OECD, ISO, ASABE, BIS, RNAM and other standardization agencies.

2.3.1 OECD test code

There is a considerable amount of international trade of agricultural tractors. The purpose of this international test code is to facilitate trade by enabling either an exporting or an importing country to accept with confidence the results of tests carried out in another country. The Standard Code for the Official Testing of Agricultural Tractors was established by the Decision of the Council of the Organisation for European Economic Co-operation (OEEC) on 21st April 1959. This Code remained in force pursuant to the decision of the Council of the Organisation for Economic Co-operation and Development (OECD) on 30th September 1961, and it was made public on 3rd May 1962 [www.oecd.org]. List of OECD standard Codes are given in Annexure 2.1.

Tests carried out under the OECD tractor codes are officially approved only after the OECD Secretariat is satisfied that the tests have been carried out in accordance with the procedures laid down in the applied code. In the case of tests on protective structures, in addition, there are pass/fail criteria, which stipulate certain minimum performance levels for the structure tested. Other performance tests carried out under internationally recognized procedures may be included / reported if the procedures are available in published form and in an official language of OECD, will be clearly marked as not submitted for OECD approval.

It is desirable that tests under the codes be carried out in the tractor's country of origin for simplification of the work, procedures and reduction in costs. This may not always be possible or convenient, for instance, the producing country does not participate in the OECD. In such cases, the importing country may carry out the tests. For tests under Code 2, verification is made with the OECD to ensure that no other country has carried out the relevant tractor tests, hence avoiding unnecessary duplication of effort and cost.

Even though, testing agencies can make whatever tests they wish, but only one OECD approval number will be issued for a given tractor or for a tractor-protective structure combination. If modification to the tested models make it necessary to retest them within the limits specified in each code, a new OECD test report is issued. Approval can be extended for modified tractors or tractor variants within the limitations of each code and it can be requested only by the testing station where the original test has been carried out. Tractors which need approval by extension may give rise to the publication of a test report or an extension report, provided that reference to the originating tractor be visible in the test or extension report and that the modifications of the specifications and results, when relevant, be clearly identified. In this case, the reports will receive the same approval number, complemented by an appropriate numerical designation [178].

2.3.2 ISO standards

ISO standard - ISO/TC 23, Tractors and machinery for agriculture and forestry comprises standardization of tractors, machines, systems, implements and other equipment used in agriculture and forestry as well as gardening, landscaping, irrigation and other related areas in which such equipment is used, including agricultural electronics [www.iso.org]. Details are given in Annexure 2.2. Agricultural machinery standards cover the internal mechanical workings, safety considerations and requirements, electronics, controls, and more for a variety of agricultural machines. As with most industries, safety is a key concern for both manufacturers and end-users, influencing everything from the initial design process to consumer usage and maintenance/repair lifecycle. Sprayers, soil working, and mowing machines are featured prominently, and standards for other machinery and their components are included as well.

a) Safety standards for agricultural machinery

Safety standards for agricultural machinery address a wide range of different types of machinery with the multi-part ISO 4254 series, with further information provided by other ISO standards. Most of the standards under this area are focused on individual agricultural machines, intended for use with more generalized standards such as ISO 4254-1 [135], to provide a detailed look into the safety considerations of each piece of agricultural machinery.

b) Mechanical standards for agricultural machinery

Mechanical Agricultural Machinery Standards, a series of ISO standards dealing with tractors and associated agricultural machinery, cover safety, mechanical specifications, and more. With a focus on stability, guard openings, drive shafts, and belts, these standards provide an important component of the overall standardization effort for agricultural machinery.

c) Soil working and mowing standards for agricultural machinery

Soil Working and Mowing Standards for agricultural machinery deal with acceptance criteria for physical specifics and test methods. Rotary (disc and drum) and flail mowers are covered, as well as equipment used for sowing, planting, and other soil working.

d) Sprayer standards for agricultural machinery

Agricultural Spraying Standards, published by ISO, split their focus between inspections of sprayers in use and environmental requirements for sprayers. These further break down into general considerations, horizontal boom sprayers, sprayers for bush and tree crops, and fixed and semi-mobile sprayers.

e) Electronics and control standards for agricultural machinery

Electronics and Control Standards for Agricultural Machinery are a varied group of standards that range from communications and electrical connections, to coding and symbols, and sensors and electromagnetic compatibility.

2.3.3 Bureau of Indian Standards (BIS)

The Indian Standards Institution (ISI) came into existence on the 6th January, 1947. During the initial years, the main focus was on standardization activity and to provide the advantages to common consumers, Certification Marks Scheme was launched in 1955-56. It enabled ISI to grant licenses to manufacturers producing goods in conformity with Indian Standards and to apply ISI Mark on their products. To govern the formulation of standards and other related work, a bill was introduced in the Indian Parliament on 26th November, 1986 for establishing Bureau of Indian standards (BIS). By setting up the national standard body, the government envisaged building a climate for quality culture and consciousness and greater participation of consumers in formulation and implementation of national standards [www.bis.org.in].

Testing and evaluation of tractors and agricultural machinery is governed by BIS standards and the same is conducted in different authorized testing centres. The list of sectional committees (FAD) is given in Annexure 2.3. Different testing systems currently prevailing in India includes the following:

i) National testing

Under this system, the tests are conducted by the Farm Machinery training and Testing Institutes under the jurisdiction of Ministry of Agriculture and Farmers Welfare, Government of India. Currently such institutes are located at Budni (MP), Hissar (Haryana), Anantpur (AP) and Vishwanath Chariali (Assam) and are fully functioning. Type of tests conducted under the national testing programme have been discussed later in this chapter. An exhaustive list of authorized testing centres is given in Annexure 2.5.

ii) Prototype testing

Prototype testing includes the testing of research prototypes as well as the testing of production prototypes. Currently, different State Agricultural Universities (SAU's) and ICAR Institutes with Engineering Divisions conduct prototype testing of their research equipment. Some of these institutions also undertake testing of production prototypes manufactured by different firms.

iii) Testing for quality marketing

A. Testing is carried out by the Bureau of Indian Standards under their Certification Marks Scheme through their Central Laboratory and other approved test laboratories located all over the country.

B. The national tests are conducted on tractors, power tillers and agricultural machinery under the control of Ministry of Agriculture and Farmers Welfare:

a) Confidential test

Confidential test is meant for providing confidential information on performance of machines which are required for commercial production or to provide any special data that may be required by a manufacturer/applicant. The following categories of machines are covered under the scope of the confidential tests:

i) a prototype model before it is ready for commercial production.

ii) an improved model prior to progressive manufacture/import on large scale.

iii) a machine under commercial production, but with modification of one or more systems for improved performance.

iv) the machines submitted for test under BIS Certification Marks Scheme.

A confidential test report cannot be used for a commercial purpose. An applicant is not permitted to publish the confidential test report in full or in abbreviated form or to divulge the test results contained therein to any person or body.

b) Commercial test

Commercial tests are conducted on the machines, which are ready for commercial production or already in production to establish their performance characteristics. The following types of commercial tests are undertaken:

i) Initial commercial : On indigenous or imported prototype machines
   test   ready for commercial production.

ii) Batch test	:	In view of the need for continuous improvement in the quality, an indigenous machine under commercial production is tested after a certain time interval initially after 3 years of initial commercial test and subsequently after every 5 years for conformity of production. At present this is applicable to tractors only.
iii) Test as per OECD Code	:	This is conducted on machines (which have already undergone initial commercial test) on specific request of the applicant/manufacturer exclusively for export purposes. The test are conducted in accordance with OECD standard test cord. Commercial test report can be published in full without any alteration or omission.
iv) Repeat test	:	This test is conducted on tractor to validate its performance for the same parameters and on the same sample under test after rectifying the defect or after replacement of defective part/assembly with the new part/assembly of the same specifications.
v) Technical extension	:	This test is conducted on tractor for certain parameters/improvements or statuary requirements after incorporating the improvements, which shall remain permanently on the tractor that had already undergone commercial test.

2.3.4 American society of agricultural and biological engineers standards

American Society of Agricultural and Biological Engineers (ASABE) is accredited by the American National Standards Institute (ANSI) to coordinate and develop the U.S. position in fourteen distinct areas of international standards development [www.asabe.org]. Each of these areas has an associated ASABE technical committee, made up of U.S. experts from a diverse range of backgrounds. These committees (called TAGs, or Technical Advisory Groups) operate in a manner that complies with all applicable ANSI and ISO/IEC procedures. The list of machinery systems committees is given in Annexure 2.4.

2.4 Testing Networks and Centers

2.4.1 European Networks for Testing of Agricultural Machines (ENTAM)

ENTAM is the network constituted by the official testing stations in those European countries which have signed an agreement on shared activities. These activities have, the implementation of standardized tests of the performance, safety and environmental aspects of agricultural machinery and tools. At the manufacturer's request, tests are carried out by specialist testing stations [www.entam.net].

The tests are based on National, European or International standards, or shared agreements (or methodologies), and can provide the manufacturer with useful information on ways in which to improve its machinery. The results may be issued as test reports published by the testing stations, which work in partnership with ENTAM. As such, applying for and obtaining the ENTAM mark on the reports is a way of confirming not only that the manufacturer works to international standards but also that all applicable regulatory requirements have been met. This allows the farmer to make an informed choice when purchasing new agricultural machinery. ENTAM works in two levels, which is shown in Fig. 2.1. ENTAM registration is the "first step" of the ENTAM procedure, which implies that the machines have been submitted for testing procedures in one of the partners' testing laboratories and are therefore certified in at least one European country.

Any manufacturer requiring European-level certification in other EU countries can apply for ENTAM recognition. The "ENTAM recognized" logo is awarded when a test report, which should be registered by the proposing member with its own number, has been sent to the ENTAM office and has obtained recognition from at least two other ENTAM member institutions.

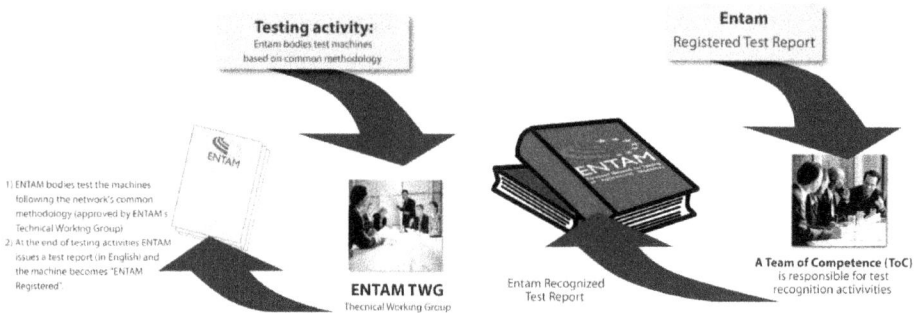

Fig. 2.1: Working of ENTAM – First and Second Level

ENTAM member countries

ENTAM is currently made up of eleven members, an honorary witness (FAO) and four observer members (INTA, AFMSPTC, CEA, VIM) respectively from Argentina, Bulgaria, Brazil, Russia). The eleven official members are based in eight European countries:

- Austria (BLT)
- France (Cemagref)
- Germany (JKI, DLG, KWF)
- Greece (N.AG.RE.F)
- Hungary (HIAE)
- Italy (ENAMA)
- Poland (PIMR)
- Spain (CMA, EMA/CENTER)

According to the ENTAM agreement, every testing system is always based on official international standards (ISO, EN, OECD) and to develop a common testing activity, the different testing stations work for a common system called '*Entam Common Methodology*'. It consist of technical instruction which are discussed, approved and updated unanimously by the specific Technical Working Group held by the ENTAM Members. The ENTAM common methodology remains confidential and is not published.

The 'Team of Competence' is made of a restricted group of experts chosen among those of the corresponding ENTAM Technical Working Group. It is looking after the ENTAM recognition of the test assuring the correspondence of the technical paper data to the related ENTAM common methodology. The introduction of the 'Team of Competence' allow the network to be more flexible and reactive in specialized mechanization sectors.

2.4.2 Asian and Pacific Network for Testing of Agricultural Machinery (ANTAM)

The Regional Network for Agricultural Machinery (RNAM) was established by the United Nations Economic and Social Commission for Asia and the Pacific (ESCAP) in cooperation with the United Nations Industrial Development Organization (UNIDO) and the Food and Agriculture Organization (FAO) in 1977 in Los Baños, Philippines. There were eight participating countries: India, Indonesia, Islamic Republic of Iran, Pakistan, Philippines, Republic of Korea, Sri Lanka and Thailand. In 2002, RNAM was upgraded to a centre known as the Asian and Pacific Centre for Agricultural Engineering and Machinery (APCAEM) located in Beijing, China. APCAEM then adopted its current name as Centre for Sustainable Agricultural Mechanization (CSAM) in 2012 [www.un-csam.org].

ANTAM was formally proposed at the Roundtable Forum for the Regional Agricultural Machinery Manufacturers/Distributors Associations held by APCAEM in Seoul, Korea in 2006 in collaboration with the Ministry of Agriculture and Forestry of the Republic of Korea and the Korea Agricultural Machinery Industry Cooperative (KAMICO). In preparation for the establishment of ANTAM, APCAEM commissioned a feasibility study in 2009 and the study suggested phased approach to:

i) develop region-wise testing codes

ii) upgrade the existing infrastructure

iii) improve the professional knowledge and skills of the technicians

The 1st ANTAM Annual Meeting in Beijing, China, in September 2014 approved the Terms of Reference (ToR) and agreed to establish a Technical Working Group (TWG) to develop mutually recognized test codes [http://antam.un-csam.org].

ANTAM member countries

The member countries (total 18, including associate member of ESCAP) decided on December, 2017 are:

- Armenia
- Bangladesh
- Cambodia
- China
- France
- Hong Kong, China
- India
- Indonesia
- Malaysia

- Nepal
- Pakistan
- Philippines
- Russia
- Republic of Korea
- Sri Lanka
- Thailand
- Turkey
- Viet Nam

Technical Working Groups (TWGs) approved the following three ANTAM standards for testing:

a) ANTAM Standard code for testing of Paddy Transplanters

b) ANTAM Standard code for testing of Powered Knapsack Mister-cum-Dusters

c) ANTAM Standard code for testing of Power Tillers

2.5 Emerging Standards

With any developing technology industry, there are emerging standards. In agriculture, many of the new technologies are not specifically related to tractors but are more related to an entire field known as 'Precision Agriculture' (PA / PAg). Precision agriculture, also known as site-specific management, addresses spatial variability within a field and managing that variability to maximize production and profitability while minimizing risk. Site-specific management may be applied to such decisions as variety selection, weed and pest management, nutrient management, and irrigation. Many of the new standards include the interoperability of systems, which provide variable rate capability for fertilizer application, herbicide application, and seed placement. Another part of precision agriculture is the use of GPS guidance and auto-steer functions.

Some of the more global standards being worked on by ASABE are in the field of aquaculture. Aquaculture is the commercial farming of seafood in a controlled environment. With earth's increasing population, aquaculture is becoming increasingly important for feeding the world. Aquaculture will need standards dealing with health and environmental concerns. Tractor standards have improved the ability of manufacturers to market their products worldwide meanwhile the same standardization is needed for implements. Many manufacturers wish to distribute their implements globally, but currently implements do not have the same standards infrastructure to support them. Finally, the latest standard development for tractors has come as a result of new emissions standards and the manufacturers have to select one of two primary methods for improving emissions. The first is the use of a Diesel Exhaust Fluid, future performance test must measure the amount of fluid consumed as a portion of the fuel efficiency test. The other method requires use of regenerating particulate filters, which must also be evaluated [131].

2.5.1 Electronic standards and communication protocols

Monitoring and control of agricultural machines, equipment, facilities and processes, and the management of farms and agricultural companies through the use of electronic and computer systems has led to new and better standards.

The agricultural equipment industry has been incorporating electronics to their products with the aim of providing more information to the operator about machine performance, registering that information for future analysis, automatically controlling the machines, freeing operator attention to other tasks, optimizing the use of the machine and avoiding unnecessary wear, and optimizing the use of inputs [111, 227]. This increased the performance and reliability of machines. Examples are:

i) on-board computers for tractors, which monitor variables of the engine, transmission and slip,

ii) monitoring and/or control systems in planters, sprayers and combines (harvesters), which compensate the influence of ground speed variation on the actual application rate of inputs (such as fertilizers or herbicides) or on the rate of harvesting,

In recent years, incompatibility between the software, hardware and the data format has become a major problem. These issues are standardized in accordance with ISO 11783 and ISOBUS (International Standardization Organization Binary Unit System).

Connected agricultural system can only be achieved when the farm machines seamlessly talk to each other and to related systems (farm management systems) and necessary standards and communication protocols are required. In this area, CEMA taking initiatives such as the AEF and AgGateway and areas of expertise is shown in Fig. 2.2 [108]. The overlapping (grey) area shows the fields in which both organizations work together to align with the development and acceptance of data exchanges standards.

a) The Agricultural Industry Electronics Foundation (AEF):

AEF [www.aef-online.org] is an independent international organization, established in 2008 by 7 agri-equipment manufacturers and 2 associations, to support the development, implementation and enhancement of standards for the increased use of electronic and electrical systems in mobile farming equipment. Currently, more than 190 members are working under AEF. Main focus was the development of ISOBUS standard (ISO 11783), however, this has been expanded to cover additional areas such as:

- Farm Management Information Systems (FMIS)
- wireless in-field communication
- high-speed ISOBUS
- electric drives
- camera systems

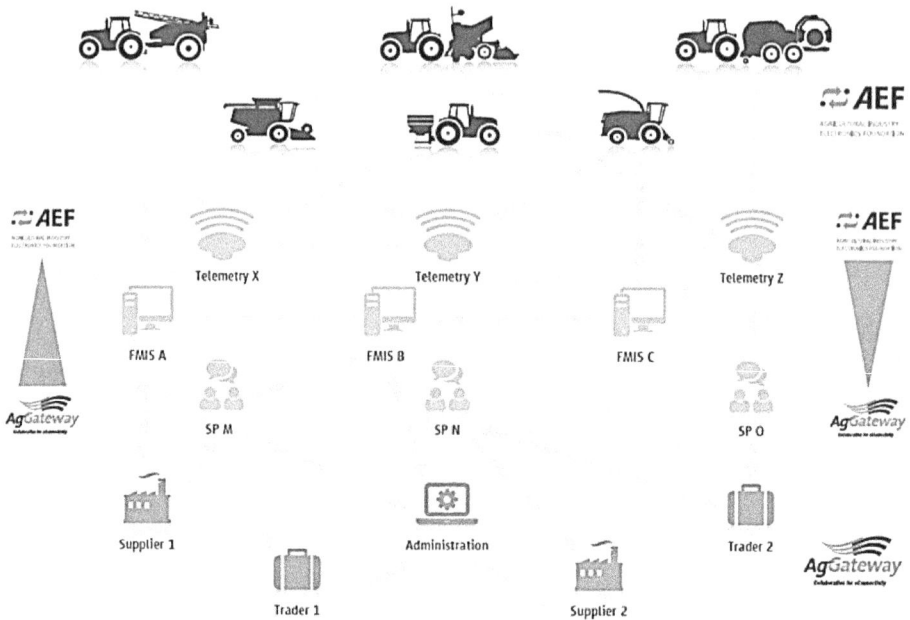

Fig. 2.2 : Respective areas of expertise of AEF and AgGateway and areas of cooperation are highlighted in grey

b) AgGateway: standardizing data exchange within the agricultural supply chain

AgGateway [http://aggatewayglobal.net] is the recognized international organization that uses the concept of industry cooperation to expand the use of e-business standards and guidelines globally and, as such, to enable the use of information and communication systems in farming. It considers regional operating practices to identify opportunities and balanced approaches to standardization. The objective is to share best practices, i.e. "what has worked" in different regions of the world so as to promote global e-business, and to collaborate on the development of the necessary standards where such specific needs exist. It also takes the initiative to develop missing standards for specific area, in cooperation with existing standardization boards.

For the standardized exchange of order-, invoice-, and dispatch-data the UN/CEFACT, GS1 or UBL standards are leading. For exchanging e.g. laboratory analysis results and crop-related data for compliance purposes, new standard UN/CEFACT messages are being developed in recent years. For data exchange, the unique identification of farmers, crop fields, inputs, etc. is very important. GS1 provides a set of worldwide implemented standards for unique identification such as GLN (Global Location number), GTIN (Global Trade Item Number) or GPC (Global Product Classification). AgGateway has become a member of AEF to make the standards for data exchange future-proof and adapt it to the needs of digital farming. The added data exchange standards between mobile farm equipment and farm management systems (or data management systems) are also standardized in ISO11783, Parts 10 & 11 [137].

Annexure 2.1

OECD Standard Codes for the Official Testing of Agricultural and Forestry Tractors - 2008

Code 1   :   Repealed - for the record.

Code 2   :   Repealed and replaced by testing of agricultural and forestry tractor performance.

Code 3   :   Testing of the strength of protective structures for agricultural and forestry tractors (dynamic test).

Code 4   :   Testing of the strength of protective structures for agricultural and forestry tractors (static test).

Code 5   :   Noise measurement at the driver's position(s).

Code 6   :   Testing of front-mounted protective structures on narrow-track wheeled agricultural and forestry tractors.

Code 7   :   Testing of the rear-mounted protective structures on narrow-track wheeled agricultural and forestry tractors.

Code 8   :   Testing of protective structures on tracklaying tractors.

Code 9   :   Protective structures for telehandlers (testing of falling-object and roll-over protective structures fitted to self-propelled variable reach all-terrain trucks for agricultural use).

Code 10   :   Testing of Falling object protective structures

Annexure 2.2

ISO Standards (ISO/TC 23)
Tractors and machinery for agriculture and forestry

Subcommittee	Subcommittee Title	Published standards	Standards under development
ISO/TC 23/SC 2	Common tests	38	2
ISO/TC 23/SC 3	Safety and comfort	14	6
ISO/TC 23/SC 4	Tractors	54	6
ISO/TC 23/SC 6	Equipment for crop protection	47	12
ISO/TC 23/SC 7	Equipment for harvesting and conservation	24	5
ISO/TC 23/SC 13	Powered lawn and garden equipment	20	6
ISO/TC 23/SC 14	Operator controls, operator symbols and other displays, operator manuals	8	2
ISO/TC 23/SC 15	Machinery for forestry	20	9
ISO/TC 23/SC 17	Manually portable forest machinery	30	6
ISO/TC 23/SC 18	Irrigation and drainage equipment and systems	38	7
ISO/TC 23/SC 19	Agricultural electronics	40	15
	Total	333	76

Annexure 2.3
BIS Standards Sectional Committees
Tractors, Farm Machinery, Irrigation, Dairy Equipment

Sectional Committee	Sectional Committee Title	Published standards
FAD 11	Agricultural Machinery and Equipment	208
FAD 17	Farm Irrigation and Drainage Systems	42
FAD 19	Dairy Products and Equipment	126
FAD 20	Agriculture & Food Processing Equipment	58
FAD 22	Agricultural Systems and Managements	10
	Total	444

Annexure 2.4
ASABE Standards Management Systems Technical Committees

S. No.	MS*	U.S. TAG for	Committee name
1.	MS-23	ISO/TC 23	Tractors and machinery for agriculture and forestry
2.	MS-23/2	ISO/TC 23/SC 2	Common tests (for agricultural equipment)
3.	MS-23/3	ISO/TC 23/SC 3	Safety & comfort (of the operator)
4.	MS-23/4	ISO/TC 23/SC 4	Tractors
5.	MS-23/6	ISO/TC 23/SC 6	Equipment for crop protection (spray equipment and drift mitigation)
6.	MS-23/7	ISO/TC 23/SC 7	Equipment for harvesting and conservation
7.	MS-23/14	ISO/TC 23/SC 14	Operator controls, operator symbols and other displays, operator manuals
8.	MS-23/19	ISO/TC 23/SC 19	Agricultural electronics (equipment, animal ID)
9.	NRES-03/2	ISO/TC 23/SC18	Irrigation and drainage equipment and systems
10.	ASE-134	ISO/TC 134	Fertilizers and Soil Conditioners
11.	PAFS-234	ISO/TC 234	Fisheries and Aquaculture
12.	ES-238	ISO/TC 238	Solid Biofuels
13.	ES-255	ISO/TC 255	Biogas
14.	PRS-293	ISO/TC 293	Feed Machinery

*MS – ASABE Machinery Systems Technical Community

Annexure 2.5

List of Authorized Testing Centres approved for
Testing and Certifying Agricultural Machineries and Equipment in India (as on 01.03.2018)

S. No.	Testing Centre	Phone	Email address
(1)	(2)	(3)	(4)
1.	Central Farm Machinery Training and Testing Institute, Budni, Madhya Pradesh	07564-234988 07564-234987	fmti-mp@nic.in
2.	Northern Region Farm Machinery Training &Testing Institute, Tractor Nagar, Sirsa Road, Hissar, Haryana	0166-2276984 99263 14603	fmti-nr@nic.in rajendranema@yahoo.com
3.	Southern Region Farm Machinery Training and Testing Institute, Anantpur, Andhra Pradesh	08551-286441	fmti-sr@nic.in
4.	North Eastern Region Farm Machinery Training and Testing Institute, Sonitpur, Biswanath Chariali, Assam	03715-222094	fmti-ner@nic.in
5.	Acharya N.G. Ranga Agricultural University Rajendra Nagar, Hyderabad, Telengana	24018277 9989625237	fimscheme@gmail.com
6.	Faculty of Agricultural Engineering, Rajendra Agricultural University, Pusa, Bihar		dean_cae@rediffmail.com , kumar.kranti@yahoo.com
7.	State level Agriculture Implement Testing Centre, Agri. Dept., Govt. of Chhattisgarh, TeliBandha, Gorav Path, Raipur, Chhattisgarh	09827151776, 0771-2430156	jdaengg.test-cg@nic.in
8.	Division of Agricultural Engineering, Indian Agriculture Research Institute, New Delhi	011-25842294, 9968190405	indramani@iari.res.in satishiari@gmail.com
9.	College of Agricultural Engineering & Technology, Junagarh Agricultural University, Junagarh, Gujarat	9898942762	gupta-r-a@jau.in
10.	College of Agricultural Engineering & Technology, CCS Agricultural University, Hissar, Haryana	9416397798 8607866633	mukeshjainhisar@ rediffmail.com
11.	Sher-e-Kashmir University of Agriculture Science & Technology, Srinagar, Jammu & Kashmir	094192-30487	sk_sharma@yahoo.com

(1)	(2)	(3)	(4)
12.	Birsa Agricultural University (BAU), Jharkhand	09431543781	dkrusia@gmail.com
13.	Jharkhand Agriculture Machinery Testing and Training Centre (JAM-TTC) Govt. of Jharkhand, Ranchi, Jharkhand	9431395373 9934156377 8986672378	singh.omkar14@yahoo.co.in jarkhandranchi@gmail.com
14.	University of Agricultural Sciences, Gandhi Krishi Vignyan Kendra, Bangalore, Karnataka	9482297957	palanimuthuv@gmail.com
15.	College of Agricultural Engineering, UAS, Raichur, Karnataka	9480163904	anantachary@gmail.com
16.	Farm Machinery Testing Centre, KAU Kelappaji CoAE&T, Tavanur, Malappuram (Dist.), Kerala	9446113559	sindhu.bhaskar@kau.in fmtctve@kau.in
17.	Central Institute of Agricultural Engineering Berasia Road, Bhopal, Madhya Pradesh	0755-2521082	kna@ciae.res.in
18.	Dr. A.S. College of Agricultural Engineering Rahuri, Distt. Ahmednagar. Maharashtra	9850612344 9049051525	diraj_karale@hotmail.com vskankal@gmail.com
19.	Farm Machinery Testing, Training and Production Centre, Department of FM&P, Dr. PDKV, Akola, Maharashtra	0724-2259403	sht1964@rediffmail.com
20.	College of Agricultural Engineering and Technology, Dr. Balasaheb Sawant Konkan Krishi Vidyapeeth, DAPOLI. Maharashtra	9423784381	shahareprashant@rediffmail.com
21.	College of Agricultural Engineering and Technology, Orissa University of Agriculture and Technology, Bhubaneswar, Orissa	9437386861	debaraj1963@rediffmail.com
22.	State Level Farm Machinery Training & Testing Centre, Agriculture Department, Government of Orissa, Bhubaneswar, Orissa	09437815544 09439727549	pkmsonu@gmail.com dasalok61@gmail.com
23.	College of Agricultural Engineering and Technology, Punjab Agricultural University, Ludhiana, Punjab	9888997292 09779906139	sksingh@pau.edu goyalrajat@pau.edu
24.	Central Institute of Post- harvest Engineering and Technology (CIPHET), PAU campus Ludhiana, Punjab	9417596894 0161-2313123	Kadam1k@gmail.com

(1)	(2)	(3)	(4)
25.	College of Technology and Agricultural Engineering, Maharana Pratap University of Agriculture and Technology, Udaipur, Rajasthan	07891781022	shashipawar5406@gmail.com
26.	Farm Implements and Machinery Testing & Training Centre, Central Workshop, Swami Keshwanand Rajasthan Agricultural University, Bikaner, Rajasthan	0941423034 0152-2250576	vladha@yahoo.com
27.	College of Agricultural Engineering and Post-Harvest Technology, Ranipool, Gangtok, Sikkim	09933469544	snyadavkpl@yahoo.com
28.	Tamil Nadu Agricultural University, Coimbatore, Tamil Nadu	0422-2457576	tnaushridhar@gmail.com
29.	State Level Farm Machinery Training and Testing Institute, Govt. of U.P., Rehmankhera, Lucknow, Uttar Pradesh	8176004522 0522-2841146	sametiup@gmail.com
30.	Sam Higginbottom Institute of Agriculture, Technology & Science (AAI), Allahabad, Uttar Pradesh	9794618648	fdean_engg@shiats.edu.in
31.	College of Technology, Gobind Ballabh Pant University of Agriculture and Technology, Pantnagar, Uttarakhand	0941051494 05944233380	jayantsingh07@gmail.com
32.	Department of Agriculture & Food Engineering, IIT, Kharagpur, West Bengal	9434704801 03222-282244	hifjur@agfe.iitkgp.ernet.in
33.	State Farm Machinery Training-cum-Testing Institute, Faculty of Agricultural Engineering, Mohanpur, Dist. Nadia, West Bengal	03473-22265	souti62@rediffmail.com
34.	College of Agriculture, University of Agricultural Sciences, Belgaum Rd, Krishi Nagar, Dharwad, Karnataka		deanacb@uasd.in
35.	Farm Machinery Testing Center, ICAR-Central Institute of Agricultural Engineering- Regional Center, regional centre, near AMRC, Coimbatore, Tamil Nadu	0422–2472624 9842955606	thasekumar@gmail.com

Annexure 2.6

List of OECD Designated Authorities and Testing Stations
(as on Feb. 2017)

AUSTRIA	HBLFA - B.L.T. Höhere Bundeslehr – und Forschungsanstalt für Landwirtschaft, Landtechnik und Lebensmittel/technologie Francisco Josephinum in Wieselburg Rottenhauser Strasse 1, A-3250 WIESELBURG	Tel: +43.74 16 52 175 39 Fax: +43.74 16 52 175 45 E-mail: ewald.luger@ josephinum.at
BELGIUM	D.G.R. DÉPARTEMENT GENIE RURAL CENTRE WALLON DE RECHERCHES AGRONOMIQUES 146, Chaussée de Namur B-5030 GEMBLOUX	Tel: +32.81.62.71.40 Fax: +32.81.61.58.47 E-mail: huyghebaert@cra. wallonie.be
CHINA	CERTIFICATION AND ACCREDITATION ADMINISTRATION OF THE PEOPLE'S REPUBLIC OF CHINA (C.N.C.A.) Department for International Cooperation 9 Madian East Road, Tower B HAIDIAN, BEIJING 100088	Tel: +86.10.822.62.669 Fax: +86.10.822.60.819 E-mail: dchj@cnca.gov.cn, Zhangxd@cnca.gov.cn
	C.O.T.T.E.C. CHINA OFFICIAL TRACTOR TEST AND EVALUATION CENTER Jianxi District, Henan Province LUOYANG 471039	Tel: +86.379.6269.0095 Fax: +86.379.6269.0350 E-mail: cottec.oecd@ vip.163.com
	C.A.M.T.C. CHINA AGRICULTURAL MACHINERY TESTING CENTRE No. 96, Dongsanhuan Nanlu, BEIJING 100122	Tel: +86.10.67.32.64.90 Fax: +86.10.67.34.37.54 E-mail: kjch@camtc.net
CZECH REPUBLIC	S.Z.Z.P.L.S. GOVERNMENT TESTING LABORATORY OF AGRICULTURAL FOOD INDUSTRY AND FORESTRY MACHINES Tranovskeho 622 / 11 CZ 163 04 PRAHA 6 REPY	Tel: +420.737.859.247 Fax: +420.235.018.226 E-mail:kulhavy@szzpls.cz

FINLAND	MTT Agrifood Research Finland Testing and Standardisation Vakolantie 55, FIN-03400 VIHTI	Tel: +358.9.224.252.14 Fax: +358.9.224.6210 E-mail: Lauri.Tuunanen@ mtt.fi
FRANCE	MINISTÈRE DE L'AGRICULTURE et de la PÊCHE S.A.F.S.L/S.D.T.P.S. 78, rue de Varenne, F-75349 PARIS 07 SP	Tel: +33.1.49.55.82.17 Fax: +33.1.49.55.47.70 Email: dominique. dufumier@ agriculture. gouv.fr
	Irstea INSTITUT NATIONAL DE RECHERCHE EN SCIENCES ET TECHNOLOGIES POUR L'ENVIRONNEMENT ET L'AGRICULTURE 1, rue Pierre Gilles de Gennes CS 10030 F-92761 ANTONY CEDEX	Tel: +33 (0)1 40 96 61 58 Fax: +33 (0)1 40 96 61 62 E-mail: thierry.langle@ irstea.fr
GERMANY	Federal Ministry of Food, Agriculture and Consumer Protection (BMEL) Rochusstrasse 1, 53123 Bonn Germany	Tel: +49 228 99 529 3480 Fax: +49 228 99 529 55 3480 Email: 514@bmel.bund.de
	D.L.G. DEUTSCHE LANDWIRTSCHAFTS- GESELLSCHAFT Testzentrum Technik und Betriebsmittel, Max-Eyth-Weg 1 D-64823 GROSS-UMSTADT	Tel: +49.69.24.78.8-640 Fax: +49.69.24.78.8-90 E-mail: A.Ai@DLG.org
ICELAND	RANNSÓKNASTOFNUN LANDBÚNADARINS BÚTAEKNIDEILD Agricultural Research Institute Technical Department Hvanneyri, 311 BORGARNES	
INDIA	CENTRAL FARM MACHINERY TRAINING AND TESTING INSTITUTE Ministry of Agriculture Department of Agriculture Cooperation & Farmers Welfare Tractor Nagar BUDNI (Madhya Pradesh) 466 445	Tel: +91.7564.2347.29 Fax: +91.7564.2347.43 E-mail: fmti-mp@nic.in, kalevn2000@yahoo.co.in

	JOINT SECRETARY, GOVT OF INDIA MINISTRY OF AGRICULTURE Mechanization and Technology Division Krishi Bhawan, NEW DEHLI 110001	Tel: +91.11.2338 9208 Fax: +91.11.2338 3040 E-mail: upma.srivastva@ gmail.com
IRELAND	TEAGASC OAK PARK RESEARCH CENTRE CARLOW	Tel: +353 503 70200 Fax: +353 503 42423 E-mail: dforristal@oakpark. teagasc.ie
ITALY	For the MINISTERO DELLE POLITICHE AGRICOLE, Direzione Generale delle Sviluppo Rurale, Infrastructure e Servizi : ENAMA (ENTE NAZIONALE MECCANIZZAZIONE AGRICOLA) Via Venafro, 5, I - 00159 ROMA	Tel: +39.06.40.86.00.30 Fax: +39.06.40.76.264 E-mail: info@enama.it, trattori.ocse@enama.it
	UNIVERSITÀ DEGLI STUDI DI BOLOGNA DEIAgra Dipartimento di Economia e Ingegneria Agraria Via Gandolfi, 19 I-40057 Cadriano, BOLOGNA	Tel: +39.051.76.66.32 Fax: +39.051.75.53.18 E-mail: valda.rondelli@ unibo.it
	UNIVERSITÀ DEGLI STUDI DI MILANO D.I.A. Dipartimento di Ingegneria Agraria Via G. Celoria, 2 I-20133 MILANO	Tel: +39.02.50.3168.76 Fax: +39.02.50.31.68.45 E-mail: domenico.pessina@ unimi.it
	I.M.A.M.O.T.E.R. Istituto per la Meccanizzazione Agricola E Movimento Terra Strada delle Cacce, 73 I-10135 TORINO	Tel: +39.011.397.72.25 Fax: +39.011.348.92.18 E-mail: e.cavallo@ imamoter.cnr.it
	CREA-ING Laboratorio di ricerca di Treviglio Via Milano, 43 I-24047 TREVIGLIO / BG	Tel: +39.03.634.96.03 Fax: +39.03.634.96.03 E-mail: ing.bg@crea.gov.it
JAPAN	I.A.M - B.R.A.I.N INSTITUTE OF AGRICULTURAL MACHINERY / BIO-ORIENTED TECHNOLOGY RESEARCH ADVANCEMENT INSTITUTION 1-40-2 Nisshin-cho, Kita-ku Saitama-shi, Saitama-Ken 331-8537	Tel: +81.48.654.7102 Fax: +81.48.654.7135 E-mail: mtakaahashi@affrc. go.jp, s.tsukamoto@affrc. go.jp

KOREA (REPUBLIC OF)	NATIONAL INSTITUTE OF AGRICULTURAL ENGINEERING (NIAE) Machinery Utilization and Testing Division 249 Seodun-dong, Suwon-si, Gyeonggi-do, R.O.K 441-100 SUWON	Tel: +82.31.290.19.53 Fax: +82 31.290.19.60 Email: agrihj@rda.go.kr
	FOUNDATION OF AGRICULTURAL TECHNOLOGY COMMERCIALIZATION AND TRANSFER (FACT) Agr. Machinery Verification Team 211-2 Seodun-don Suwon-si, Gyeonggi-do KOREA (REPUBLIC OF) 441-857	Tel: +82.31.290.19.52 Fax: +82 31.290.19.65 Email: jeongsr@efact.or.kr
LUXEMBOURG	MINISTÈRE DE L'AGRICULTURE 3, rue de la Congrégation L-LUXEMBOURG	Tel: +352 247 82500 Fax: +352 46 40 27 E-mail: info@sip.etat.lu
NORWAY	AGRICULTURAL UNIVERSITY OF NORWAY DEPT. OF AGRICULTURAL ENGINEERING, P. O. Box 65, N-1432 ÅS	Tel: +47.6494.8692 Fax: +47 6494 8820
POLAND	ITP INSTITUTE OF TECHNOLOGY AND LIFE SCIENCE Falentu, Al. Hrabska, 05-090 Raszyn, Poland	Tel: +48.22.720.05.31 Fax: +48.22.628.37.63 E-mail: itep@itep.edu.pl, p.pasyniuk@itep.edu.pl
PORTUGAL	Ministério da Agricultura, do Desenvolvimento Rural e das Pescas DGADR/ Director-Geral da Dircção- Geral de Agricultura e Desenvolvimento Rural Tapada da Ajuda – Edifico 1 1349-018 LISBOA	Tel: +351.213.613.298 Fax: +351.213.613.222 E-mail: mfunenga@dagdr. pt
RUSSIAN FEDERATION	MINISTRY OF AGRICULTURE Association of Testers of agricultural machinery and technology 82, Shosseynaya str., pos. Ust-Kinelskiy Kinel, Samara oblast, 446442	Tel: +7 846 63 46 1 43 E-mail: vadim_pronin@ mail.ru

SERBIA	IPM - ASSOCIATION OF MANUFACTURERS OF IPM – Association of Manufacturers of Tractors and Agricultural Machinery in Serbia Makenzijeva str. 79/II 11000 Belgrade	Tel: +381 11 2 457 135 Fax: +381 11 2 458 844 Email: udruipm@bitsyu.net
	LABORATORY FOR POWER MACHINES AND TRACTORS Faculty of Agriculture University of Novi Sad Dositeja Obradovica Sq. 8, 21000 Novi Sad, Serbia	Tel: 381.21.48.53.301 Fax: 381.21.459.989 Email: savlaz@polj.ns.ac.yu
SLOVAK REPUBLIC	MINISTRY OF TRANSPORT, CONSTRUCTION AND REGIONAL DEVELOPMENT OF THE SLOVAK REPUBLIC Section Road Transport and Roads Department of Road Transport and Regulation of Road Transport Námestie Slobody č.6, P.O.BOX 100, 810 05 Bratislava	Tel.: 02/ 594 94 709 Fax: 02/ 524 92 720 Email: Dusan.Stofik@ mindop.sk
SPAIN	MINISTERIO DE MEDIO AMBIENTE Y MEDIO RURAL Y MARINO Dirección General de Recursos Agrícolas y Ganaderos Subdirección General de Medios de Producción C/ Alfonso XII, 62, E-28014 MADRID	Tel: + 34.91 347 66 06 Tel: + 34.91 34740 58 Fax: +34.91 347 40 87 E-mail: sgmpagri@mapa.es
	E.M.A. ESTACIÓN DE MECANICA AGRICOLA Carrera de Madrid-Toledo, km 6.8 E-28916 LEGANES (MADRID)	Tel: +34.91.341.90.14 Fax: +34.91.341.82.95 E-mail: vmontema@ magrama.es
SWEDEN	SMP Svensk Maskinprovningar AB Fyrisborgsgatan 3 S-754 50 UPPSALA	Tel: +46 18 56 15 00 Fax: +46 18 12 72 44
	STATENS MASKINPROVNINGAR Box 56 S-230 53 ALNARP	
	STATENS MASKINPROVNINGAR Box 5053 S-900 05 UMEA	
SWITZERLAND	F.A.T. Swiss Federal Research Station for Agricultural Economics and Engineering CH-8356 TÄNIKON / AADORF	Tel: +41.52.368.31.31 Fax: +41.52.365.11.90 Email: info@art.admin.ch

TURKEY	T.C. TARIM VE KÖYISLERI BAKANLIGI TARIM ALET VE MAKINALARI TEST MERKEZI MÜDÜRLÜGÜ *(Directorate Testing Center of Agricultural Equipment and Machinery)* P.K. 22 TR- 06 170 YENIMAHALLE / ANKARA	Tel: +90.312. 315.65.74 / 315.56.85 Fax: +90.312. 315.04.66 E-mail: info@tamtest.gov.tr
UNITED KINGDOM	VEHICLE CERTIFICATION AGENCY Midlands Centre Watling Street, CV10 0UA Nuneaton	Tel: +44.247.632.84.21 Fax: +44.247.632.92.76 E-mail: Derek.Lawlor@vca. gov.uk
UNITED STATES	A.E.M. ASSOCIATION OF EQUIPMENT MANUFACTURERS 6737 West Washington Street Suite 2400, Milwaukee, WI 53214-5647 USA	Tel: +1.414.298.4128 Fax: +1.414.272.1170 E-mail: mpankonin@aem. org
	NEBRASKA TRACTOR TEST LABORATORY Biological Systems Engineering Department 245 L.W. Chase Hall P.O. Box 830726 LINCOLN, NE 68583-07262 USA	Tel: +1.402.472.0956 Fax: +1.402.472.6338 E-mail: rhoy2@unl.edu

Chapter 3

Testing and Evaluation of Agricultural Tractors

It has become inevitable now to conform to the National and International standards of performance as well as to meet the customer needs of comfort, safety, durability and cost economics due to fast-growing competition and technological advancements in the manufacturing of quality tractors. Agricultural tractors are being increasingly used worldwide to mechanize farm operations. The tractor industry has developed as one of the major engineering industries in India since manufacture of tractors started in the year 1960. Tractor is the largest segment in the agricultural equipment sector with an annual sale of 600,000-700,000 units and more than 15 firms engaged in tractor manufacturing in India. This has necessitated testing and evaluation of their performance on a uniform and rationalized basis to carry out different operations, primarily for agricultural purpose. These include:

- Carrying, pulling or propelling agricultural implements and machinery and, where necessary, supply power to operate them with the tractor in motion or stationary.
- Transporting the produce and inputs by means of trailers.

3.1 Terminology

a) *Length:* The distance between the two vertical planes at right angles to the median plane of the tractor and touching its front and rear extremities. This is true when all parts of the tractors and in particular the components projecting at the front or rear are contained between these two planes. The removable hitch components at the front and rear are not included in the length.

b) *Width:* The distance between two vertical planes parallel to the median plane of the tractor, each plane touching the outermost point of the tractor on its respective side. This is true when all parts of the tractor, in particular all fixed components projecting laterally are contained between these two planes. The removable attachments like cage wheels etc. are not included in the width.

c) **Height:** The distance between the supporting surface and the horizontal plane touching the uppermost part of the tractor.

d) **Track width:** The distance between the median planes of wheels on the same axle measured at the point of ground contact.

e) **Wheel base:** Horizontal distance between front and rear wheels measured at the centre of their point of ground contact.

f) **Ground Clearance:** The distance between the firm horizontal supporting surface and the lowest point of the tractor.

g) **Position of centre of gravity:** The position of the centre of gravity is defined by:

 • Height above the supporting surface;

 • Distance to right or left of the median plane of the tractor;

 • Distance from the vertical plane passing through the line representing the track of the rear wheels, or sprockets.

h) **Turning space diameter:** The diameter of the smallest circle described by the outermost point of the tractor while executing its sharpest practical turn.

i) **Turning diameter:** The diameter of the circle described by the center of tyre contact with the ground of the outermost wheel of the tractor while executing its sharpest practical turn.

j) **Rated speed:** The engine speed specified by the manufacturer for continuous operation at full load.

k) **Engine Power:** The power measured at the flywheel or the crankshaft, when the engine is running at its rated speed.

l) **Power take-off power (PTO power):** The power measured at any shaft designed by the tractor manufacturer to be used for PTO work.

m) **Drawbar power:** The power available at the drawbar, sustainable over a distance of at least 20 meters.

n) **Maximum drawbar pull:** The mean maximum sustained pull which the tractor can maintain at the drawbar over a given distance, the pull being exerted horizontally and in the vertical plane containing the longitudinal axis of the tractor.

o) **Specific fuel consumption (SFC):** It is defined as the amount of fuel consumed per unit of power developed per hour. It is a clear indication of the efficiency with which the engine develops power from fuel. This parameter is widely used to compare the performance of different engines.

$$SFC(g/kW-h) = \frac{Fuel\ consumed\ (g/h)}{Horsepower\ developed}$$

p) **Specific energy:** Work per unit volume of fuel consumed.

q) **Three-Point linkage:** A combination of one upper link and two lower links each articulated to the tractor and the implement at their ends to connect the implement to the tractor.

3.2 Scope of Test

The tractor shall be tested as per BIS test code or OECD/country specific test code with the following objectives towards quality control:

i) To assess evaluative requirements applicable to qualify minimum performance characteristics (Standard) of the agricultural tractors.

ii) To assess tolerances of the values declared by the manufacturer and statutory requirements under the relevant act of the agricultural tractor.

iii) To assess criteria for determining variants and new model of tractors for the purpose of testing and certification.

iv) To assess criteria for providing Administrative Extension and Technical Extension to earlier tested tractor model.

3.2.1 Selection of sample for test

The tractor submitted for commercial test shall be taken at random from series production preferably by the representative of the testing institute or centre in consultation with the applicant and as per the requirement of the test code. The sample tractor shall be a production model in all respects, strictly conforming to the description and specification sheet submitted by the manufacturer/applicant. The test report shall include the mode of selection of tractor.

3.2.2 Manufacturer's instructions

Once the test has started, the sample tractor shall not be operated in a way that is not in accordance with the manufacturer's published instructions, which may be in the form of Operator Manual, Repair and Maintenance Manual or Service Manual etc.

3.2.3 Preparation for test

The sample tractor shall be well run in for the hours as prescribed by the applicant. The fuel tank, coolant and lubricant shall be filled to the specified level. Tyre inflation shall be in accordance to the pressure prescribed by the applicant in Operation Manual.

3.2.4 Specification sheet

The tractor manufacturer/applicant shall furnish specifications of the tractor before the start of testing. These shall be checked and verified by the testing centre/institute and included in the test report highlighting deviations, if any.

3.2.5 Fuels and lubricants

Fuels and lubricants shall be selected from the range of products commercially available in the country where the equipment is tested. It shall conform to the minimum standards approved by the tractor manufacturer. If the fuel or lubricant conforms to a national or international standard, it shall be mentioned in the report.

3.2.6 Auxiliary equipment

For all the tests, accessories such as air compressor may be disconnected, only if it is advisable for the operator to do so in normal practice of working, otherwise they should remain connected and operated at minimum load.

3.2.7 Measuring instruments

Measuring instruments shall be inspected and calibrated before use.

3.2.8 Adjustments of the engine during the tests

The engine shall be thoroughly adjusted conforming to the values given in the manual/specifications sheet before the first test. These adjustments shall not be changed throughout the test.

3.2.9 Operating conditions

No corrections shall be made to the test results for atmospheric conditions or other factors. Atmospheric pressure shall not be less than 96.6 kPa. If this is not possible because of conditions of altitude, a modified fuel injection pump setting may have to be used, details of which shall be included in the report. The pressure will be stated for each reading in the report. Stable operating conditions must have been attained at each load setting before beginning of test measurements.

3.2.10 Ballasting of tractor

Ballasting weights may be fitted to the tractor having pneumatic tyres. Liquid ballast in the tyres may also be used. The overall static weight on each tyre (including liquid ballast in the tyres and a 75 kg weight representing the driver), and the inflation pressure shall be within the limits as specified by the tyre manufacturer, except as specified for the 5 hour drawbar test. Inflation pressure shall be measured with the tyre pressure value in the lowest position.

3.2.11 Repairs during tests

All repairs made during the test together with comments shall be reported for any practical defects or shortcomings.

3.2.12 Fuel consumption

a) When consumption is measured by mass

To obtain hourly fuel consumption by volume and the work performance per unit volume of fuel, a conversion of unit of mass to unit volume shall be made using the density value at 15°C.

b) When consumption is measured by volume

The mass of fuel per unit of work shall be calculated using the density corresponding to the fuel temperature at which the measurements were made. This value shall then be used to obtain hourly consumption by volume and the work performed per unit volume of fuel, using the density value at 15°C for conversion from unit of mass to unit of volume.

3.2.13. Suspension of test
The test shall be suspended/stopped in any of the following case:

a) When any break down or abnormality occurs during the test and makes the normal execution of the subsequent test impossible.

b) When the applicant proposes to stop the test due to certain specific reasons.

3.2.14 Atmospheric conditions

a) Temperature
The ambient temperature for PTO performance test and engine test shall be 27±7°C. The PTO test under high ambient condition shall be conducted at 43 ±2°C.

b) Atmospheric pressure
Minimum 96.6 kPa is required during testing. If it is not possible, modified fuel pump setting may be used and reported.

Permissible measurement tolerances

Rotational speed (rpm)	: ±0.5%
Time (s)	: ±0.2
Distance (m or mm)	: ±0.5%
Force (N)	: ±1.0%
Mass (kg)	: ±0.5%
Atmospheric pressure (kPa)	: ±0.2
Tyre pressure (kPa or kg/cm^2)	: ±5.0%
Hydraulic system pressure (kPa)	: ±2.0%
Temperature of fuel (°C)	: ±2.0%
Wet & dry bulb thermometers (°C)	: ±0.5
Angle (°)	: +0.5
Fuel consumption - Drawbar test (kg)	: ±2.0%
Fuel consumption - PTO (kg)	: ±1.0%

3.2.16 Referred standards

a) OECD Standards Codes [178]

b) ISO 789 Part 1-13 (1982-2018) [136]

c) IS 5994: 19981 [75]

d) IS 9939: 1981 [106]

e) IS 9253: 2013 [96]

f) IS 14414: 1996 [62]

g) IS 13581: 1993 [61]

h) IS 13548: 1992 [60]

i) IS 12226: 1995 [54]

j) IS 12224: 1987 [53]

k) IS 12207: 2014 [52]

l) IS 12180: 1987 [51]

m) IS 12061: 1994 [50]

n) IS 12036: 1995 [49]

o) IS 11859: 2004 [48]

p) IS 11442: 1996 [45]

q) IS 10743: 1983 [42]

r) IS 10274: 1993 [40]

s) IS 10273: 1987 [39]

3.3 Basic Measurements

The basic measurements undertaken to evaluate the performance of an engine are:

3.3.1 Speed

A wide variety of speed measuring devices are available in the market. These range from a mechanical tachometer to digital and triggered electrical tachometers.

3.3.2 Fuel consumption

The fuel consumed by an engine can be measured by determining the volume of flow of the fuel in a given time interval and multiplying it by the specific gravity of the fuel which should be measured occasionally to obtain accurate information.

Another method is gravimetric method and in this method, the time to consume a given weight of the fuel is measured. These measurements can be made automatic by the addition of suitable devices.

Continuous flow meters like Flotron are also used to measure fuel consumption which give instantaneous readings. Such flow meters are very useful especially in testing of high horse power engines.

3.3.3 Smoke density

All the widely used smoke meters are carbon density (g/m^3) measuring devices. The meter readings are a function of carbon mass in a given volume of exhaust gas. A Bosch smoke meter is very popular and it has a calibrated chart for defining smoke density with *Bosch Number*. The basic principle in such a smoke meter is that a fixed quantity of exhaust gas is passed through a fixed filter paper and the density of the smoke stains on the paper is evaluated optically.

Von brand smoke meter can give continuous readings of smoke density. A filter tape is continuously running at a uniform rate to which the exhaust from the engine is fed. The smoke stains developed on the filter paper are sensed by a recording head. The signal obtained from the reading head is calibrated to give smoke density.

3.3.4 Power measurement

Power measurement is one of the most important parameter in the test schedule of engine. It involves the determination of torque and angular speed of the output shaft.

a) Observed power

The power obtained at the dynamometer without any correction for atmospheric temperature, pressure, or vapour pressure.

b) Corrected power

The power obtained by correcting observed power to standard conditions of sea level pressure (101.325 kPa), temperature (15.5°C) and zero vapour pressure.

3.4 Methods of Tests

3.4.1 Specifications checking

3.4.2 Fuels and lubricants

The recommendation of fuel, lubricants (engine oil, Air cleaner oil, Transmission, Differential, final drive, hydraulic system & steering housing oil and grease) by the manufacturer and actually used by testing institutes shall be reported.

3.4.3 PTO performance test

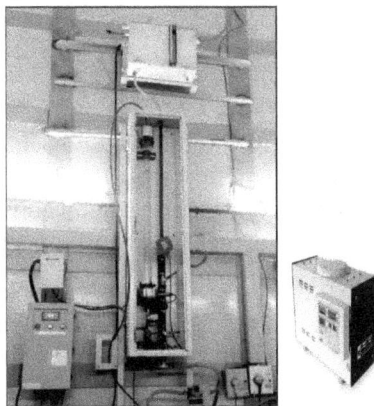

The test shall be carried out with the help of a dynamometer on tractor's main power-take-off (PTO) shaft as shown in Fig. 3.1(a) and 3.1(b) illustrates SE500 kW eddy current dynamometer [SAJ, Pune] on an adjustable test bed, adaptable to the required height of tractor PTO. The angle of the connection of the shaft connecting the PTO to the dynamometer shall not exceed 2". The specific fuel consumption is calculated by measuring the hourly fuel consumption using the gravimetric fuel measurement setup and fuel consumption meter as shown in Fig. 3.2. Also refer section 3.3.2.

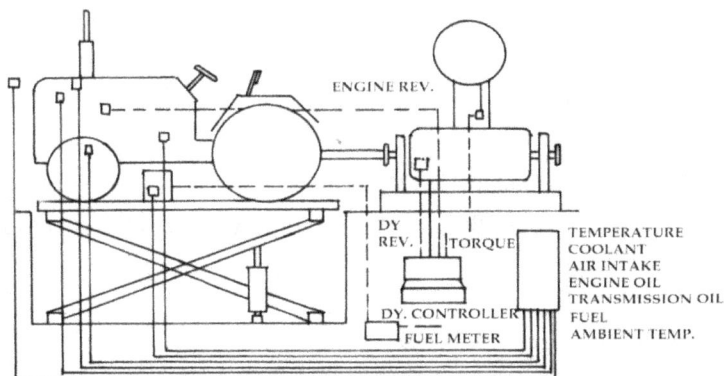

Fig. 3.1(a): Typical testbed for PTO performance test

Courtesy: SAJ, India

Fig. 3.1 (b): PTO performance test   **Fig. 3.2:** Fuel consumption meter

The torque and power values in the test report shall be obtained from the dynamometer without correction for losses in power transmission between the power-take-off and the dynamometer. The shaft connecting the power-take-off to the dynamometer shall not have any appreciable angularity at the universal joints and in any case not more than two degree.

3.4.3.1 Under normal ambient temperature (27 ± 7°C) condition

i) Test at varying speed

ii) Two-hour test at maximum power

iii) Power at rated engine speed

iv) Power at standard PTO speed

v) Test at varying load but at rated engine speed

    a) Torque corresponding to maximum power available at rated engine speed

    b) 85% of the torque obtained in (a)

    c) 75% of the torque obtained in (b)

    d) 50% of the torque obtained in (b)

    e) 25% of the torque obtained in (b)

    f) Unloaded

vi) Varying load test at standard PTO speed

    a) Torque corresponding to maximum power available at standard PTO speed

    b) 85% of the torque obtained in (a)

    c) 75% of the torque obtained in (b)

    d) 50% of the torque obtained in (b)

    e) 25% of the torque obtained in (b)

    f) Unloaded

3.4.3.2 Two-hour test at maximum power under high ambient temperature condition (43 ± 2°C)

The test given under section 3.4.3.1 (i), (ii), and (iii) shall be conducted under high ambient temperature of 43±2°C. The power, torque and fuel consumption shall be reported. The lubricating oil consumption on mass basis during maximum power test shall be measured and reported.

3.4.3.3 Belt or Pulley shaft Tests (Optional)

At the manufacturer's request, the power available at the belt or pulley shaft of tractors, if fitted, may be measured. If the rated engine speed does not correspond to a standard belt speed, measure the performance of the engine at the speed corresponding to the standard belt speed of 15.75±0.25 m/s.

The test report shall include presentation of the following curves made for the full range of engine speed tested.

a) Power and equivalent crankshaft torque as function of speed – Fig. 3.3

b) Specific fuel consumption and fuel consumption as a function of speed – Fig. 3.4

c) Specific fuel consumption as a function of power – Fig. 3.5

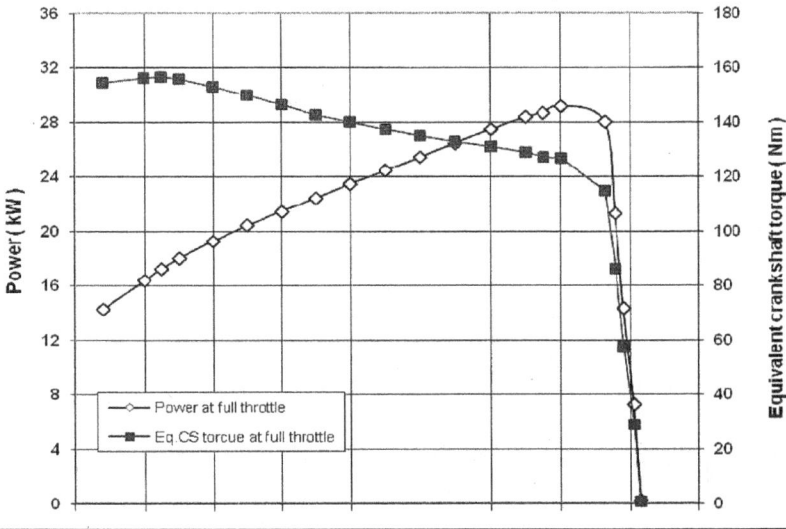

Fig. 3.3: Engine speed (rpm)

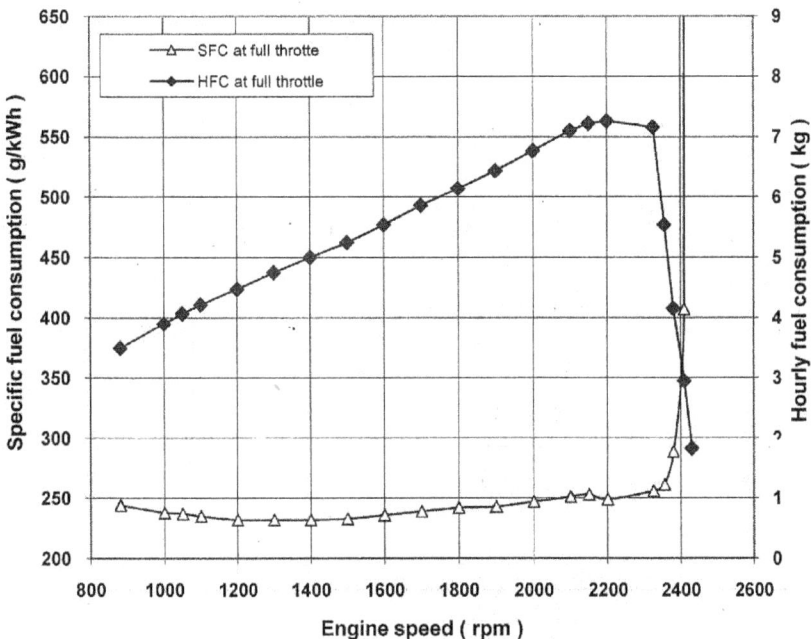

Fig. 3.4: Engine speed (rpm)

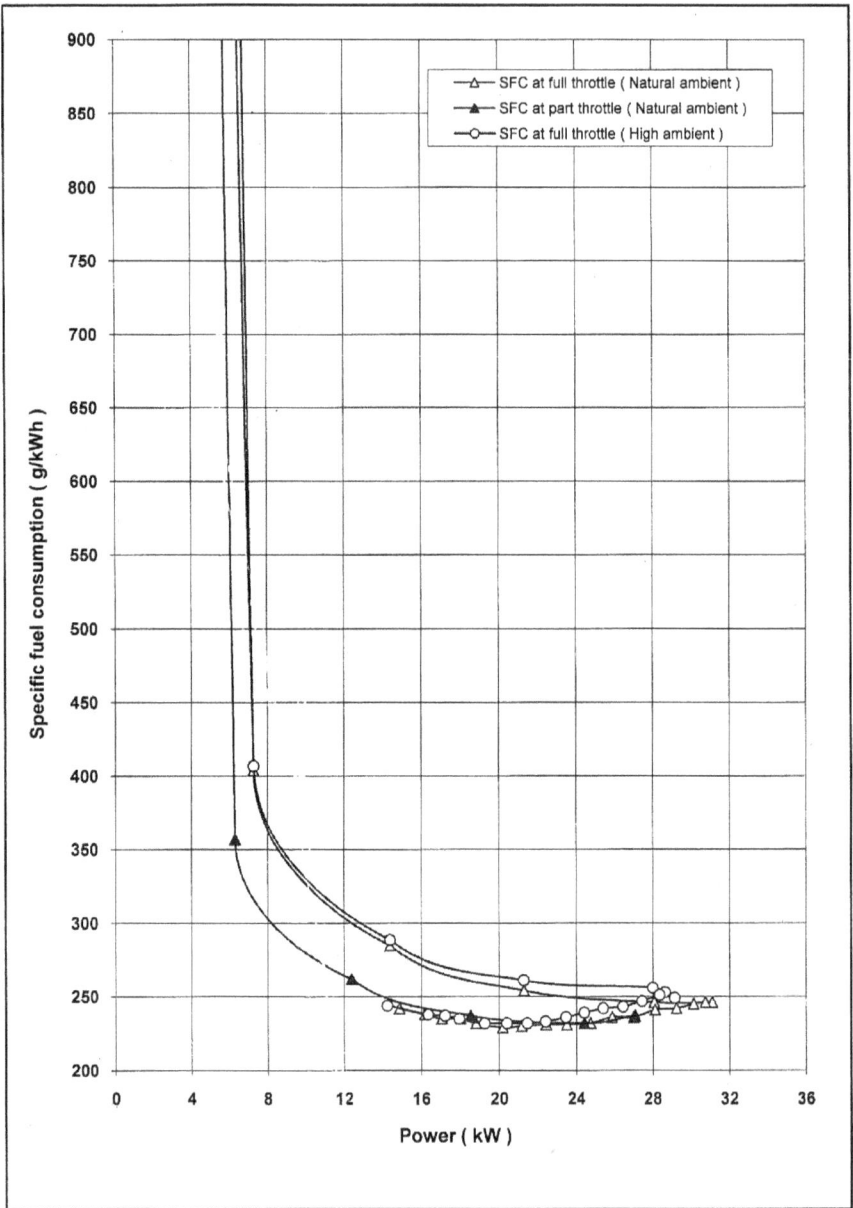

Fig. 3.5: Typical curve showing specific fuel consumption vs power

3.4.3.4 Drawbar performance test

This test is conducted with the loading car on a standard concrete test track as shown in Fig. 3.6. The following tests shall be conducted:

a) Maximum power test (tractor un-ballasted)

b) Maximum power test (tractor ballasted)

c) Five hours test at 75% of pull obtained at maximum power (ballasted wheel tractor)

d) Five hour test at pull corresponding to 15% wheel slip (ballasted wheel tractor)

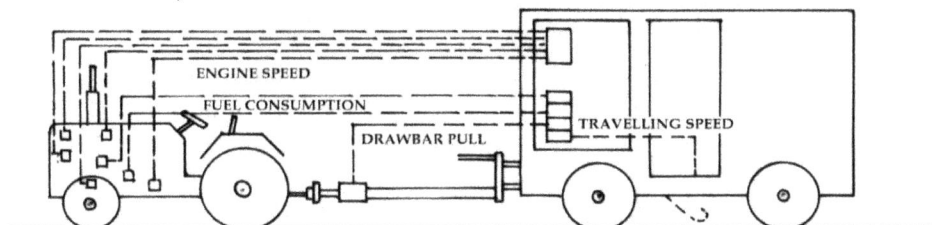

Fig. 3.6: Drawbar Performance Test

The test report shall include presentation of following curves for each gear tested:

a) Drawbar power and Wheel slip as a function of drawbar pull - Fig. 3.7

b) Forward speed as a function of drawbar pull - Fig. 3.8

c) Specific fuel consumption as a function of drawbar pull - Fig. 3.9.

3.4.3.5 Power lift and Hydraulic pump performance test

3.4.3.6 Hydraulic power test

The governor control lever shall be set for maximum power at rated engine speed. At start of each test, the temperature of the hydraulic fluid in the tank shall be 65±5°C. If this cannot be achieved owning to the presence of an oil cooler, the temperature measured during the test shall be stated in the test report. A pressure gauge shall be fitted immediately next to the external tapping of the tractor. To ensure the lifting capacity of the hydraulic lift to be adequate for effective practical use and also to allow for variation in the performance, the measured maximum performance should be reported, at 90% of the pressure relief valve setting of hydraulic fluid pressure. The schematic diagram of hydraulic pump inline tester is shown in Fig. 3.10.

$$\text{Hydraulic power (PS)} = \frac{\text{Pump delivery rate (lpm) ñ pressure (kgf/cm}^2)}{450}$$

3.4.3.7 Hydraulic lift test

The unballasted tractor shall be secured in the horizontal position in such a way that the tyres are not deflected by the reactive force of the power lift. The lifting force shall be measured at the hitch points and on a frame attached to the three-point linkage. The frame shall have a mast height appropriate to the linkage category of the tractor.

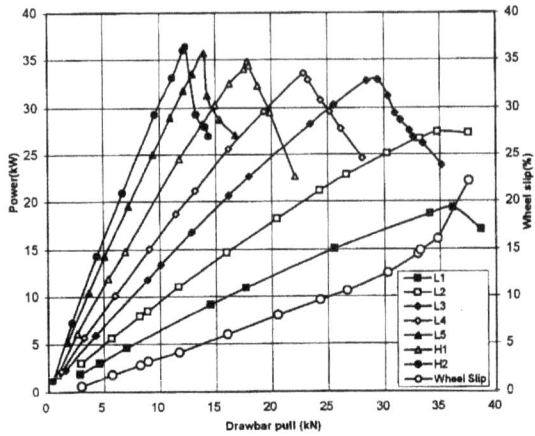

Fig. 3.7: Drawbar power and Wheel slip as a function of drawbar pull

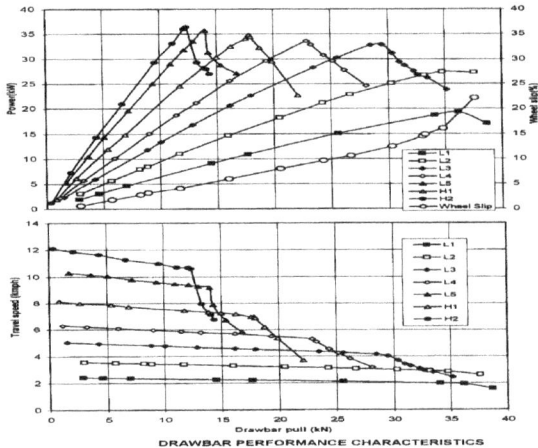

Fig. 3.8: Forward speed as a function of drawbar pull

source: CFMTTI, Budni Tractor Test Report

3.4.3.8 Maintenance of lift load test

A vertical force to the frame equal to 90% of the maximum force exerted throughout the full range of movement shall be applied at its centre of gravity. When the hydraulic lift is at the uppermost position and the control lever in the raise position, the engine shall be stopped and following measurements shall be made and reported.

a) The force applied to the frame.

b) The decrease in the vertical height at 5 minutes interval over a period of 30 minutes.

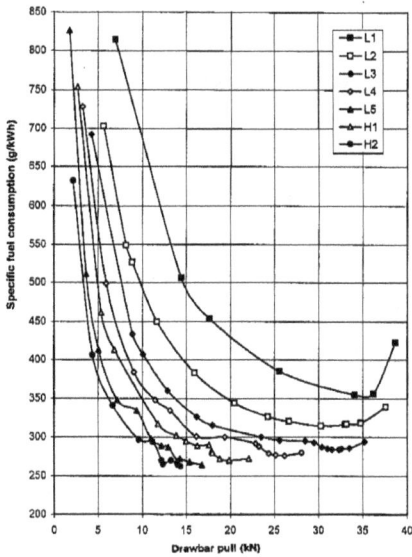

Fig. 3.9: Specific fuel consumption as a function of drawbar pull

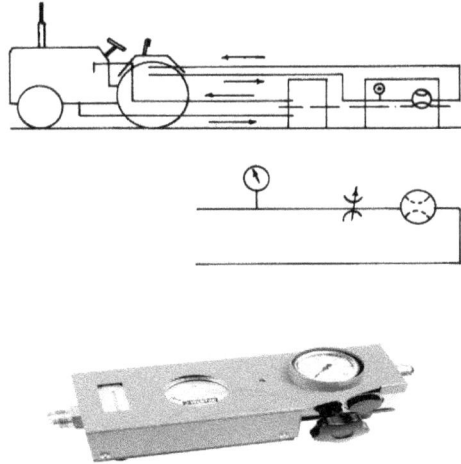

Fig. 3.10: Hydraulic inline tester

3.4.4 Brake test (with standard specified ballast) on standard concrete test track.

3.4.4.1 Service Brake

The test to be conducted at both maximum attainable speed and at 25 km/h travel speed to find out the force applied (N), mean deceleration (ms⁻²) and stopping distance (m).

a) Used brake test

b) Brake fade test

3.4.4.2 Parking brake test

Parking brake test shall be conducted on 18% slope and 12% slope with facing up & facing down to test the efficacy of parking brake.

3.4.5 Noise measurement

The test is conducted on standard concrete test track, at drawbar pull at which the tractor develops the maximum noise level in all gears corresponding to the nominal travelling speed nearest to 7.5 km/h [Fig. 3.11(a) and (b)].

a) In all gears at travelling speed before acceleration.

b) Noise at operator's ear level.

3.4.6 Mechanical vibration test

This test shall be conducted by parking the tractor on standard concrete test track at no load and at load corresponding to 85% of maximum PTO power at the following point:

a)	Foot rest	i)	Gear shifting lever
b)	Steering wheel	j)	Accelerator Lever
c)	Seat	k)	Brake pedal
d)	Mudguard	l)	Clutch pedal
e)	Head light	m)	Main hydraulic control lever
f)	Battery base (centre)	n)	PTO engaging lever
g)	Tail light	o)	Differential lock lever
h)	Plough light	p)	Any other

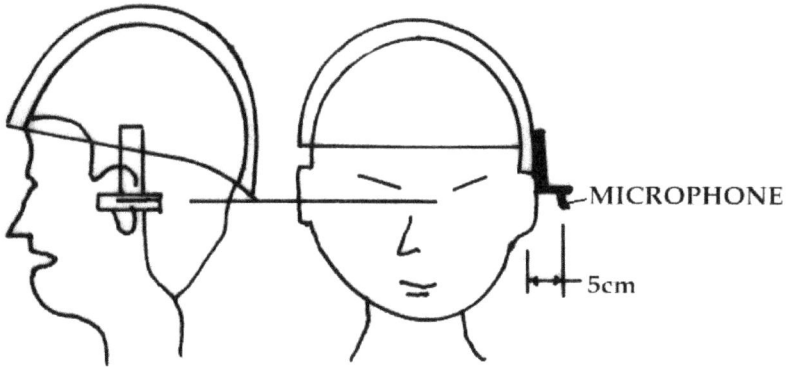

Fig. 3.11 (a): Arrangement for noise measurement at operator's ear level

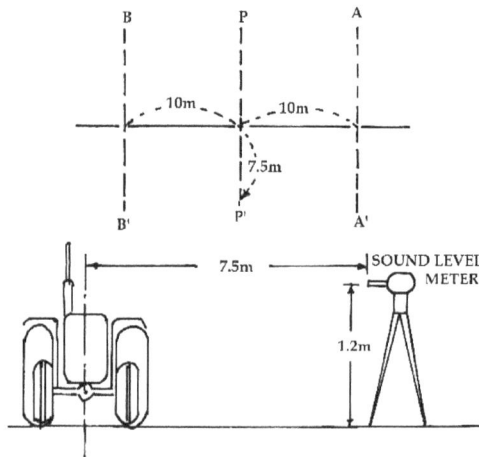

Fig. 3.11 (b): Noise measurement emitted by tractor

3.4.7 Position of centre of gravity

The test will be conducted with standard ballast condition of tractor, but with all the liquid reservoirs full and the operator is replaced by 75 kg mass of the seat.

3.4.8 Air - cleaner oil pull-over test

The tractor shall be placed on a level ground. The air-cleaner shall be cleaned and filled with oil of viscosity recommended by the manufacturer, up to the marked level. The engine shall then be operated at full governor speed for 15 minutes. This shall be followed by sudden acceleration and deceleration made after every 30 seconds for a period of 15 minutes. The air-cleaner assembly shall be weighed before and after the test. The loss of mass of oil, shall be calculated and reported.

If there is no oil pull over with the tractor in the level position the following additional testes shall be carried out.

In case of wheeled tractor, the tests shall be repeated with a tilt angle of 15° to both side and then 15° forward and backward in relation to the direction of travel of the tractor. In case of crawler tractor, the forward and backward tilt angle shall be 30°.

3.4.9 Turning ability test

The tractor shall be tested either on a test track as the drawbar test or on a horizontal and compact area. The track setting shall be the one commonly used for general purpose and be stated. The tractor shall be unballasted and moving slowly at a speed of 1.5 to 2 km/h. The test shall be conducted by turning the tractor to right and left, with and without using the steering brakes. After each turn (right or left), the following shall be measured and recorded.

  a) Minimum turning diameter

  b) Minimum clearance diameter or turning space

3.4.10 Visibility test

The unballasted tractor shall be parked on a level and horizontal surface. An artificial light, i.e. an electric bulb will be placed at a height of 720 mm from the centre of driver seat referred as an operator's vision and 75 kg weight will be placed on driver's seat referred as weight of operator. The line separating the visible area from the non-visible area be marked, measured and a graph may be drawn as shown in Fig. 3.12 (a) and (b).

For ensuring safety of operator w.r.t. driver's field of vision, a standard (AIS-107) for measurement of visibility parameters, was formulated with Tractor Manufacturers Association (TMA) / AISC / CMVR-TSC. This standard applies to the 180-degree forward field of vision of the drivers of agricultural tractors [20].

3.4.11 Field test

The field test with commonly used implements namely disc plough, mould board plough, cultivator and disc harrow shall be conducted to assess practical performance of the test tractor. A test duration of 50 h with each of these implements is recommended. In addition, puddling test of 50 h may also be conducted, if the tractor has been declared suitable for wet land operations. Various performance parameters like rate of work, fuel consumption, quality of work, etc. shall be evaluated based upon observations recorded during such testes.

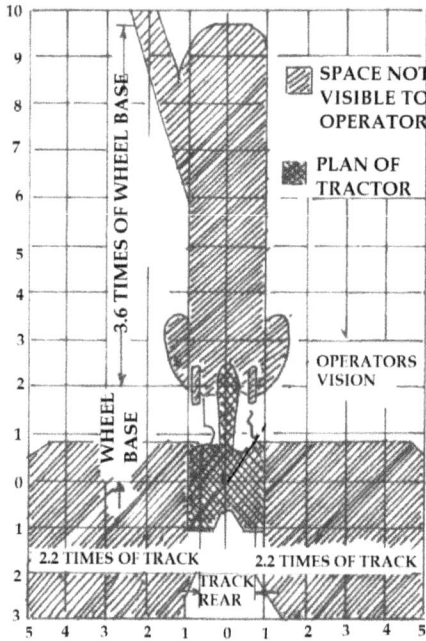

Fig. 3.12 (a): Visibility test

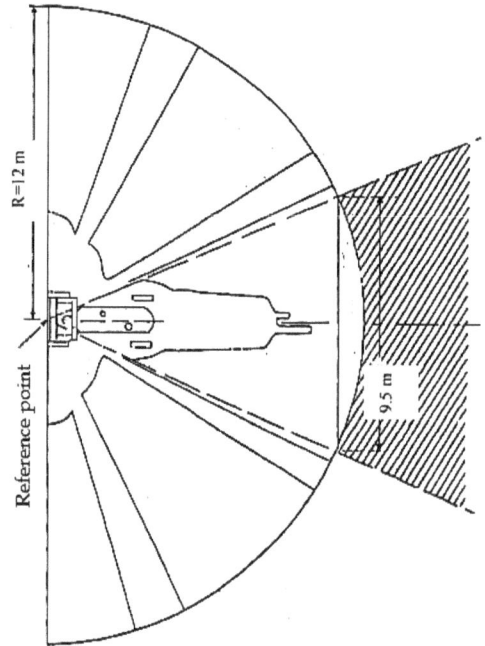

Fig. 3.12 (b): Semi circle of vision [20]

3.4.12 Haulage test

The haulage test shall be carried out with two-wheel and/or four-wheel trailers at the gross trailer loads recommended by the manufacturer, on level tar macadam track and having gradients not exceeding 8 percent at certain short lengths. The tractor shall be ballasted in accordance with the recommendations of the manufacturer / applicant.

The test consisting of equal test run in both the directions shall be conducted under identical conditions. The total distance covered in both the directions of test run shall be minimum of 40 km. Minimum two trials each with two-wheel and four-wheel trailers shall be conducted and range of the average test observations shall be reported. In case, the tractor is not capable of four-wheel trailer, it shall be tested with two-wheel trailer only. The gears selected shall be appropriate for the satisfactory and safe operation. The speed can be reduced by reducing the engine speed as and when necessary while conducting the test.

3.4.13 Water proof test

The test shall be carried out on the tractor suitable for wet land cultivation. The objective is to ascertain the dust and water proofing, mainly of wheel and brakes. The tractor shall be operated at a speed of about 6 km/h on the test bed for a period of 5 hours (Fig. 3.13). The water level in the bed shall be adjusted to the centre of front wheel. After the test, axle and brakes shall be dismantled and inspected for any entry of mud or water.

3.4.14 Assessment of power drop and wear test

a) Power drop test

After conducting all laboratory and field tests, duration of which shall not be less than 300 hours, a two-hour maximum power test shall be carried out to see any drop in power. While conducting this test, no special adjustment or change of part shall be made except normal adjustment by the manufacturer.

b) Wear test

The rate of wear shall be calculated based on the initial average value when the engine is new and the maximum permissible wear indicated by the tractor manufacturer in the instruction manual.

i)   *Cylinder bore:* The cylinder bore shall be measured on the thrust side and perpendicular to it at the top, middle and at the bottom position of the liner. The wear (%) shall be calculated and reported.

ii) *Piston diameter:* The piston diameter shall be measured on the thrust side and perpendicular to it at the top above the gudgeon pin and at the skirt. The wear (%) shall be calculated and reported.

iii) *Ring end gap:* The ring end gap for all compression and oil rings shall be measured at the top, middle and at the bottom position of the liner. The wear (%) shall be calculated and reported.

iv) *Ring groove clearance:* The ring groove clearance shall be measured with piston and oil ring. The wear (%) shall be calculated and reported.

v)   *Clearance of main and big end bearing:* The radial and axial clearance of main and big end bearings shall be measured. The radial clearance shall be measured after tightening the crack shaft bolts with the torque specified by the manufacturer. The wear (%) shall be calculated and reported.

vi) *Valves, guides and tappets:* The valve shall be inspected for carbon deposition on the stem overheating sign of pitting of the seats if any. The timing gear cover shall be opened and backlash between each pair of the meshing gears shall be measured calculated and reported.

vii) *Clutch:* The clutch shall be opened and inspected for wear of the lining and pressure plate, condition of the clutch-release bearing, pilot bearings, springs and fingers. The clutch housing shall be inspected for the entry of dust, water etc.

viii) *Gear box:* The top cover of the gear box shall be opened and inspected for visual wear and damage to the gear teeth. The backlash between each pair of the meshing gear shall be measured. The wear percent of backlash shall be calculated and reported. If the tractor has been used for puddling test, the entry of water and mud inside the gear box housing shall also be checked.

ix) *Brakes:* The brake housing shall be opened and inspected. The wear of brake lining shall be determined by measuring the thickness of lining and wear (%) of the lining shall be calculated and reported. if the tractor has been used for puddling test the entry of water and mud in the housing, shall also be inspected and reported

x) *Front axle:* The king pin and stub axle shall be dismantled and inspected for wear of king pin and bushes. The condition of thrust bearings, bearings and seals for stub axle and king pin shall also be examined for entry of dust, water, etc. For track type tractors, wear of sprocket, pin, grouser plate, idler, etc., shall be measured and reported.

xi) *Dynamo and starter:* These shall be dismantled and inspected for entry of water and dust. The condition of bearings shall also be examined and reported.

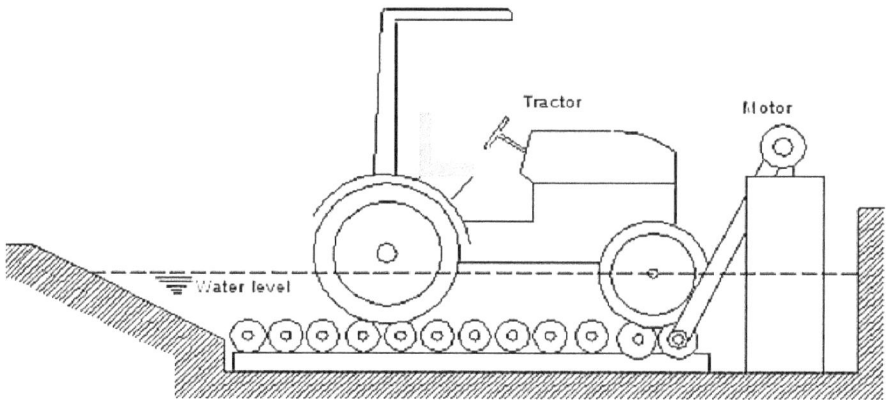

Fig. 3.13: Water proof test

3.5 Interpretation of Tractor Test Results

PTO performance: measured in the laboratory with the PTO of the tractor coupled to a dynamometer.

Maximum power: It is the maximum power, measured during two hours maximum power test at natural ambient temperature of 27±7 °C, available at the tractor PTO with the governor control fully open.

Power at standard PTO speed: It is the maximum power available for PTO work at the standard PTO speed of 540/1000 rpm at natural ambient temperature of 27±7 °C.

Torque back up ratio: It gives an indication of how much the engine torque increases to maximum (ratio of maximum torque and torque at maximum power) and hence the ability of the machine to keep working as the engine speed pulls down due to increased load requirements over and above the maximum power.

Specific fuel consumption (SFC) at maximum power: It is the mass of fuel consumed per unit work measured during two hours maximum power test (at natural ambient temperature of 27±7 °C) indicates the fuel efficiency at that power level.

Drawbar performance: It is measured on a specially constructed concrete test track under both unballasted and ballasted conditions with the governor control fully open.

Maximum pull: The maximum pull available at the drawbar at 15% wheel slip of the tractor or refers to pull corresponding to rated engine speed in case of tractors, where 15% wheel slip level is not reached because of their high weight/power ratio.

Hydraulic performance: It is measured in the laboratory on a special designed test rig.

Maximum hydraulic power: It is the power available to drive the hydraulic ram of the tractor to lift load or to drive external hydraulic rams on implements and machines with the governor control fully open.

Flow rate at maximum power: It indicates the speed at which the hydraulic ram of the tractor or implement and machines can be operated.

Lift capacity at standard frame: determines the force that can be applied by ram on an implement or a machine.

Brake performance: It is measured on the standard concrete test track.

Maximum stopping distance: The stopping distance obtained at an applied force of 600 N (61 kg approximate) on the brake pedals and the tractor initially running at its maximum achievable speed.

Force for deceleration of 2.5m/s: to check the effort to produce the braking performance to be within 600 N.

Parking performance: the effectiveness of the parking braking device (at force not exceeding 600 N at brake pedal and 400 N at hand brake respectively) to hold the tractor stationary when facing up or down under unballasted conditions at 12% gradient with trailer having gross mass equal to the mass of tractor of 3 tonnes whichever is minimum and at 18% gradient under ballasted conditions.

Noise level: It is measured in the open on a concrete test track, both at the bystander's position and at the operator's ear level position and should not exceed 90 dB(A) as specified by ILO (International Labour Organization).

Mechanical vibration: It is the amplitude of average mechanical vibration of assemblies and components measured on the tractor, which are functionally important with the engine running at its rated speed both on load and unloaded conditions and should not exceed 100 microns.

Field performance: It is carried out in the field for a duration of atleast 50 hours with each implement. It should be noted that the performance and fuel consumption varied over the operating range of the engine should be noted and not a single value be taken as representative for all conditions.

Performance parameter for evaluation of tractor and Nebraska tractor test report has been shown in annexure 3.1 and 3.2 respectively.

Illustrative Example for Calculation of Wear:

Manufacturer prescribes the piston ring gap in the initial setting as 0.3 to 0.5 mm and recommends that the piston rings should be changed when the gap reaches 1.5 mm. The measured value after the test was found to be 1.2 mm. Calculate wear percentage?

Solution:

$$Average\ initial\ wear = \frac{(0.3+0.5)}{2} = 0.4mm$$

$$Observed\ wear = \frac{(1.2-0.4)}{(1.5-0.4)} \times 100 = 72.2\%$$

Illustrative Example for Calculation of Tractor Performance Parameters:

A tractor was tested on a firm surface and the following data was obtained.

Rear wheel weight, W_r = 1463 kg	Distance, no-load, m_o = 47.8 m
Engine power, Q_e = 41.74 kW	Time, t = 22.6 s
Drawbar pull, P = 17.9 kN	Distance, load, m = 40.2 m
Fuel consumed, F = 74 g	

Determine the wheel slip, travel speed, drawbar power, tractive efficiency, fuel consumption and specific fuel consumption.

Solution:

$$Wheelsip,\ i = \frac{(m_o - m)}{m} \times 100 = \frac{(47.8-40.2)}{47.8} \times 100 = 15.9\%$$

$$travel\ speed,\ V = \frac{m}{t} = \frac{40.2}{22.6} = 1.78\,m/s = 6.40km/h$$

$$Drawbar\ power,\ Qd = PV = 17.9 \times 1.78 = 31.86\ kW$$

$$Assuming\ transmission\ efficiency\ \eta_r = 0.9$$

$$Wheel\ power,\ Q_w = 0.9 \times 41.74 = 37.57\ kW$$

$$tractive\ efficiency, = \frac{Q_d}{Q_w} = \frac{31.86}{37.57} \times 100 = 84.8\%$$

$$Fuel\ consumption\ rate,\ FC = \frac{F}{t} = \frac{74}{22.6} = 3.27g/s = 11.79kg/h$$

$$Specific\ fuel\ consumption,\ SFC \wp \frac{FC}{Q_d} \quad \frac{1179}{31.86} \quad 370\,g/kWh$$

Annexure 3.1

Evaluative (mandatory)/ Non-evaluative (Non-mandatory) parameters applicable for qualifying minimum performance criteria as per IS Standard for acceptance of the tractor

Sr. No.	Characteristics	Category (Evaluative/Non Evaluative)	Requirements as per Standard	Values declared by the applicant(D)/ Requirement (I)	As Observed	Whether meets the requirements (Yes/No)
1	2	3	4	5	6	7
1.1 PTO Performance						
a	Maximum Power under 2h test, (kW) Natural ambient condition)		PTO power >26kW: -5/+10% PTO power ≤26kW: -7.5/+10% Engine power >26kW: -5/+10% Engine power ≤26kW: -7.5/+10%			
b	Power at related engine speed, (kW)					
c	Specific fuel consumption corresponding to max. power (g/kWh)		+5%			
d	Maximum equivalent crankshaft torque, (Nm)		±8%			
e	Back-up torque, %		10%, min			
f	Maximum Operating Temperature (°C)					
	1) Engine Oil					
	2) Coolant		Should not exceed boiling temp.			
g	Engine Oil consumption, (g/kWh)		Not exceeding 1% of SFC at max. power under HAC			
h	Smoke level		Max. LAC 3.25/m or eq. BOSCH No. 5.2 or 75 Halridge value (as per CMVR)			
1.2 Drawbar Performance						
a	Max. drawbar pull with ballast corresponding to 15% wheel slip, (kN)		Min. 65% of static mass with ballast			
b	Max. drawbar pull without ballast corresponding to 15% wheel slip, (kN)		Min. 65% of static mass without ballast			
c	Max. drawbar power without ballast, (kW)		Min 80% and 75% of PTO power for tractors having static mass greater than and less than 1500kg resp.			
d	Max. transmission oil temperature (°C)		Should not exceed boiling temp.			

1.3 Power lift and hydraulic pump performance

a	Maximum lifting capacity throughout the range of lift, (kN):	
	1) At hitch point	-10%
	2) With standard frame	Atleast 24kg/PTO kW
b	Maximum drop in the height of the point of application of the force after each 5 minutes interval for a total duration of 30 min (mm)	Should not exceed 50mm

1.4 Brake performance at 26 kmph

a	Maximum stopping distance at a force, equal to or less than 600N on brake pedal with road ballast, (m)	
	1) Cold brake	10
	2) Hot brake	10
b	Maximum force exerted on the brake pedal to achieve a deceleration of 2.m/s^2 (N)	600
c	Whether parking brake if effective at a force of 600N at foor pedal(s) or 400N at hand lever	Yes / No

1.5 Noise measurement

a	Maximum ambient noise emitted by the tractor dB(A)	As per CMVR
b	Maximum noise at operator's ear level	

1.6 Amplitude of mechanical vibrations

1)	Left foot rest	
	Right foot rest	100 microns (max.)
2) Seat (with driver seated)		
3) Steering wheel		

1.7 Air cleaner oil pull-over test

Maximum percentage of oil pull-over	0.25% (max.)

1.8 Haulage requirements (for both two wheel and four wheel trailers)

a	Gross mass of trailers (tones)	--
b	Distance travelled/litre of fuel consumption (km/l)	--
c	Fuel consumption (ml/km/tone)	--

1.9 Wetland cultivation: Tractor is not recommended for wetland cultivation

Sealing for Clutch assembly, Brake housings, Front axle hubs, Engine oil, Transmission oil	As per IS:11082

1.10 Safety features

a	Guards against moving and hot parts	As per CMVR
b	Lighting arrangements	
c	Seating requirement (Tractors having more than 1150 mm rear track width)	As per IS:4931
d	Technical requirements for PTO shaft	As per IS:4931
e	Dimension of three point linkage	As per IS:4468(1)
f	Specification of linkage and swinging drawbars	As per IS: 12953 & 12362(3)

1.11 Labelling of tractors (Provision of labeling plate)

1) Make

2) Model

3) Year of Manufacture

4) Engine Serial Number

5) Chassis Serial Number

6) Declaration of PTO power, kW

7) SFC, g/kwh

Should conform to CMVR and along with declared value of PTO hp

1.12 Discard limit

a Cylinder bore diameter, (mm)

b Clearance between piston & cylinder liner at skirt, (mm)

c Ring and gap (mm):

1st comp. ring.

2nd comp. ring.

Oil ring

d Ring groove clearance (mm)

1st comp. ring.

2nd comp. ring.

Oil ring

e Clearance of main bearings (mm)

Diametrical clearance

Crankshaft endpoint

f Clearance of big end bearings, (mm)

Diametrical

Axial

g Clearance between king pin and bush, (mm)

h Clearance between center pin and bush, (mm)

To be specified by the manufacturer

1.13 Literature (Submission to test agency)

a   Operator manual

b   Parts catalogue

c   Workshop/ Service manual

1.14 Category of breakdowns/ defects

a   Critical

b   Major   Not one Max. 2 (not repetitive) Max. 5

c   Minor   (freq. of each not more than 2) Max. 5

d   Total breakdowns (major+minor)

1.15 Optional requirements

a   Fitment of ROPS   As per IS: 11821

b   Accessories

Annexure 3.2
Nebraska Tractor Test Report (for reference)

NEBRASKA TRACTOR TEST 1872
JOHN DEERE 5303 DIESEL
9 SPEED

POWER TAKE-OFF PERFORMANCE

Power HP (kW)	Crank shaft speed rpm	Gal/hr (l/h)	lb/hp.hr (kg/kW.h)	Hp.hr/gal (kW.h/l)	Mean Atmospheric Conditions
MAXIMUM POWER AND FUEL CONSUMPTION					
Rated Engine Speed—(PTO speed—545 rpm)					
55.98 (41.74)	2400	3.57 (13.50)	0.445 (0.271)	15.69 (3.09)	
Maximum Power (1 hour)					
56.42 (42.07)	2151	3.42 (12.95)	0.424 (0.258)	16.49 (3.25)	
VARYING POWER AND FUEL CONSUMPTION					
55.98 (41.74)	2400	3.57 (13.50)	0.445 (0.271)	15.69 (3.09)	Air temperature
49.56 (36.96)	2507	3.35 (12.67)	0.472 (0.287)	14.81 (2.92)	77°F (25°C)
38.03 (28.36)	2549	2.86 (10.83)	0.526 (0.320)	13.29 (2.62)	Relative humidity
25.66 (19.14)	2578	2.23 (8.45)	0.608 (0.370)	11.50 (2.27)	15%
12.96 (9.66)	2611	1.43 (5.42)	0.772 (0.469)	9.06 (1.78)	Barometer
0.57 (0.42)	2628	0.96 (3.63)	11.789 (7.171)	0.59 (0.12)	28.71"Hg (98.22 kPa)

Maximum Torque 164 lb.-ft. *(223 Nm)* at 1402 rpm
Maximum Torque Rise - 34.1%
Torque rise at 1900 rpm - 21%

TRACTOR SOUND LEVEL WITHOUT CAB	dB(A)
At no load in 5th(B2) gear	92.0
Transport speed - no load - 9th(C3) gear	93.3
Bystander in 9th (C3) gear	82.6

TIRES AND WEIGHT

	Tested without ballast
Rear Tires–No., size, ply & psi*(kPa)*	Two 16.9-28; 8; 12 *(85)*
Front Tires–No., size, ply & psi*(kPa)*	Two 7.50-16; 6; 28 *(195)*
Height of Drawbar	17.5 in *(445 mm)*
Static Weight with operator– Rear	3225 lb *(1463 kg)*
– Front	1800 lb *(816 kg)*
– Total	5025 lb *(2279 kg)*

Location of tests: Nebraska Tractor Test Laboratory, University of Nebraska, Lincoln Nebraska 68583-0832

Dates of tests: March 15 - 30, 2006

Manufacturer: John Deere Commercial Products Inc., 700 Horizon South Parkway, Grovetown Ga. USA, 30813

FUEL, OIL and TIME: Fuel No. 2 Diesel **Specific gravity converted to 60°/60° F** *(15°/15°C)* 0.8395 **Fuel weight** 6.990 lbs/gal *(0.838 kg/l)* **Oil SAE** 15W40 **API service classification** CG-4 **Transmission and hydraulic lubricant** John Deere Hy-Gard Fluid **Total time engine was operated** 9.0 hours

ENGINE: Make John Deere Diesel **Type** three cylinder vertical with turbocharger **Serial No.** *PY3029T105719* **Crankshaft** lengthwise **Rated engine speed** 2400 **Bore and stroke** 4.19" x 4.33" *(106.4 mm x 110.0 mm)* **Compression ratio** 17.8 to 1 **Displacement** 179 cu in *(2934 ml)* **Starting system** 12 volt **Lubrication** pressure **Air cleaner** one paper element and one polyester felt element **Oil filter** one full flow cartridge **Fuel filter** one paper element **Muffler** underhood **Exhaust** vertical **Cooling medium temperature control** one thermostat

ENGINE OPERATING PARAMETERS: Fuel rate: 23.5 - 25.9 lb/h *(10.7 - 11.7 kg/h)* **High idle:** 2575 - 2650 rpm **Turbo boost:** nominal 10.9 - 13.8 psi *(75 - 95 kPa)* as measured 12.6 psi *(87 kPa)*

CHASSIS: Type standard **Serial No.** *PY5303U005246* **Tread width** rear 55.7" *(1415 mm)* to 71.5" *(1815 mm)* front 56.3" *(1430 mm)* to 80.7" *(2050 mm)* **Wheelbase** 80.3" *(2040 mm)* **Hydraulic control system** direct engine drive **Transmission** selective gear fixed ratio **Nominal travel speeds mph** *(km/h)* first 1.39 *(2.23)* second 2.01 *(3.23)* third 3.02 *(4.85)* fourth 3.89 *(6.26)* fifth 5.62 *(9.05)* sixth 8.46 *(13.62)* seventh 9.01 *(14.50)* eighth 13.03 *(20.98)* ninth 19.62 *(31.57)* reverse 2.33 *(3.75)*, 6.55 *(10.54)*, 15.13 *(24.35)* **Clutch** single dry disc operated by foot pedal **Brakes** single wet disc mechanically operated by two foot pedals which can be locked together **Steering** hydrostatic **Power take-off** 540 rpm at 2376 engine rpm **Unladen tractor mass** 4900 lb *(2223 kg)*

THREE POINT HITCH PERFORMANCE (OECD Static Test)

CATEGORY: II
Quick Attach: None

Maximum force exerted through whole range:	3591 lbs	*(16.0 kN)*
i) Opening pressure of relief valve:	NA	
Sustained pressure of the open relief valve:	2796 psi	*(193 bar)*
ii) Pump delivery rate at minimum pressure and rated engine speed:	12.4 GPM	*(46.9 l/min)*
iii) Pump delivery rate at maximum hydraulic power:	11.4 GPM	*(43.2 l/min)*
Delivery pressure:	2444 psi	*(169 bar)*
Power:	16.3 HP	*(12.1 kW)*

THREE POINT HITCH PERFORMANCE

Observed maximum pressure psi. *(bar)*	2796 *(193)*
Location:	remote outlet
Hydraulic oil temperature: °F *(°C)*	185 *(85)*
Location:	hydraulic sump
Category:	II
Quick attach:	none

SAE Static Test—System pressure 2480 psi *(171 Bar)*

Hitch point distance to ground level in. *(mm)*	8.0 *(203)*	15.0 *(381)*	22.0 *(559)*	29.0 *(737)*	36.0 *(914)*
Lift force on frame lb	6633	5486	5067	4734	3978
" " " " " " *(kN)*	*(29.5)*	*(24.4)*	*(22.5)*	*(21.1)*	*(17.7)*

REPAIRS AND ADJUSTMENTS: No repairs or adjustments.

REMARKS: All test results were determined from observed data obtained in accordance with official OECD, SAE and Nebraska test procedures. For the maximum power tests, the fuel temperature at the injection pump inlet was maintained at 125°F *(52°C)*.

We, the undersigned, certify that this is a true and correct report of official Tractor Test No. **1872,** May, 22, 2006.

Leonard L. Bashford
Director

    M.F. Kocher
    V.I. Adamchuk
    J.A. Smith
    Board of Tractor Test Engineers

	SAE Test		OECD Test	
	inch	*mm*	inch	*mm*
A	23.3	*590*	23.5	*597*
B	11.0	*280*	11.0	*280*
C	13.7	*347*	13.7	*347*
D	11.8	*300*	11.8	*300*
E	13.2	*335*	13.2	*335*
F	6.9	*175*	6.9	*175*
G	26.4	*670*	26.4	*670*
H	0.4	*10*	0.4	*10*
I	15.7	*397*	15.7	*397*
J	19.5	*495*	19.5	*495*
K	16.1	*410*	16.1	*410*
L	38.6	*980*	38.6	*980*
M	21.7	*550*	21.7	*550*
N	32.6	*830*	32.6	*830*
O	8.0	*203*	8.0	*203*
P	38.6	*980*	43.5	*1105*
Q	32.5	*825*	32.5	*825*
R	21.2	*540*	21.2	*540*

HITCH DIMENSIONS AS TESTED - NO LOAD

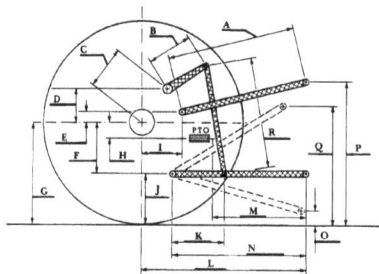

John Deere 5303 Diesel

Institute of Agriculture and Natural Resources
University of Nebraska–Lincoln

Chapter 4

Testing and Evaluation of Power Tillers

Mechanization of farming has made remarkable progress in India and other developing countries and as a result, the productivity has increased manifold. Power tiller is a small, composite and multipurpose agricultural machine by which almost all operations in agriculture can be done as in the case of a tractor. Power tiller normally comes in the range of 8 to 12 hp and is suitable for both wet and dry land operations. Earlier, power tillers were being manufactured as only walk behind type, but ridding type (Fig.4.1) are also being manufactured by mounting a seat on it. There are different types of power tillers: a) Track-Laying Power Tiller - a power tiller propelled by tracks passing around the drive sprocket, b) Wheeled Power Tiller - a power tiller propelled by wheels fitted with pneumatic tyres. In the wheeled power tiller, it can be single wheel power tiller and two wheel power tiller.

4.1 Terminology

Engine Power: Power measured at the crankshaft of the engine with the governor control lever in the position recommended by the manufacturer.

Operational Mass: The mass of the power tiller without operator in normal working condition with fuel tank and radiator (if fitted) full and lubricants filled to the specified levels.

Ground Clearance of Power Tiller: The height of the lowest point of the power tiller chassis from a firm horizontal supporting surface.

Height of Power Tiller: The distance between a firm horizontal supporting surface and the horizontal plane touching the uppermost part of the power tiller with the engine in horizontal position.

Tyre Rolling Radius: The effective radius corresponding to the average distance travelled by the power tiller in one rotation of the driving wheels (i.e. distance divided by 2). When the power tiller is driven without drawbar, load at a speed of approximately 2 km/h.

Courtesy: VST Tillers Tractors, India
Fig. 4.1: Riding type power tiller

4.2 Scope of Test

4.2.1 Laboratory test

i)   Specification checking

ii) Engine performance test

iii) Rotary shaft performance test

iv) Drawbar performance test

v)   Brake performance test

vi) Noise level test

vii) Air cleaner oil pull over test

viii) Mechanical vibration test

ix) Turning ability test

4.2.2 Field performance test

i)   Rotavator in dry land

ii) Puddling test in wet land

iii) Haulage test

4.2.3 Hardness test of blade

The blades shall be heat treated, quenched and tempered. The hardness in edge portion shall be 56±3 HRC and in shank portion shall be 37 to 45 HRC.

4.2.4 Chemical composition of blade

Typical material used for manufacturing of blades are carbon steel, silico manganese steel [T 70 Mn 65, T 75, T 80 Mn 65 and 55 Si 2 Mn 90].

4.2.5 Wear & tear test of components

Blade is an important soil-engaging component of the power tiller. It wears out earlier than other components causing its replacement very often. Therefore wear of blade shall be calculated in percentage.

4.2.6 Measuring Tolerances

Refer to Section 3.2.15

4.2.7 Referred Standards

The testing of power tiller will be done as per following test codes.

a) IS: 6690 - 1981 [83]

b) IS: 9980 - 2004 [107]

c) IS: 13539 - 2008 [59]

d) SNI: 0738 - 2014 [176]

e) IS: 9935 - 2012 [105]

f) IS: 9939 - 1981 [106]

g) GB/T: 6229 - 2007 [224]

h) ANTAM: 2015 [19]

4.3 Test Methods

4.3.1 Laboratory test

a) Specification checking

The specification of power tiller will be checked and compared with the specifications as supplied by the manufacturer and reported.

b) Fuel and lubricants

The fuel and lubricants as recommended by the manufacturer and as actually used by the testing institute shall be reported in the test report.

c) Engine performance test

The angle of connection of the shaft connecting the crankshaft to the dynamometer shall not have any appreciable angularity and any case not more than 2°. The governor control shall be set for maximum power. The engine test shall be conducted under normal ambient (27±7°C) and under high ambient (43±2°C) conditions:

I. Under normal ambient temperature (27±7°C) condition:

i) Varying speed test
- Maximum power and fuel consumption
- Power at rated engine speed
- Maximum torque test

ii) Two hours maximum power test

iii) Part load test
   a) Torque corresponding to maximum power available at rated engine speed
   b) 85% of torque obtained in (a)
   c) 75% of torque defined in (b)

d) 50% of torque defined in (b)

e) 25% of torque defined in (b)

f) Unloaded

II. Under high ambient temperature (43±2°C) condition:

i) Varying speed test

- Maximum power and fuel consumption
- Power at rated speed
- Maximum torque test

ii) Two hours maximum power test

The engine performance characteristics shall be reported in tabular form and in graphical form. The typical curves (Fig. 4.2 to 4.4) of the following shall be given in the report:

i) Engine power and crankshaft torque as a function of engine speed

ii) Specific fuel consumption and hourly fuel consumption as a function of engine speed

iii) Specific fuel consumption as a function of engine power

The curves (i) and (ii) shall be reported separately for tests under normal ambient conditions and high ambient conditions.

d) Rotary shaft performance test

The following test shall be conducted with a suitable dynamometer:

i) Maximum power test

ii) Rated power test

iii) Maximum torque test

iv) Five hour rating test

- Four hours at 90% of torque corresponding to maximum power
- Fifth hour at maximum power

The rotary shaft performance characteristics shall be reported in tabular form and in graphical form. The typical curves (Fig. 4.4 to 4.7) of the following shall be given in the report:

i) Rotary shaft power and shaft torque as a function of engine speed

ii) Specific fuel consumption and hourly fuel consumption as a function of engine speed

iii) Specific fuel consumption as a function of rotary shaft power

e) Drawbar performance test

The power tiller shall be fitted with pneumatic wheels and the test shall be conducted on a clean, horizontal and dry concrete test track with minimum number of joints. During the test, the line of pull shall be maintained horizontal. The height of the drawbar shall remain fixed in relation to the power tiller. The test shall be conducted for at least 20 m continuously without varying atmospheric or track conditions significantly. If the manufacturer recommends ballasting of power tiller, the test shall be conducted both at ballasted and unballasted conditions and the results shall be reported separately.

   i)   Maximum power test

   ii) Ten hour test at 75% of pull obtained at maximum power.

Results of the drawbar performance characteristics, a) drawbar power and wheel slip (Fig. 4.8), b) forward speed (Fig. 4.9), and c) specific fuel consumption, as a function of drawbar pull, shall be reported. Datasheets has been given in annexure 4.1 and 4.2.

f) Brake performance test

Test will be conducted on concrete standard test track.

   i)   Cold brake

      Un-ballasted power tiller with trailer of same mass of power tiller or one tonne whichever is less.

   ii) Hot brake

      Un-ballasted power tiller with trailer of same mass of power tiller or one tonne whichever is less

   iii) Parking brake test: this test will be carried out at:

      • 18% slope up & down with trailer of same mass of power tiller or one tonne whichever is less.

      • 12% slope up & down with trailer of same mass of power tiller or one tonne whichever is less.

g) Noise level test

   i)   Maximum noise level at bystander's position at rated speed (stationary conditions)

   ii) Noise at operator ear level

h) Air cleaner oil pull over test

Air cleaner oil pullover test will be carried out at each slope for a period of 15 minutes followed by sudden acceleration and deceleration made after every 30 second for a period of 15 minutes.

   i)   Level ground

   ii) 15° laterally on RHS

   iii) 15° laterally on LHS

   iv) 15° longitudinally with front-up

   v) 15° longitudinally with rear end up

i) Mechanical vibration test

The test shall be conducted by parking the power tiller on standard concrete test track with no load and at load corresponding to 85% or maximum PTO power at all important components / assemblies.

j) Turning ability test

The test area shall be a horizontal, compacted or paved surface having good tyre adhesion and capable of displaying legible marking. The power tiller shall be tested with all liquid reservoirs filled to the specified level but without ballast, mounted implements and any other specified components. At the beginning of the test, the height of the tyre tread bars shall not be less than 65% of their height when new. The inflation pressure in the tyres shall be maintained as recommended for the road work by the manufacturer. The test shall be carried out, using minimum travel speed, on the power tiller by turning it to the right and the left side by the use of steering clutch till a 360 degree turn is completed (Fig. 4.10). During the test, diameter of the minimum turning circle, and diameter of minimum turning space shall be recorded.

4.3.2 Field performance test

a) Dry land operation

The land selected for test shall preferably be agricultural land on which a crop was harvested within the last one year. The field operation should be made, as far as possible, in test plots where furrow length of minimum 50 m would be available. Field operations with each implement shall be carried out for a duration of at least 25 hours. The total duration of field operations with different implements during initial commercial and batch testing of the power tiller, shall be 150 and 75 h, respectively.

    i)   Ploughing with mould board plough/disc plough,

    ii) Tilling with harrow/cultivator,

    iii) Tilling with rotary tiller, and

    iv) Any other (to be specified by the manufacturer)

b) Wet land operation

In case of puddling test, the furrow length may be a minimum of 25 m. The field should have an average initial water depth of at least 100 mm. Operational speed should be selected to obtain proper tilth taking into account the manufacturer's/applicant's recommendations.

c) Stationary operations, if recommended by the manufacturer

The power tiller shall be tested for running a centrifugal pump and sprayer by the testing station on the request of the applicant.

d) Haulage test

The power tiller shall be tested for haulage with a trailer, on a level road/track with a gradient not exceeding 10%, of the capacity recommended by the manufacturer. The total distance covered in both the directions of one test run shall be up to 50 km. During the initial commercial testing, three test runs and during batch testing, only one test run would be covered. The appropriate gears may be selected for satisfactory and safe operation. Also, the speed can be reduced by reducing the engine speed as and when necessary while conducting the test.

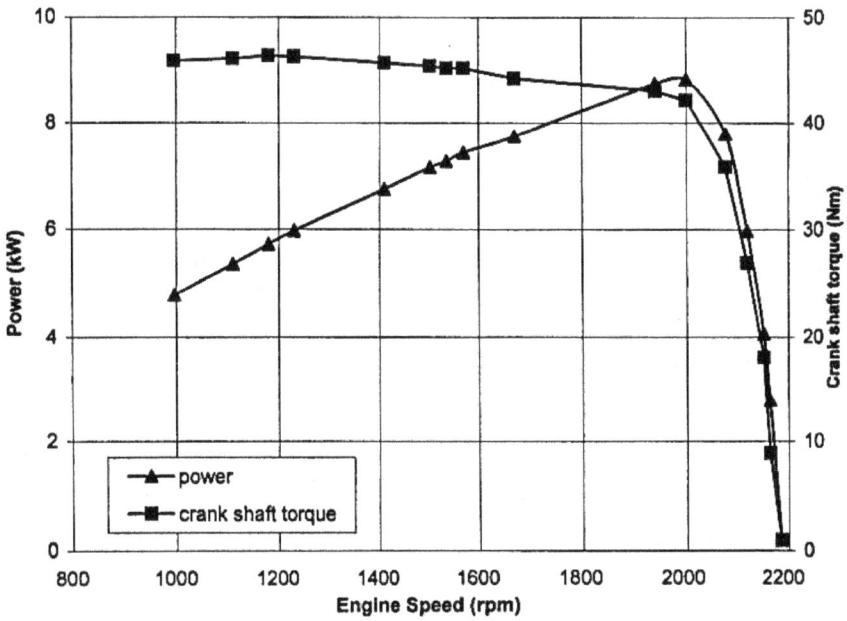

Fig. 4.2: Typical curve showing engine power and crankshaft torque

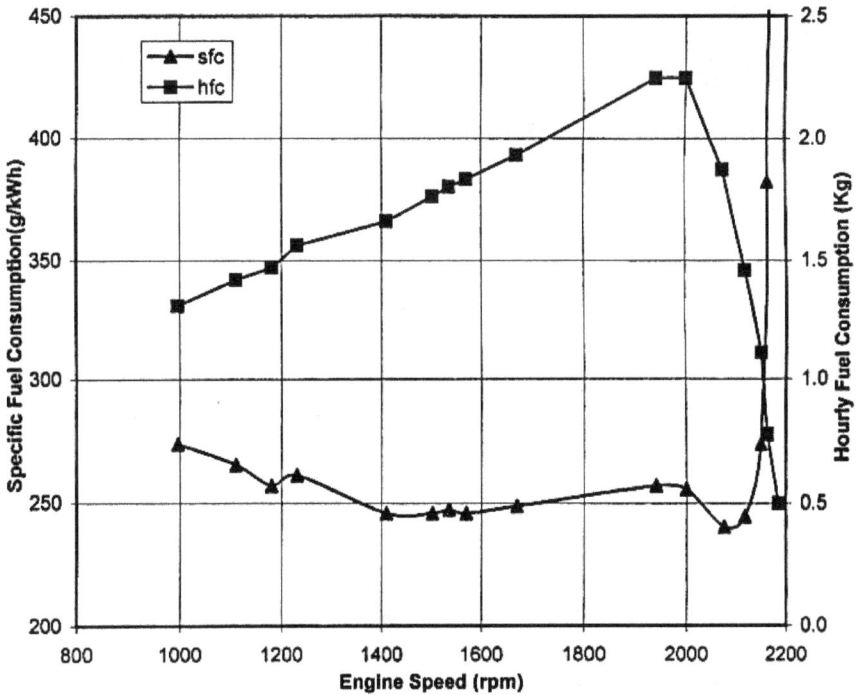

Fig. 4.3: Typical curve showing specific fuel consumption (SFC) and hourly fuel consumption (HFC)

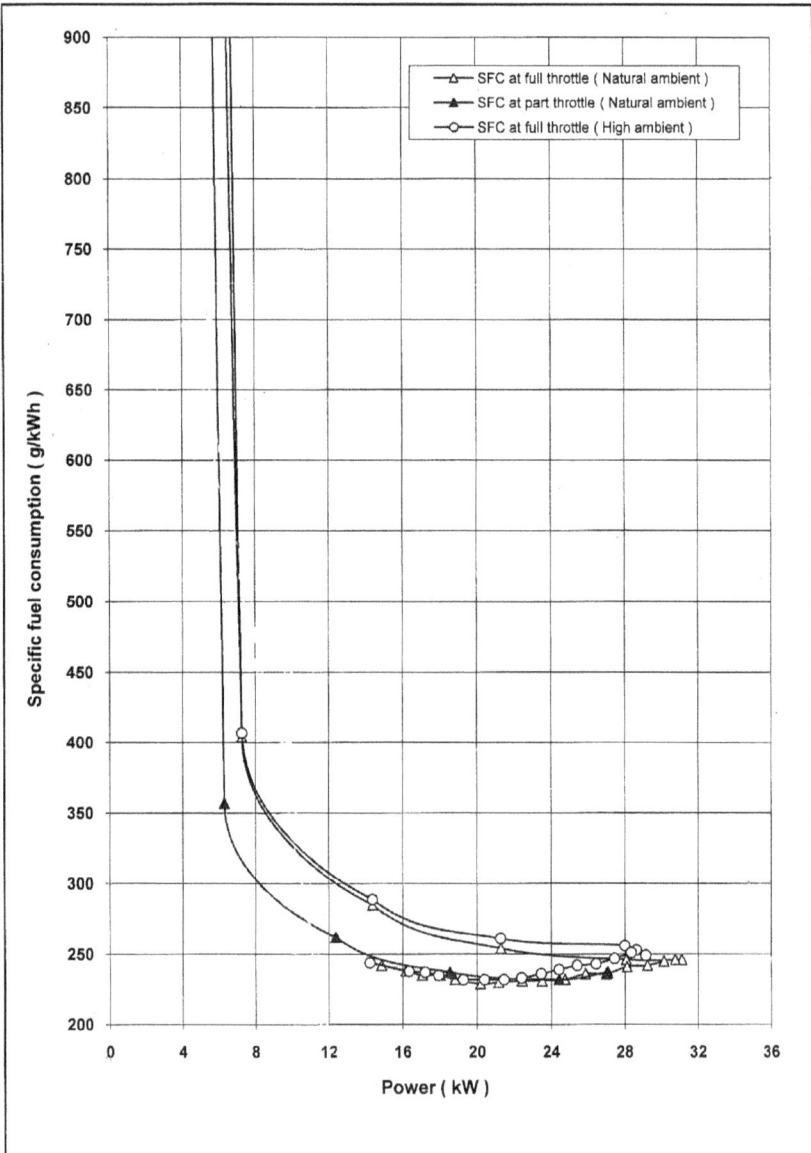

Fig. 4.4: Typical curve showing specific fuel consumption vs power

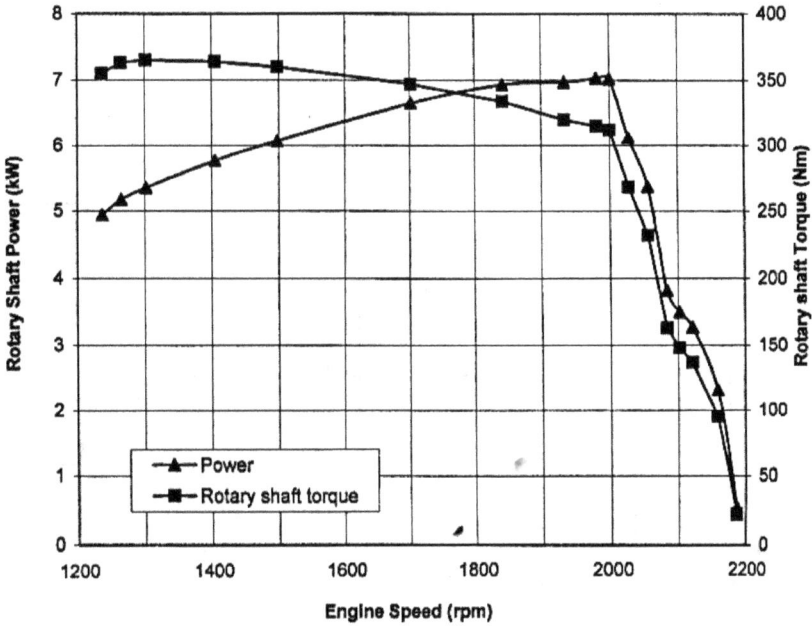

Fig. 4.5: Typical curve showing rotary shaft power and torque

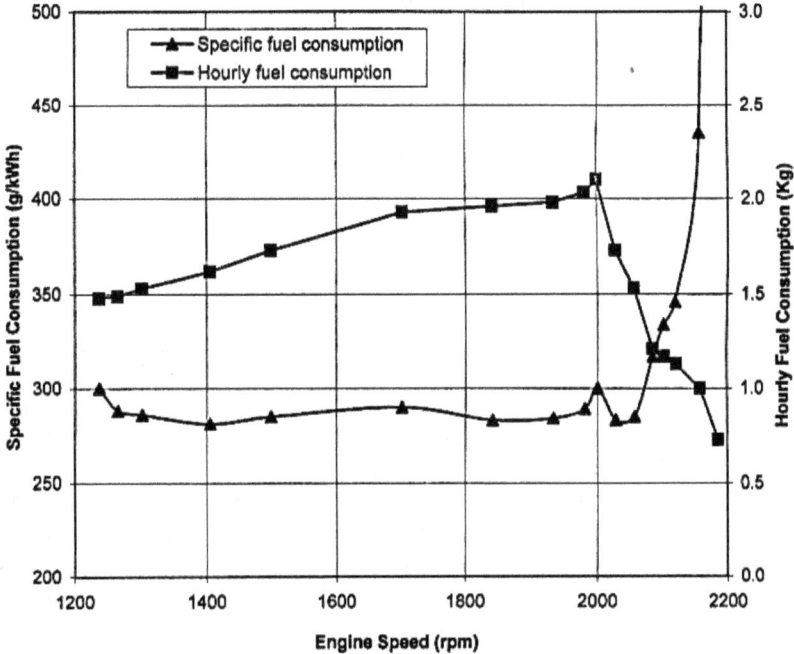

Fig. 4.6: Typical curve showing specific fuel consumption (SFC) and hourly fuel consumption (HFC)

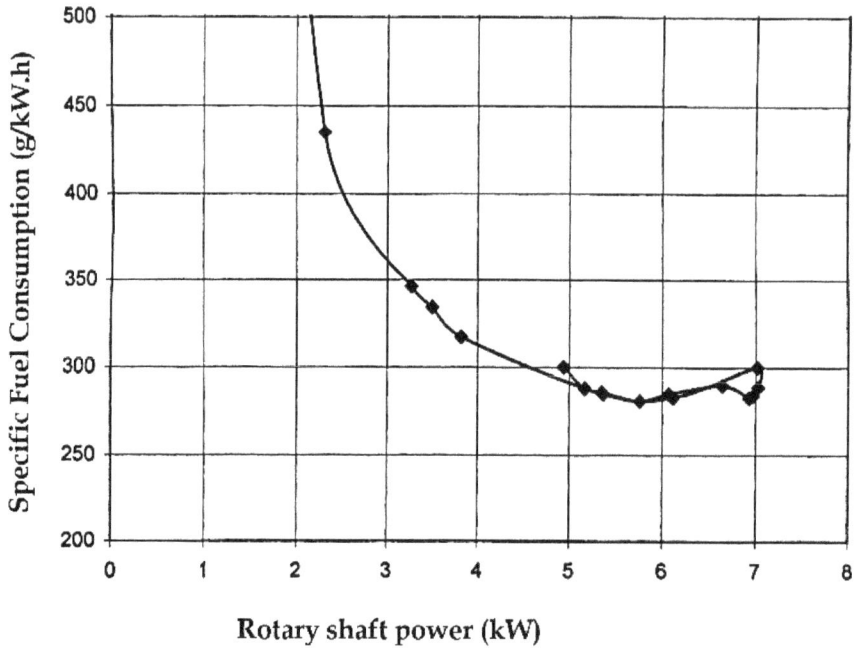

Fig. 4.7: Typical curve showing specific fuel consumption

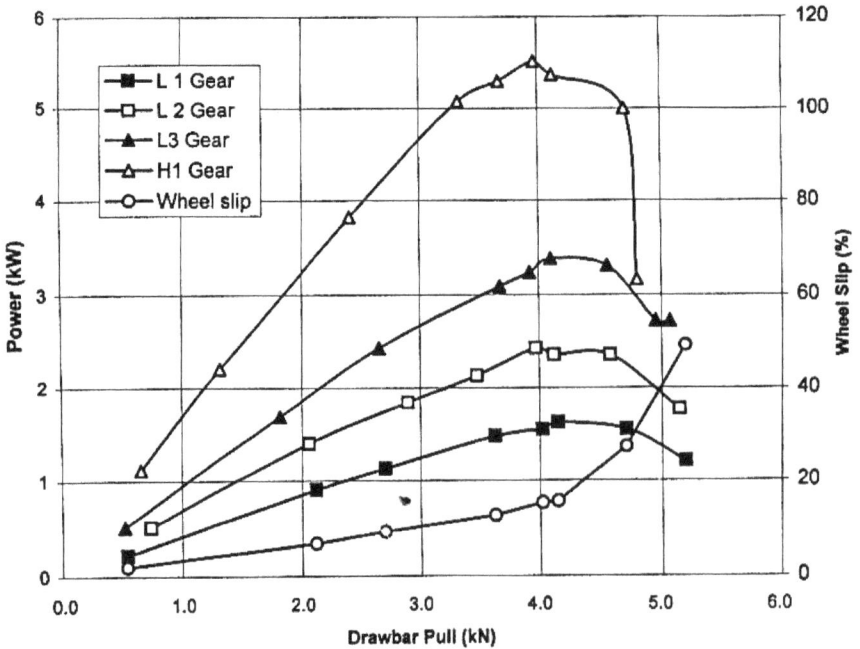

Fig. 4.8: Typical curve showing power and wheel slip as a function of drawbar pull

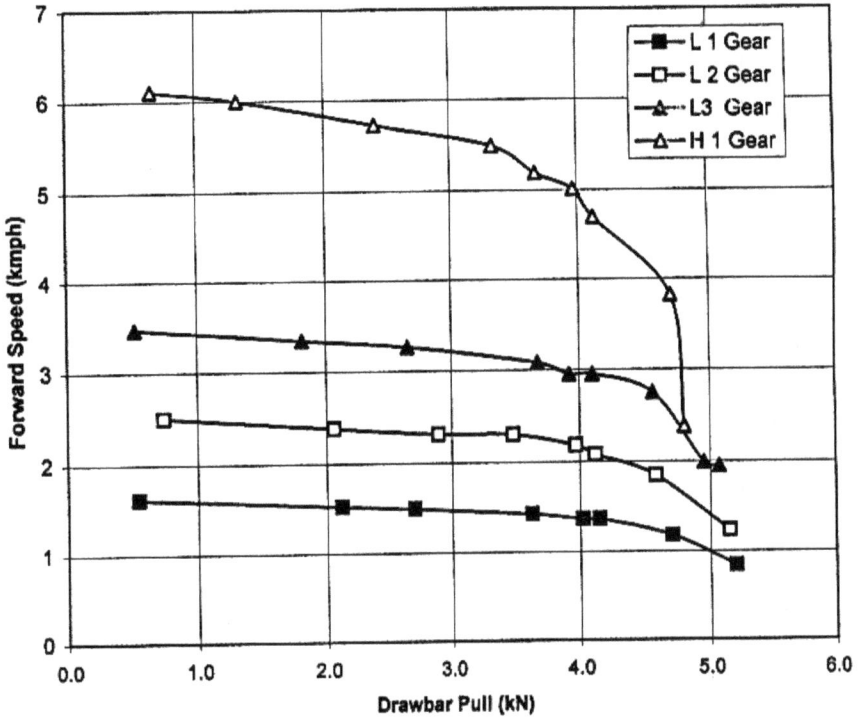

Fig. 4.9: Typical curve showing forward speed and drawbar pull

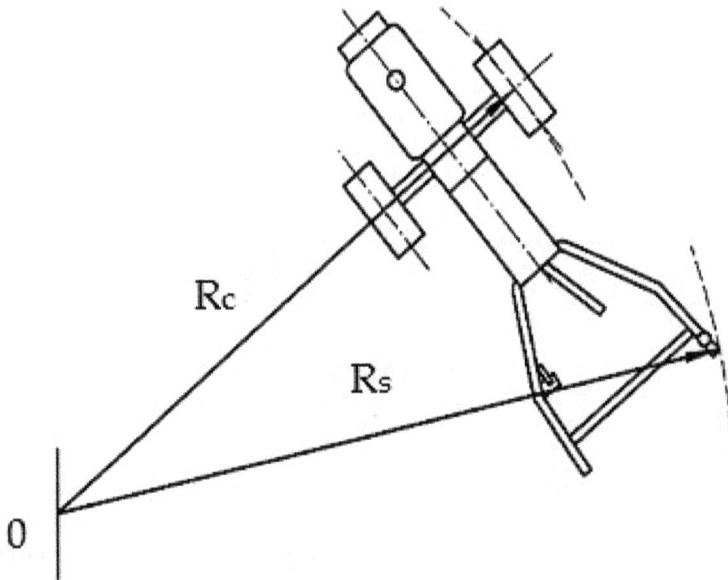

Fig. 4.10: Radius of turning circle & turning space

R_c – turning circle radius; R_s – turning space radius

Annexure 4.1: Data Sheet – Engine Performance Test

Test conditions	Test	Power (kW)	C. Shaft torque (Nm)	Engine speed (rpm)	Fuel consumption		Specific energy (kW.h/l)	Temperature (°C)			Atmospheric conditions		
					Hourly (kg/h)	SFC (g/ kW.h)		Fuel	Engine oil	Engine coolant	T (°C)	RH (%)	P (kPa)
Normal ambient conditions	Max. power test												
	Power at rated engine speed												
	Varying engine speed at full load												
	i.												
	ii.												
	...												
	Rated engine speed at varying load												
	i.												
	ii.												
	...												
5 Hour Test													
Normal ambient conditions	At load corresponding to 80% of max. power (4 hour)												
	i.												
	ii.												
	...												
	At load corresponding to max. power												
	i.												
	ii.												
	...												

Annexure 4.2: Data Sheet – Rotary Shaft Performance Test & Drawbar Performance Test

Test conditions	Test	Rotary shaft power (kW)	Rotary shaft torque (Nm)	Engine speed (rpm)	Fuel consumption		Specific energy (kW.h/l)	Temperature (°C)						Atmospheric conditions			Rotary shaft chain speed, rpm	Rotary shaft oil temp., °C
					Hourly (kg/h)	SFC (g/ kW.H)		Intake air Fuel	Engine oil	Intake air	exhaust	coolant	T (°C)	RH (%)	P (kPa)		
Normal ambient conditions	Varying engine speed at full load i. ...																	
	5 Hour test at rated power of rotary shaft																	
	At load corresponding to 90% of max. power (4 hour) i. ...																	
	At load corresponding to min. power i. ...																	

Drawbar Performance Test

Test	Gear number used	Travel speed (km/h)	Drawbar pull (kN)	Drawbar power (kW)	Wheel slip (%)	Engine speed (rpm)	Fuel consumption			Atmos. conditions		
							kg/h	g/ kWh	kWh/l	T (°C)	RH (%)	P (kPa)
Max. power test (unballasted) i. ...												

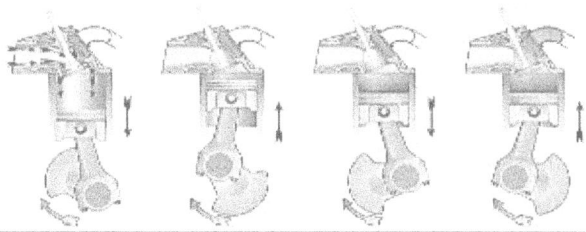

Chapter 5

Testing and Evaluation of Internal Combustion Engines

Internal combustion (IC) engine is defined where combustion takes place inside the combustion chamber of the engine. The engine is a contrivance in which, fuel and air are filled and start burning. Air-fuel mixture expand rapidly while burning and pushes outward. This push can be used to move a part of the engine and transmitted to carry useful work through suitable mechanisms. The IC engines are employed as prime movers for stationary work or mobile applications in agricultural operations. The reciprocating, rotary and turbines are the most popular designs, which are commercially available. However, the reciprocating type IC engines are the most preferred design, which are available in smaller and large power output range. The IC engines are classified further based on the method of ignition, a) spark ignition (SI) engine, and b) compression ignition (CI) engine. In SI engines, a mixture of air and fuel is injected during the suction stroke via the carburetor that controls the quantity and the quality of the mixture. On the other hand, in CI engines, fuel is injected into the combustion chamber towards the end of the compression stroke and it starts burning instantly due to the high pressure. To inject fuel in SI engines, a fuel pump, carburetor and spark plug are required. In CI engines, the quantity of fuel to be injected is controlled but the quantity of air to be injected is not controlled. The compression ratio of the fuel is in the range of 6 to 10 depending on the size of the SI engine and the power to be produced. In contrast, higher compression ratio of 16 to 20 for air results in high temperatures, which ensures the self-ignition of fuel.

In Indian agriculture, mostly the reciprocating IC engines are used for different operations. However, these engines may be four stroke or two stroke type. On the basis of fuel, these may be classified as petrol, diesel, CNG engines, biodiesel, and bioethanol. Cross sectional view of a typical diesel engine is shown in Fig. 5.1.

5.1 Terminology

e) *Compression ignition engine:* An engine in which ignition occurs by the temperature of the cylinder contents, resulting solely from their compression, delivering shaft power through one or more crankshafts.

f) *Spark ignition engine:* An engine in which ignition occurs by means of an electric spark.

g) *Engine speed:* Mean rotational speed of its crankshaft or shafts in revolutions per minute (rpm).

h) *Power:* It is a quantity proportional to the mean torque calculated or measured, and to the mean rotational speed of the shaft or shafts transmitting this torque, declared in kW.

i) *Brake power:* The power or sum of the powers measured at the driving shaft or shafts.

j) *Rated power:* The power obtained on the test bed at the rated speed specified by the manufacturer.

k) *Compression ratio:* The numerical value of the cylinder volume divided by the numerical value of the combustion space volume.

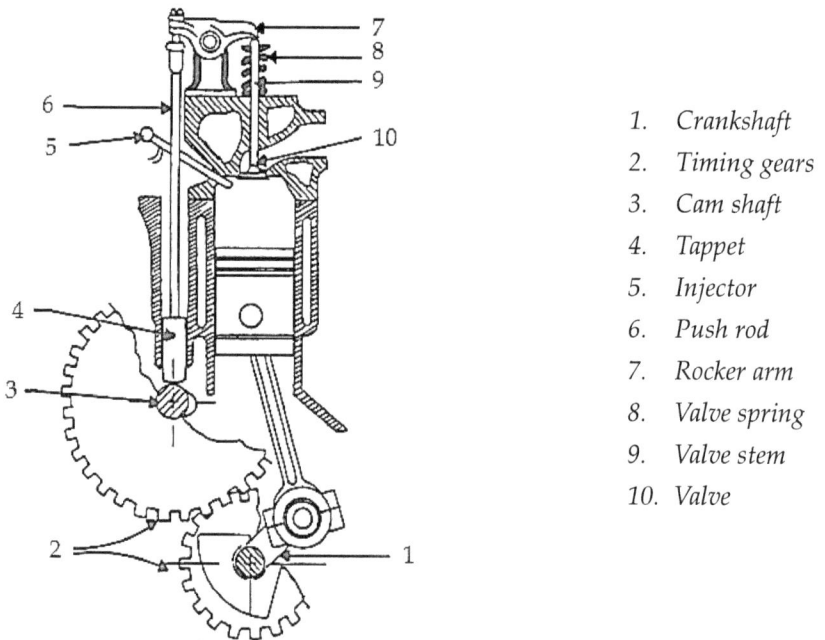

1. *Crankshaft*
2. *Timing gears*
3. *Cam shaft*
4. *Tappet*
5. *Injector*
6. *Push rod*
7. *Rocker arm*
8. *Valve spring*
9. *Valve stem*
10. *Valve*

Fig. 5.1: Cross section of a typical diesel engine

5.2 Scope of Test

a) Verification of specifications and other data furnished by the applicant

b) Engine performance test

c) Nature of breakdowns and repairs

d) Wear assessment of various critical components

5.2.1 Referred standards

a) IS: 10000 (1) - 1980. [27]

b) IS: 10000 (2) - 1980. [30]

c) IS: 10000 (3) - 1980. [31]

d) IS: 10000 (4) - 1980. [32]

e) IS: 10000 (5) - 1980. [33]

f) IS: 10000 (6) - 1980. [34]

g) IS: 10000 (8) - 1980. [35]

h) IS: 10000 (9) - 1980. [36]

i) IS: 10000 (12) - 1980. [28]

j) IS: 10000 (13) - 1980. [29]

k) IS: 11170 - 1985. [43]

l) ISO: 3046 (3) - 2006. [132]

5.3 Testing of IC Engine

5.3.1 Specification checking

The specification of engine shall be measured and compared with the specifications supplied by manufacturer and shall be reported.

5.3.2 Fuel and lubricants

The fuel and lubricant as recommended by the manufacturer and actually used in testing will be reported in the test report.

5.3.3 Engine performance test

The Engine test shall be conducted under normal and high ambient temperature conditions. Prior to test running in of the engine will be done as per recommendations of the manufacturer.

I. Under normal ambient temperature (27 ± 7^0C) condition:

a) Varying speed test

i) Maximum power and fuel consumptions

ii) Power at rated speed

iii) Maximum torque test

b) Two hours maximum power test:

c) Part load test

i) Torque corresponding to maximum power

ii) 85% of torque obtained in (i)

iii) 75% of torque obtained in (ii)

iv) 50% of torque obtained in (ii)

v) 25% of torque obtained in (ii)

vi) Unloaded

II. Under high ambient temperature ($43\pm2°$C) condition:

a) Varying speed test

i) Maximum power and fuel consumption

ii) Power at rated engine speed

iii) Maximum torque test

b) Two hours maximum power test

c) Five hours rating test

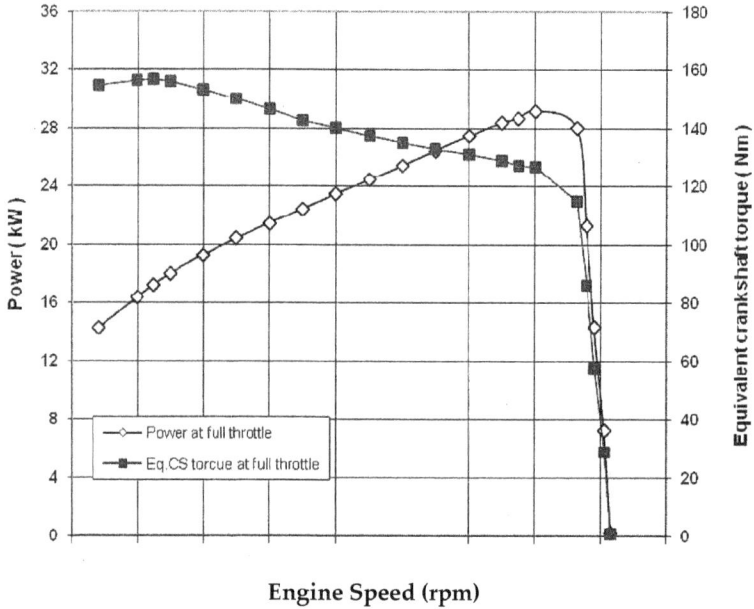

Fig. 5.2: Typical curve showing power and crankshaft torque vs engine speed

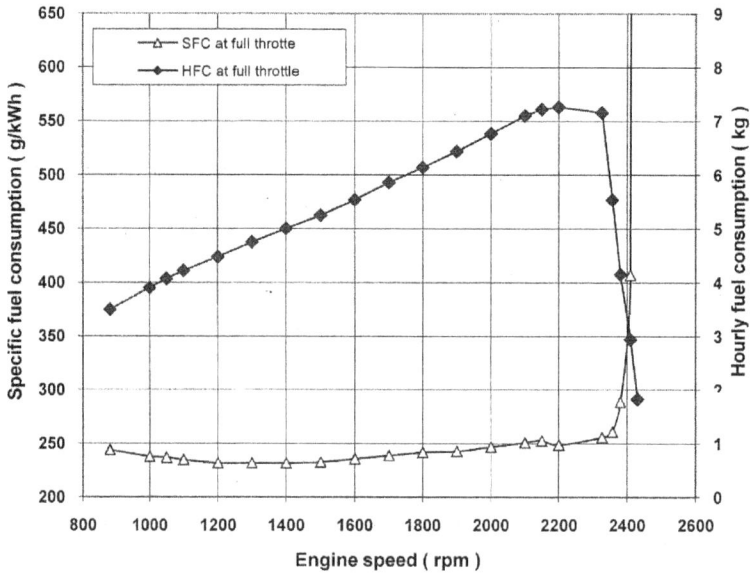

Fig. 5.3: Typical curve showing fuel consumption and specific fuel consumption vs engine speed

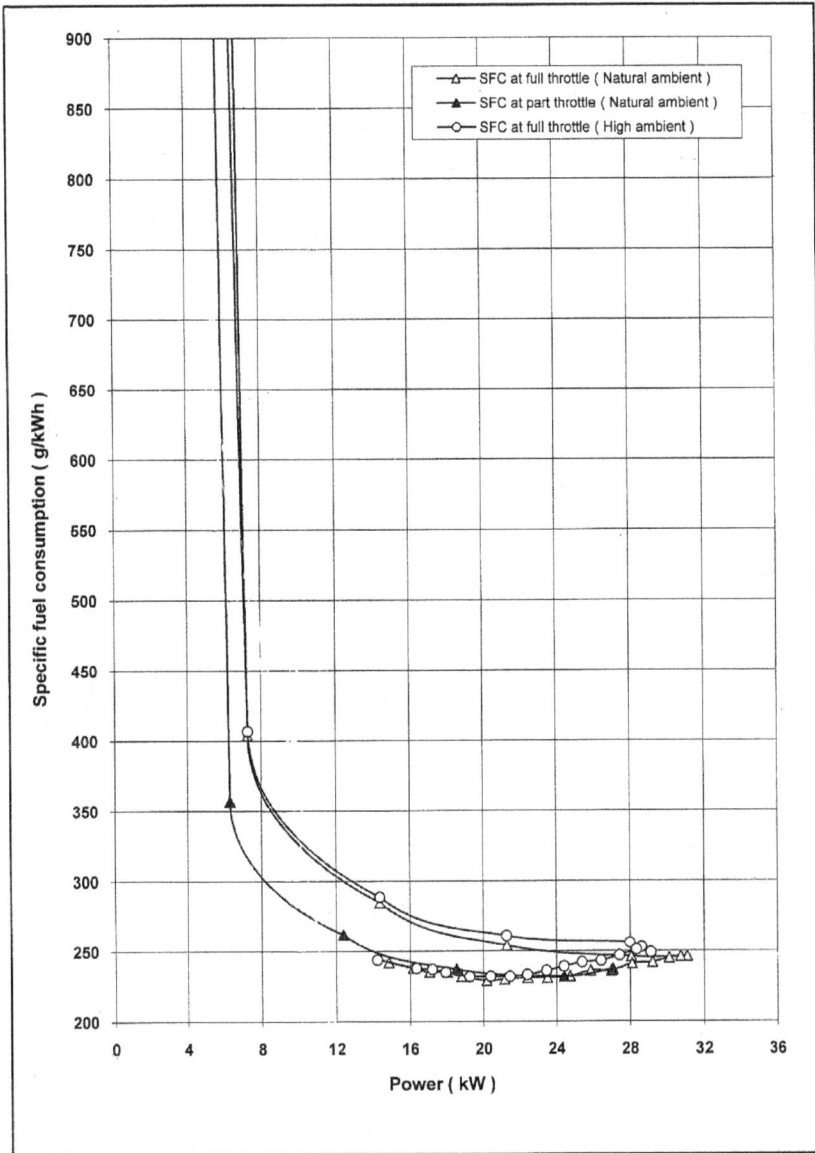

Fig. 5.4: Typical curve showing specific fuel consumption Vs power

Fig. 5.2 shows the variation of engine power and crankshaft torque as a function of engine speed. The power rises nearly linearly with torque in the governor-controlled range because the speed varies little. Torque varies only modestly in the load-controlled range, so power falls in nearly direct proportion to the fall in speed. For example, at the high setting, governor's maximum speed is at 2200 rpm and that would be the rated speed of the engine.

Fig. 5.3 shows the specific fuel consumption and hourly fuel consumption as function of engine speed. This helps to find out best engine speed where we will have minimum/optimum fuel consumption.

Specific fuel consumption vs engine power under normal and high ambient temperature condition is shown in fig. 5.4. It indicates that as the power increases, the specific fuel consumption decreases. This happens upto the maximum power of the engine.

5.4 Nature of Critical Breakdowns and Limits

a) Critical Break-Downs/defects

b) No critical breakdown should occur during the course of test.

Code No.	Aggregate	Critical Defects	Sub-assembly / Part
(1)	(2)	(3)	(4)
C-1	Engine	Engine seizure	Piston liner
C-2	-do-	-do-	Main/Big end bearing
C-3	-do-	Breakage of	Piston
C-4	-do-	-do-	Connecting rod
C-5	-do-	-do-	Crankshaft
C-6	-do-	-do-	Lubricating oil pump
C-7	-do-	-do-	Fuel injection pump
C-8	-do-	-do-	Governor
C-9	Engine	Breaking of	Cylinder blocks
C-10	-do-	-do-	Cylinder head
C-11	-do-	-do-	Timing gears
C-12	-do-	-do-	Valves
C-13	-do-	-do-	Cam shaft
C-14	Transmission	-do-	Clutch housing
C-15	-do-	-do-	Gearbox housing/ Transmission gears
C-16	-do-	-do-	Axle housing
C-17	Steering system	-do-	Steering ram cylinder
C-18	-do-	-do-	Steering shaft
C-19	-do-	-do-	Steering wheel
C-20	Brake system	-do-	Actuating linkage parts
C-21	-do-	-do-	Brake pedal
C-22	Front axle	-do-	Front axle
C-23	Rear axle	-do-	Rear axle

(1)	(2)	(3)	(4)
C-24	-do-	-do-	Rear axle support
C-25	-do-	-do-	Pivot pin and k lock
C-26	-do-	-do-	Hydraulic cylinder
C-27	-do-	-do-	Radius rod
C-28	Wheel equipment	-do-	Wheel rim
C-29	-do-	-do-	Wheel disc
C-30	-do-	-do-	Wheel bolts

5.5 Major Break-Downs/Defects and Limits

The total major breakdown should not be more than three in numbers and neither of them should be of repetitive in nature.

Code No.	Aggregate	Major Defects	Sub-assembly / Part
Mj-1	Engine	Breakage/ Crackage of	Fan blade
Mj-2	-do-	-do-	Oil pump
Mj-3	-do-	-do-	Water pump
Mj-4	-do-	-do-	Fuel tank
Mj-5	-do-	-do-	Radiator

5.6 Minor Break-Downs/Defects and Limits

The minor break down should not be more than five and should not be repetitive (more than three times during the course of test).

Code No.	Aggregate	Minor Defects	Sub-assembly/Part
Min-1	Engine	Leakage from	Radiator joints
Min-2	-do-	-do-	Gaskets
Min-3	-do-	-do-	Seals
Min-4	-do-	-do-	O rings
Min-5	-do-	Burst/Cracked	High pressure pipe
Min-6	-do-	Multifunctioning	Fuel injector

5.7 Wear Assessment of Critical Components

The engine after the test will be dismantled and wear of major parts like cylinder bore, piston diameter, ring end gap, ring side clearance, main and big-end bearings will be measured and compared with permissible limits & reported. Valves, guides and timing gear will be inspected for any sign of overheating, pitting and damage & reported.

5.8. Comparison with Minimum Performance Standard (MPS)

Standard (MPS) given in following table (IS 15806-2008):

Characteristics	Requirements	Tolerance in percent	Remarks
(1)	(2)	(3)	(4)
Prime Mover Performance:			
a) Maximum power (absolute)- Average maximum power observed during two hours, Max power test under natural ambient conditions, kW	To be declared by the manufacturer	-5	-
b) Maximum power observed during the test after adjusting the no load engine speed as per recommendation of the manufacturer for field work, kW		-5	-
c) Power at rated engine speed, kW		-5	IS 8122 (Part 2)
d) Specific fuel consumption corresponding to average maximum power under 2 h maximum power test, g/kW.h.	-do-	±5	-
e) Maximum smoke density at 80 percent load between the speed at maximum power and 55 percent of speed at maximum power or 1000 rpm whichever is higher	As per CMVR rules	-	-
f) Maximum crank shaft torque observed during the test after no load engine speed is adjusted as per manufacturers recommendation for field works, Nm	To be declared by the manufacturer	-8	-
g) Back-up torque, percent	7 percent, min	-	-
h) Maximum operating temperatures ^{0}C: Engine oil Coolant	To be declared by the manufacturer	-	The declared value should not exceed the maximum safe value specified by the oil company
i) Lubricating oil consumption	1 percent of special fuel consumption at maximum power (high ambient)	+10 percent	The value would be based on the test conducted under high ambient conditions (5 h rating test)

Chapter 6

Testing and Evaluation of Hand Tools for Agriculture

The common hand tools used in agricultural operations are khurpi, spade, hoes, cono weeders, and single row paddy weeders. Spade may be of narrow blade or wide blade as per the requirement and field conditions. These hand tools are small in size, and their best use is for cultivation of vegetables, flowers, medicinal and aromatic plants in small plots and for general purpose in agriculture.

Khurpi is a small multipurpose tool used for loosening of soil, breaking of clods, cut and uproot the weeds. It consists of a metallic blade and a wooden or PVC handle. With khurpi, a person can cover an area of about 0.025 ha in a day.

Spade is a general purpose hand tool used for a number of agricultural operations like digging, inter-culture, ridging, and making bunds or furrows for irrigation. It consists of a blade, made of high carbon steel and a wooden handle.

Hand hoes is the most frequently used tool for weed control and soil tilling operations. The shape of the working blade varies with the type of hoe. It is also used for digging of tuber crops. The different types of hoes are hacks, v-blade hand hoe, naini hand hoe, three tine hand hoe, rippers, grubber weeder and wheel hoe. Wheel hand hoe is used in intercultural operation in crops with more than 240 mm row spacing and it is of two types, a) single wheel hoe and b) twin wheel hoe.

For weeding operation in between the rows of paddy crop in standing water, Cono weeder is the best suited hand tool. It has two conical rotors mounted in tandem with opposite orientation for effective weeding. Serrated pegs mounted on the rotor uproot and bury the weeds. The work is accomplished by alternatively pushing and pulling the cono weeder.

The above mentioned hand tools need to be evaluated for ease of operation, quality of work and most importantly the ergonomical aspects. The comparison between various make of hand tools can be made by the test engineer based on the above aspects.

6.1 Scope of Test

This procedure is applicable for the evaluation of various types of hand hoes. The procedure gives general terms and prescribes measurements and assessments to be made to establish rate and quality of work, and to evaluate ergonomics aspects of hoe use.

It will be the responsibility of the Test Engineer to decide which measurements should be made to best judge the work output and suitability of the implement.

6.2 Test Methods

6.2.1 Selection of sample

The manufacturer shall supply the hoe complete and fully assembled from the series production for commercial use together with instructions for its use. Specifications giving dimensional data and materials used in the construction shall also be provided.

6.2.2 Laboratory test

Prior to any field tests, the manufacturer's specification shall be checked. The hoe shall be examined for quality of materials and construction, especially of blade attachment and the condition of the handle(s) (free from splinters and cracks, securely attached and offering a good grip).

The hardness of the cutting element of the blade(s) should be measured and compared with the specification.

6.2.3 Field test

6.2.3.1 Test conditions

Operators: Operators selected for the tests should be representative of potential users. For example, if it is women who usually perform hoeing tasks, they will be used as operators. The operators chosen must be prepared to allow their physical dimensions to be measured and where necessary to cooperate with the recording of heart rate, respiration and subjective responses during work periods.

Fields: The rates of work achieved with a hand hoe will vary with the type and condition of the soil, the crop layout and the weed population. The test report should give soil details and weed population details. The land chosen for the tests should be typical for the region.

6.2.3.2 Preliminary trials

Trials should be carried out adjacent to the test plots to allow operators to familiarize themselves with the test implement and give some indication of the expected work rate. A plot size consistent with a total work time of approximately 3 hours should then be determined.

6.2.3.3 Performance tests

The main objectives of the performance tests are to obtain the sustainable work rate and to evaluate the demands made of the operator during the work period. Tests should be made on plots having a range of typical crop, soil and weed conditions. All operators should work on all the conditions included for testing. When the test plots have been marked out, soil samples should be taken and weed counts made.

Each plot should be weeded in one session and the following measurements should be recorded:

i) Total operating time

ii) Time taken for stoppages apart from essential rest periods

iii) Depth of work

iv) Total area of work

v) Weeding efficiency by weed count method

vi) Wear assessment

If the appropriate equipment and expertise is available, measurements may be made which indicate the energy expended by the operator during tests and the results used to rate the implement. If the instrumentation is not available, a subjective assessment of workload and the physical discomfort associated with the use of the hoe may be made by the operators.

At the end of test, subjects may be informally interviewed to gain their overall opinions of the hoe. Soil and weed measurements should be repeated to establish clod size, inversion and mixing and final weed population.

6.2.3.4 Comparative tests

Where possible, the hoe should be tested against a standard "reference" hoe (hoe traditionally used in the region), or in a group test with a number of other hoes. It is then possible rank the hoes against the criteria specified (e.g: rate of work and quality; energy requirement; subjective assessments; quality of manufacturing etc.).

6.2.3.5 On-farm durability tests

A series of trials may be undertaken on farmers' fields to enable the hoe to be evaluated in more varied field and soil conditions. They should be carried out over a complete weeding season to obtain more accurate measurements of blade wear and highlight any problems of operation or robustness. All the conditions specified in the performance tests shall be applied for testing and evaluation and reported in the test report. A sample datasheet has been given in annexure 6.1.

Annexure 6.1
Summary Sheet for Testing of Hand Tools

Diagram /Photograph		

Specification of hand tool		
Make:	*Model:*	*Serial No:*

Manufacturers name and address:

Blade

Width of cut:	*Material:*	*Hardness (Standard if applicable):*

Blade mounting

Material:	*Throat width:*	*Method of fixing blade:*	*Angle of fixing with handle*

Handle

Material:	*Length:*	*Size:*	*Working height:*	*Method of attachment:*

Support wheel (if applicable)

Material:	*Diameter:*	*Width:*

Total weight:

Details of operator

Sex:	*Age:*	*Weight:*	*Height:*

Summary of Test Results

Sl. No.	Description / Parameters	1	2	3	4	5
A. Pre-test parameters						
	Date					
	Location					
	Plot size (m x m)					
	Topography					
	Soil description					
	Previous cultivation Crop					
	Weed count before test (number/m^2)					
	Soil moisture (DB %)					
	Dry bulk density (g/cm^3)					
	Penetrometer reading (kPa)					
	Shear strength (kPa)					

B. After test parameters

> Working depth (cm)
>
> Working width (cm)
>
> Time taken to complete operation (min)
>
> Weed count after test (number/m^2)
>
> Field efficiency (%)
>
> Soil inversion (%)
>
> Weeding efficiency (%)
>
> Evenness of cultivating

Chapter 7

Testing and Evaluation of Tillage Implements and Rotavator

Tillage is the first and foremost operation in crop production. Different types of tillage include primary and secondary tillage, like ploughing, harrowing, tilling in order to get desired soil condition for sowing the seed. In recent years, rotary tillage utilizing both tractive and rotary power has emerged as a prominent tillage practice. Rotary tiller or rotavator has become most desired tillage implement in India, as it can do the job of a harrow, cultivator, and planking in single pass.

7.1 Types of Tillage

Primary tillage: Primary tillage is the first operation done with heavy implements like disc plough, mould board plough for deep ploughing

Secondary tillage: Secondary tillage utilizes lighter implements like disc harrow and cultivator to obtain the desired degree of soil breakup and pulverization. Technically, this operation is done after primary tillage operation. However, as per farm requirement, the selection of tillage implement is made.

Rotavator: It is used for dry as well as wet soil operation (puddling). Rotavator helps to prepare the seedbed in one or two passes, if soil has proper moisture content. Thus it serves the purpose of both primary and secondary tillage operations.

There is a growing interest in minimum tillage system with a view to reduce production cost, improving soil conditions and saving in energy input. Some important primary and secondary tillage implements are shown in Fig. 7.1 (a), 7.1 (b) and 7.1 (c).

Main objectives of tillage include:

- To minimize the mechanical power and labour requirements.
- To conserve the moisture and reduce soil erosion.
- To perform the operations necessary to optimize soil tilth.
- To maximize the field efficiency.

Therefore, testing and evaluation of tillage implements under the laboratory and actual field conditions is the only way of determining their quality and output.

This will also help in standardization of critical components and saving the energy through adoption of energy efficient implements.

Fig. 7.1 (a): Primary tillage implements (Reversible MB plough, Disc plough)

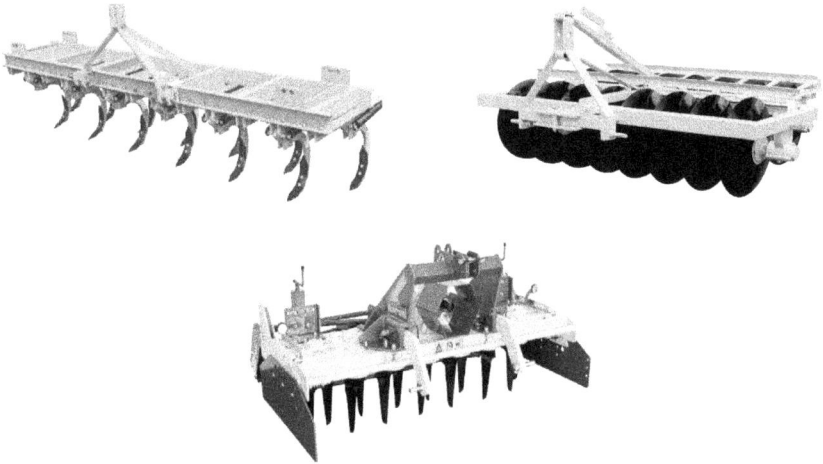

Fig. 7.1 (b): Secondary tillage implements (Cultivator, Disc harrow, Power harrow)

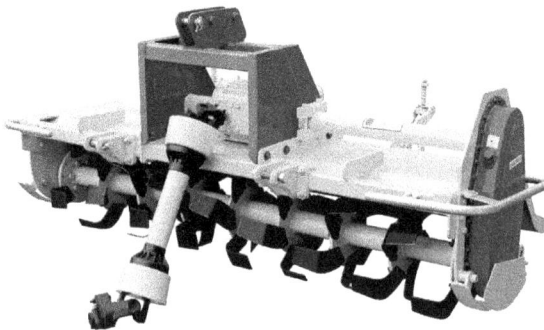

Fig. 7.1 (c): Rotavator

7.2 Terminology

a) *Chiseling:* A tillage operation in which a narrow tool is used to break up hard pan in the soil. It is usually performed at a depth greater than the normal ploughing depth.

b) *Conventional Tillage:* The combined primary and secondary tillage operations normally performed in preparing a seed bed.

c) *Furrow:* The trench formed by a tool in the soil during operation (see Fig. 7.2); Furrow Crown: The peak of the turned furrow slice; Furrow Slice: The soil mass cut and turned by tool; Furrow Sole: The bottom surface of the furrow; Furrow Wall: The undisturbed side of furrow.

d) *Minimum Tillage:* The minimum soil manipulation necessary to meet tillage requirements for crop production.

e) *Mulch Tillage:* Preparation of soil in such a way that plant residues or other mulching materials are specially left on or near the surface.

f) *Primary Tillage:* Tillage operations which constitute the initial major soil-working operation. It is normally designed to, reduce soil strength, cover plant materials and re-arrange aggregates.

g) *Puddling:* The mechanical manipulation of soil in the presence of standing water in the field to create and impervious hard pan below the puddle zone so as to prevent loss of water through leaching and facilitate transplanting of paddy seedlings by making the soil softer.

h) *Pulverization:* The general fragmentation of a soil mass resulting from the action of tillage forces.

i) Secondary Tillage: Tillage operations, following primary tillage, which are performed to create proper soil tilth for seeding and planting.

j) *Sub-soiling:* Chiseling at depth greater than 40 cm.

k) *Tilt angle:* The angle at which the plane of the cutting edge of the disc is inclined to a vertical line (Fig. 7.3 (a)).

l) *Disc angle:* The angle at which the plane of the cutting edge of the disc is inclined to the direction of travel (Fig. 7.3 (b)).

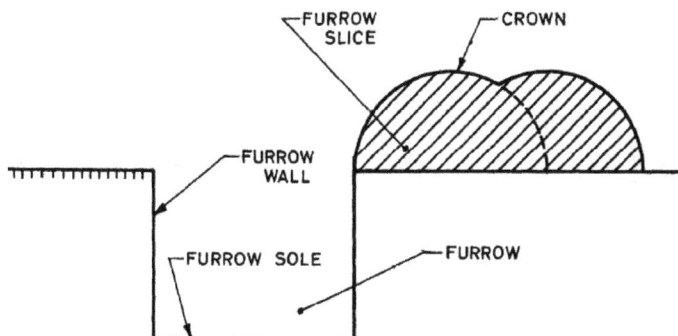

Fig. 7.2: Diagram showing furrow and its parts

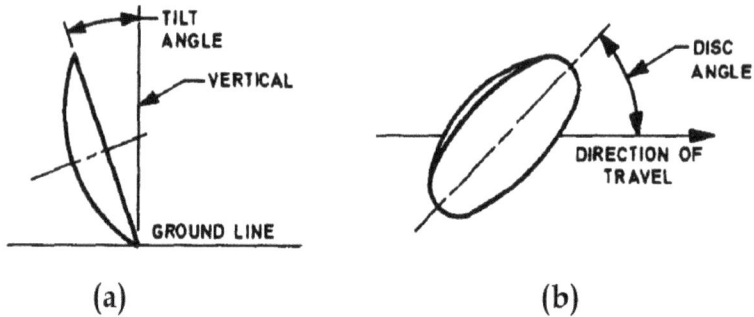

Fig. 7.3: Disc angle and tilt angle of a disc plough

m) *Horizontal clearance (side clearance):* The maximum clearance between land side and a horizontal plane touching point of share at its gunnel side and heel of landside. It is also known as horizontal suction (Fig. 7.4 (a)).

n) *Vertical clearance:* The maximum clearance under the land side and horizontal surface when the plough is resting on a horizontal surface in the working position. It is also known as vertical suction (Fig. 7.4 (b)).

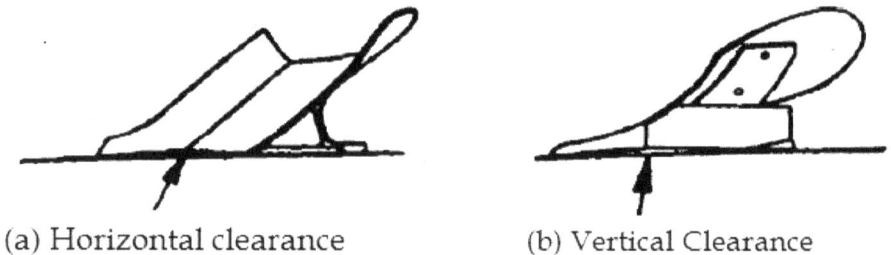

(a) Horizontal clearance   (b) Vertical Clearance

Fig. 7.4: Horizontal and vertical clearance of a mould board plough

7.3 Scope of Test

7.3.1 Laboratory tests

a) Specifications checking of implement

b) Hardness of critical components

c) Chemical analysis of soil engaging components

d) Wear analysis of soil engaging components

7.3.2 Field tests

a) Rate of work

    i) Width of cut

    ii) Effective field capacity

    iii) Field efficiency

b) Quality of work
   i)   Depth of cut
   ii) Soil inversion
   iii) Soil pulverization
c) Draft measurement
d) Fuel consumption
e) Soundness of construction
f) Ease of operation and adjustment

7.3.3 Referred standards and test codes
a) IS 6288: 1999 [77]
b) IS 6635: 2001 [81]
c) IS 6638: 2001 [82]
d) IS 6690: 1996 [83]
e) IS 7640: 1999 [87]
f) IS 9217: 2001 [95]
g) IS 10233: 2001 [38]
h) IS 10691: 2001 [41]
i) IS 4366-1: 2001 [72]
j) IS 4366-2: 2001 [73]

7.4 General Conditions

7.4.1 Selection of test sample
The test sample, which is in commercial production, shall be selected at random from the works of the manufacturer by a representative of testing institute. However, a prototype machine may be submitted directly to the testing institute. The method of selection shall be given in the test report.

7.4.2 Acquaintance with the test sample
The testing team should get fully acquainted with the construction and operational features of the equipment/machine to assess its real function and performance. The technical literature/information/manufacturing drawings etc. should be thoroughly studied. All adjustments should be made as per the manufacturer's recommendations.

7.4.3 Instrumentation
The reliability of testing data, to a great extent depends upon the accuracy of instrumentation for proper evaluation of test sample. It is necessary that various testing institutes should test machine at a common yardstick. Therefore, the measuring instruments should have accuracy as specified below:

Time (s)	±0.2
Distance (m)	±0.5%

Force (kgf)	±2.0%
Mass (kg)	±0.5%
Rotational speed (rpm)	±0.5%

7.4.4 Selection of test plot

A Tillage equipment/machine gives optimum performance in a rectangular field. The test plot should be rectangular having dimensions in the ratio of 2:1 as far as possible. If the field is irregular, then a rectangular test plot should be marked for conducting the test. The other portion of the field can be used for initial setting/adjustments of the equipment. The test plot selected should not have any previous tillage treatment after the last crop harvested.

7.4.5 Field operational pattern

Field capacity and field efficiency of an implement are affected by field operational pattern which is closely related to the size and shape of the field, the kind and size of implement. The downtime (non-working time) should be eliminated/minmised with adoption of appropriate field operational pattern. Common field operational patterns for rectangular field are shown in Fig. 7.5.

7.4.6 Duration of test

The test sample shall be operated under different soil and surface conditions i.e. dry & wet conditions for a minimum period of 50 h to establish its performance. Each test should be of minimum 3 hours.

7.4.7 Field parameters to be recorded before test

Various field parameters to define soil characteristics and surface conditions of the test plot as specified below, should be observed and recorded:

a) Location of test plot

b) Size of test plot

c) Last crop grown

d) Detail of previous tillage operation, if any

e) Topography of field

f) Type of soil

g) Bulk density of soil

h) Cone index of soil

Moisture content of the soil is computed on dry weight basis. For measurement of soil moisture, take core samples of wet soil from at least three different locations randomly selected in the test plot. Weigh the sample and record the weight. Place the sample in a hot air oven maintained at 105°C for at least 8 hours. At the end of 8 hours, cool the sample in desiccator and weigh again. Calculate the soil moisture using the following formula:

$$Soil\ moisture\ (\%\ dry\ weight\ basis) = \frac{w_1 - w_2}{w_2}$$

Where,

w$_1$ - *weight of wet soil sample (g)*

w$_2$ - *weight of dry soil sample (g)*

Bulk density of soil is defined as the mass after oven drying of soil per unit volume. For measurement of bulk density of soil, take cylindrical core sample from at least three different locations, selected randomly in the test plot. Measure the diameter and length of cylindrical soil sample. Keep the core sample in the oven. Maintained at 105°C for at least 8 hours. At the end of 8 hours, take out the sample from the oven, cool it in a desiccator, and weigh again.

$$Bulk\ density\ of\ soil\ (g/cc) = \frac{4m}{\pi d^2 l}$$

Where,

m   - mass of oven dry core sample (g)

d   - diameter of cylindrical core sample (cm)

l   - length of cylindrical core sample (cm)

Cone index also provides an indication of soil resistance and is expressed as force per square centimeter required for a cone of standard base area to penetrate into soil to different depths. Cone index for the same soil varies with the cone apex angle, area of cone base and depth of penetration. Apex angle and area or diameter of cone used should be given in the report.

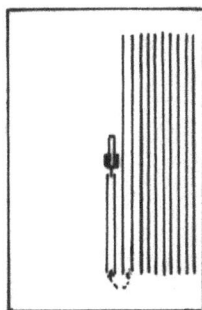

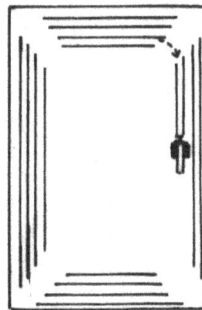

Continuous pattern strip turn   Circuitous pattern   Circuitous pattern turn strips
at each end   rounded corners   at corner diagonals

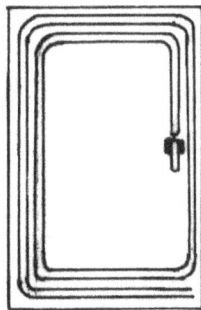

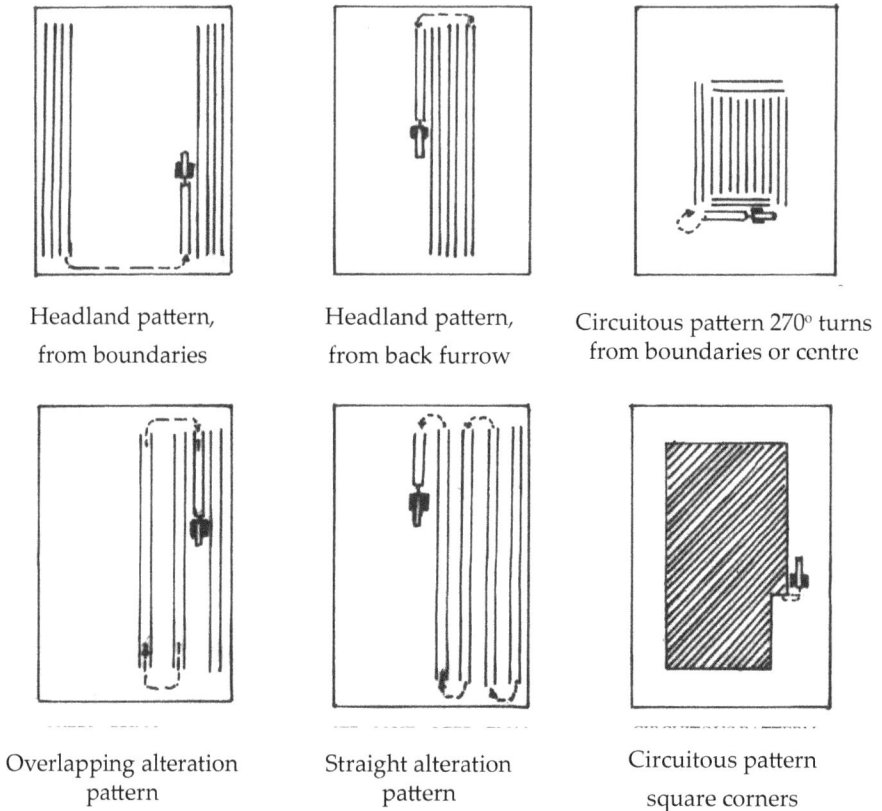

| Headland pattern, from boundaries | Headland pattern, from back furrow | Circuitous pattern 270° turns from boundaries or centre |

| Overlapping alteration pattern | Straight alteration pattern | Circuitous pattern square corners |

Fig. 7.5: Field operational patterns

7.5 Testing of Tillage Implements

7.5.1 Laboratory test

a) *Specification of implement:* The specifications of the implement and of main components should be checked and verified as against values furnished by the manufacturer. The variation if any, should be highlighted in the report.

b) *Hardness:* The hardness of various critical components should be measured and compared with the relevant Indian standards.

c) *Chemical analysis:* The chemical composition of critical soil engaging components, e.g. blade of rotavator, furrow opener of cultivator etc., should be determined and reported

d) *Wear analysis test:* The mass of critical soil engaging components should be determined before and after the field tests to assess the wear rate of soil engaging parts and reported.

7.5.2 Rate of work

a) Width of cut

For determining width of cut, average of five runs should be taken. The measurement of composite width should be taken at minimum five equidistant places in the direction of travel and average working width should be determined. A graduated rule as shown in Fig. 7.6 may be used for measuring furrow width and depth.

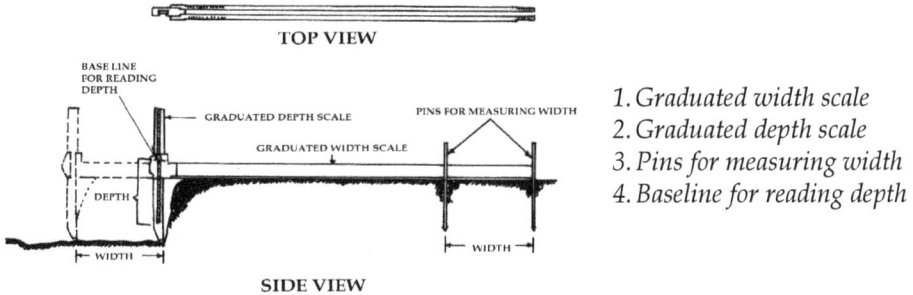

TOP VIEW

SIDE VIEW

1. *Graduated width scale*
2. *Graduated depth scale*
3. *Pins for measuring width*
4. *Baseline for reading depth*

Fig. 7.6: Graduated rule for measurement of furrow width and depth [191]

b) Speed of operation

To calculate the speed of operation, two poles 20 m apart are placed approximately in the middle of the test run. On the opposite side, two poles are placed in similar position and 20 m apart so that all four poles form corners of a rectangle. The speed will be calculated from the time required for the machine to travel the distance of 20 meter between the assumed line connecting two poles on opposite sides. The easily visible point of the machine should be selected for measuring the time. Layout of fixing poles for speed measurement is illustrated in Fig. 7.7.

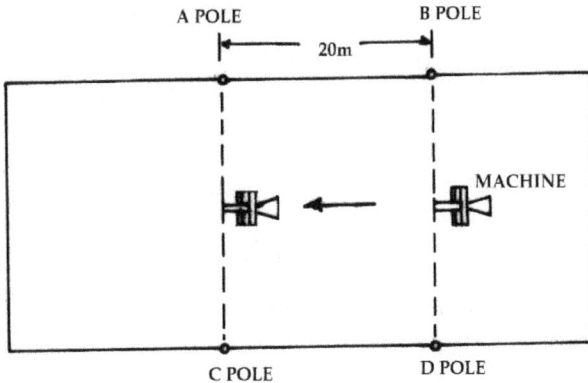

Fig. 7.7: Field measurement of forward speed

c) Wheel slip

Tractor pneumatic tyres or drive wheels experience slippage under different field operations. The distance covered by the tractor in a given number of drive wheel revolutions decreases with the wheel slip. Therefore, in case of tractor or power tiller operated implement, the wheel slip will affect the speed of operation and thereby the effective field capacity of the implement. The wheel slip is determined as under and expressed in percentage.

$$Wheel\ slip\ (\%) = \frac{N - N_1}{N} \times 100$$

Where,

N - total no. of revolutions at load

N_1 - total no. of revolutions at no load (for the marked test run)

d) Effective field capacity

The actual output in terms of area covered per hour is expressed as the effective field capacity. In calculating the effective field capacity, the time consumed for real work, in turning and adjustment etc. should be taken. Time for refueling should be deleted because usually filling up fuel, before starting the test, can make refueling unnecessary except for especially large field. Therefore:

$$Fe = \frac{A}{P + NP}$$

Where,

Fe   - effective field capacity (ha/h)

A   - area covered (ha)

P   - productive time (h)

NP   - non-productive time (h)

e) Field efficiency

The field efficiency is the ratio of effective field capacity to the theoretical field capacity expressed as percentage.

$$F_t = \frac{ws}{10}$$

Where,

F_t - theoretical field capacity, ha/h

w   - theoretical width of implement, m

s   - speed of operation, km/h

Field efficiency is calculated by the formula:

$$Field\ eficiency\ (\eta) = \frac{F_e}{F_t}$$

Field efficiencies of tillage operations ranges between 75 - 90% [Annexure 20.3].

7.5.3 Quality of work

a) Depth of cut

The vertical distance between furrow sole and ground level is referred as depth of cut. The depth of cut should be measured at minimum 10 places and its average taken. Refer Fig. 7.2.

b) The soil inversion

It is the process through which the furrow slice is inverted. The inversion characteristic can be measured by weed count method. In this method, the number of weeds/stubbles present are counted before and after the test with the help of placing a wooden square of 30 cm or 50 cm at random in the field and at minimum 5 locations. The soil inversion is expressed on percentage basis.

$$Soil\ inversion\ (\%) = \frac{W_a - W_b}{W_a} \times 100$$

Where,

$\quad\quad$ W_a $\quad$ - No. of weeds present per unit area before the operation

$\quad\quad$ W_b $\quad$ - No. of weeds present per unit area after the operation

c) Soil pulverization

Soil pulverization is the process of breaking up of soil into small aggregates resulting from the action of tillage forces. The mean mass diameter (MMD) of the soil aggregated is considered as index of soil pulverization and can be determined by the sieve analysis of the soil sample through a set of standard test sieves (IS: 460-1982). Sieving provides a simple means for measuring the range of clod size and relative amount of soil in each size class.

Method: Pass the soil sample through a set of sieves. Weigh the soil retained on the largest aperture sieve, soil passed through each sieve and retained on the next sieve and passed through the smallest aperture sieve. Mean mass diameter can be calculated as shown in table 7.1.

d) Puddling index

The mechanical manipulation of soil in presence of standing water in the field to create a least pervious layer that prevents the loss of water through percolation and facilitate the transplanting of paddy seedlings by making the soil softer is called puddling. The quality of puddling is measured by calculating puddling index, which is the ratio between volume of settled soil and total volume of sample and it is given by the formula:

$$Puddling\ Index = \frac{V_s}{V} \times 100$$

Where,

V_s - volume of settled soil,

V   - total volume of sample

Method: The samples of soil-water suspension shall be taken by immersing a glass tube to a depth of about 100 ml from a number of points and shall be collected in measuring cylinders. These shall be kept undisturbed for 18 hours to allow the soil to settle. The volume of settled soil (V_s) shall be noted and puddling index shall be calculated[46].

Table 7.1 Computing the mean mass diameter (MMD) of clods

Size of aperture (mm)	Diameter of soil passing through the upper sieve and retained on the next small aperture sieve (mm)	Representative diameter of soil (mm)	Weight of soil (kg)
2.0	< 2.0	1	A
2.8	2.0 – 2.8	2.4	B
4.0	2.8 – 4.0	3.4	C
5.6	4.0 – 5.6	4.8	D
8.0	5.6 – 8.0	6.8	E
11.2	– 11.2	9.6	F
--	> 11.2	x	G

$$MMD = \frac{1}{W}\,(A + 2.4B + 3.4C + 4.8D + 6.8E + 9.6F + xG)$$

Where,

W = A+B+C+D+E+F+G

x = Mean of measured diameter of soil clods retained in the largest aperture sieve

7.5.4 Draft measurement

a) Trailed type implement

The draft required for pulling the implement can be measured by inserting a spring/hydraulic or strain gauge type dynamometer between the hitch of the implement and power source. If the line of pull through the dynamometer is not horizontal. Then angle of inclination with horizontal plane should be recorded and draft is calculated as follows:

$$D = P\ Cos\theta$$

Where,

D   - draft, kgf

P   - pull measured by a dynamometer, kgf

θ   - angle between the line of pull and the horizontal, degree

b) Mounted type implement

Draft of a mounted type implement can be measured by two tractors and a pull type dynamometer. An implement is mounted on tractor A and this tractor is towed by another tractor B thorough a pull type dynamometer as shown in Fig. 7.8. Measure the draft when the tractor A is in neutral gear condition, but implement in operating condition. The procedure is repeated again with the implement in a lifted position. The difference between two readings gives the draft requirement of implement.

Draft power requirement for the operation of implement can be calculated by the following formula:

$$Power \ (PS) = \frac{Draft \ (kgf) \ \acute{n} \ operating \ speed \ (m/s)}{75}$$

7.5.5 Fuel consumption

Use of digital fuel flow meters in combination with data acquisition system can be used to measure the fuel consumption. Analog type fuel flow meters also can be used necessitating the need of the operation to take reading before and after each test.

Alternative method: The tank is filled to full capacity before and after the test. Amount of refueling after the test is the fuel consumption for the test. While filling up the tank, careful attention should be paid to keep the tank horizontal and not to leave empty space in the tank. The fuel consumption will give an idea of energy requirement by the implement for the operation.

7.5.6 Soundness of construction

During the entire period of testing, a complete detail of the defects and breakdowns should be recorded and reported in the test report. The wear of critical components should also be assessed during the test. These would reflect largely the soundness of construction of the implement under test.

7.5.7 Ease of operation and adjustments

A complete record of provisions for various adjustments to cover a wide range of operating conditions and ease of carrying out the adjustments, should be maintained during the test. Apart from this, the ease of operations of the implement should be observed critically and reported. Sample data sheet for specification and test result etc. has been shown in Annexure 7.1 & 7.2.

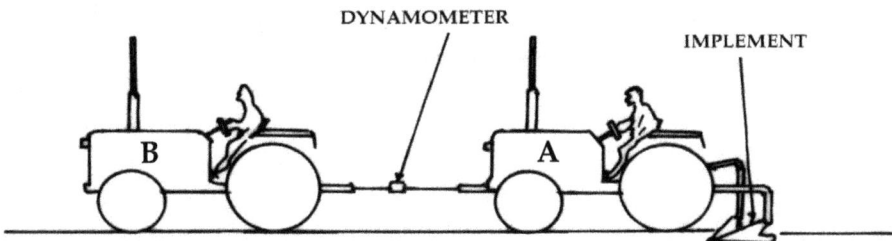

Fig. 7.8: Measurement of draft of a tractor mounted implement [191]

7.6 Interpretation of Test Results

Although evaluation of implements on the basis of their performance tests is a complex job because of number of variables affecting the performance of the implement directly or indirectly during the operation. These are soil type, moisture content, presence of weeds/stubbles, power source, provision of adjustments and skill of operator, etc. It is difficult to maintain identical field conditions to facilitate comparison of field results. However, careful evaluation of the test report can give realistic information, provided, the testing has been conducted following standard code and procedure. An implement requiring the minimum draft for operation is considered to be the best, provided the quality of work done is also satisfactory. In case of tractor/power tiller operated implements, the maximum fuel economy will be achieved at minimum draft requirement. For determining the fuel economy, comparison of fuel consumption per unit volume of soil worked will provide a realistic data to evaluate the implement. Lesser the fuel consumption per unit volume of soil worked, better is the economy of implement.

The rate of work is a function of size of implement, speed of operation, size of plot and other field parameters. Thus, the evaluation of implement on the basis of their rate of work is done by comparing the result with performance results of similar implements. However, an implement giving higher output without sacrificing the quality of work is considered better.

Soundness of construction, ease of operation and adjustment and rate of wear of critical components of the implements are also important factors to be given due weightage in their evaluation. An implement of a simple design and robust construction should be preferred over sophisticated implement having weak construction. Operating cost of the implement is another important point which is to be considered in evaluation of implement. However, quality of work cannot be sacrificed for low operational cost. The availability of efficient after sales service facilities including warranty, cost and availability of spares, use of standard hardware and other fast moving components are also important considerations.

During the process of evaluation, it may be possible that the implement may not have all the features as described in the test procedure. But it is expected that, the test results should conform to the relevant Indian standard, quality of work done is satisfactory, economical in operation, sturdy construction, adequate provisions for adjustments to cover a wide range of working conditions and safety aspects have been provided in the implement. Moreover, an overall performance index be calculated on the basis of various performance results during the testing and compared with other similar products.

Annexure 7.1

Specifications Sheet of the Tractor used during Field Test

1.	Make, model and type	:
2.	Number of cylinders	:
3.	Maximum PTO power, kW	:
4.	Engine speed corresponding to standard PTO speed (rpm)	:
5.	Rated engine speed (rpm)	:
6.	Number & size of tyre Front Rear	:
7.	Inflation pressure of tyre Front Rear	:
8.	Standard track width (mm) Front Rear	:
9.	Wheel base (mm)	:
10.	Total operational Mass (kg)	:

Annexure 7.2
Data Sheet – Test Conditions and Results

Date: ....   Location: ....   Plot size (m X m): ....

Topography: ....   Soil type: ....   Previous crop: ....

Description	Test No.				
	1	2	3	4	5
Crop residue, %					
Weed count per unit area (number/m²)					
Cultivation after last harvest					
Soil moisture, % DB					
Dry bulk density, g/cm³					
Penetrometer reading, kPa					
Shear strength, kPa					
Operational pattern					
Working depth, cm					
Working width, cm					
Draft, kgf					
Operating/forward speed, m/s					
Power, kW					
Tractor wheel slip, %					
Time taken to complete test plot, min					
Field capacity, ha/h					
Field efficiency, %					
Soil inversion, %					
Puddling index, % (for rotavator)					

Chapter 8

Testing and Evaluation of Tractor Operated Laser Land Leveller

Land leveling is a prerequisite for better agronomic, soil and crop management practices as it has a direct bearing on all the farming operations. All major farming operations from land preparation to seedbed preparation, seed placement and germination require an optimal soil moisture condition. The non-uniformity or unevenness of soil surface adversely affects efficiency of utilization of irrigation water. Higher degree of uniformity of irrigation water is ensured by precise land leveling. The fully self leveling lasers are the easiest to use and can achieve accuracy upto 2.5 mm at 100 m thereby increases the productivity [151, 144].

The accurate method to level or grade the field is to use the laser-guided leveling equipment. Laser leveling is a process of smoothening the land surface (± 2 cm) from its average elevation by using laser guided drag buckets to achieve precision in land leveling. Precision land leveling involves altering the fields in such a way as to create a constant gradient of 0 to 0.2%. This practice makes use of tractors of 40 hp or more and soil movers that are equipped with laser-guided instrumentation and/or global positioning system (GPS) so that the soil can be moved by either cutting or filling to create the desired slope or level of field [257].

Laser leveling provides a fairly accurate, smooth and graded field which allows ideal control of water distribution with negligible water losses. Laser leveling improves irrigation efficiency and reduces the potential for nutrient loss through better irrigation and runoff control. Laser land leveling facilitates uniformity in the placement of seedlings by rice transplanter, which helps in achieving higher yield levels. The precisely leveled surface leads to uniform soil moisture distribution which results in good seed germination, enhanced input use efficiency and crop yields [155]. Perfectly leveled field leads to uniform distribution of nitrogen, which further results in better fertilizer-use efficiency and higher yields. Laser leveling improves water distribution which contributes :

- increases grain yield and improves grain quality,
- results in better crop stands,
- reduces weed problems and results in uniform crop maturity

8.1 Components of Laser Land Leveler

A laser leveler involves the use of laser (transmitter), that emits a rapidly rotating beam parallel to the required field plane, which is intercepted up by a sensor (receiver) fitted to the tractor carrying the scraper unit. The signal received is converted into cut and fill level adjustment and the corresponding changes in the scraper level are carried out automatically by a hydraulic control system. The scraper guidance is fully automatic, thus eliminating the element of operator error, allowing consistently accurate land leveling. The laser transmitter, which is mounted on a high podium /platform, rotates rapidly, sending the laser light in a circle like a lighthouse except that the light is a laser, so it remains in a very narrow spectrum (beam). The mounting has an automatic leveler built into it, so when it is set to all zeros, the laser's circle of light is perfectly level. Some laser transmitters have the ability to operate over graded slopes ranging from 0.01% to 15% and apply dual controlled slope in the field. The laser receiver is a multi-directional receiver that detects the position of the laser reference plane and transmits this signal to the control box. The receiver is mounted on a manual or electric mast attached to the drag bucket. The operator can adjust the settings on the receiver, and also can override the receiver wherever necessary. A laser-controlled land leveling system consists of the following five major elements (Fig. 8.1 and 8.2):

a) Toe hook

b) Bucket frame

c) Hydrolic cylinder

d) Beam

e) Tyre axle system

f) Mast

g) Receiver column

h) Directional valve

i) Bucket angle adjusting link

j) Receiver height adjusting lever

k) Parking stand

8.2. Terminology

a) *Earth work:* It involves the total volume of soil to be shifted in a field (plot) through cut and fill operations involved in land leveling.

b) *Grid:* The grid is defined by identifying one of its corners (lower left usually), the distance between nodes in both the X and Y directions, the number of nodes in both the X and Y directions, and the grid orientation.

c) *Land uniformity co-efficient (LUC):* It represents the magnitude as well as frequency of occurrence of successively larger undulations in the field.

d) *Leveling Index (LI):* Leveling index is an indicator for the levelness of the land surface, which is the difference in desired and achieved levelness of the field.

e) *Leveling:* The tillage operation in which the soil is moved to establish a desired soil elevation slope.

f) *Water-application efficiency:* Water application efficiency is the measure of efficiency with which water delivered to the field is stored in the root zone.

g) ***Water-distribution efficiency:*** The efficiency associated with the uniform distribution of water in an irrigation system. Or, it is defined as the percent of difference from unity of the ratio between the average numerical deviations from the average depth stored during the irrigation.

h) ***Water-use efficiency:*** The amount of dry matter that can be produced from a given quantity of water.

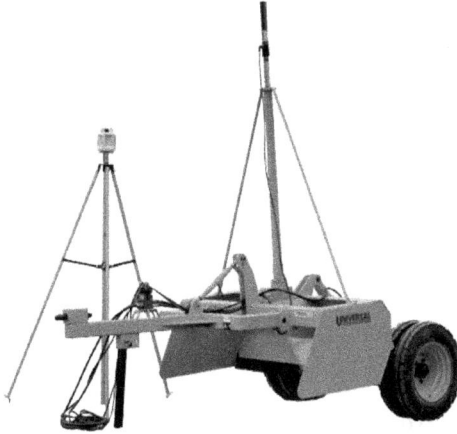

Fig. 8.1: Laser land leveler for agriculture

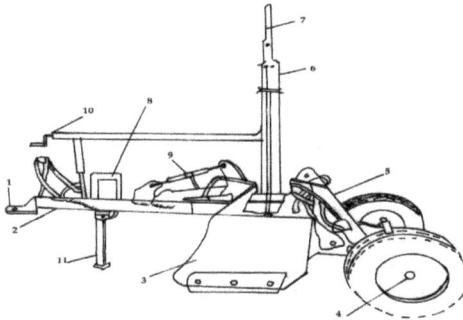

1. Toe hook 2. Beam 3. Bucket frame 4. Tyre axle system 5. Hydrolic cylinder 6. Mast 7. Receiver column 8. Directional valve 9. Bucket angle adjusting link 10. Receiver height adjusting lever 11. Parking stand

Fig. 8.2 : Schematic diagram and components of tractor operated laser land leveler

8.3 Scope of Test

8.3.1 Laboratory tests

a) Specification checking

b) Calibration of laser emitter

c) Determination of strength of laser beam

d) Effect of vibration on accuracy of measurement

e)   Effect of temperature on accuracy of measurement

f)   Water resistance test

g)   Determination of working range of laser leveler\

h)   Hardness and chemical composition of critical part

i)   Leakage test for seal and joints in the hydrolic circuit

j)   Endurance test for laser emitter and receiver

8.3.2 Field performance assessment

a)   Assessment of total volume of cut

b)   Assessment of total volume of fill

c)   Rate of work

d)   Quality of work

e)   Fuel consumption

f)   Field capacity

g)   Ease of operation, maintenance & adjustments

h)   Soundness of construction

8.3.3 Referred standards

As BIS test code is not yet available for testing of laser land levelers, the following BIS test codes / standards have been referred for this test procedure.

a)   IS 9813: 2002 [102]

b)   IS 7353: 1974 [85]

c)   IS 12362: 1994 [56]

8.4 Test Method

8.4.1 Calibration of laser emitter

The laser transmitter should be periodically checked for accuracy. Most laser transmitters have two horizontal level adjustment screws that allow minor adjustments to be made along the two axes of the horizontal plane. The axes are usually labeled 'X' and 'Y'. All checking and calibration procedures are done at the zero slope reading. The following items are required to check the accuracy of the transmitter.

- A suitable tripod that allows rotating the transmitter in 90 degree increments.

- A minimum 65 meter range that is unobstructed and as close to flat as possible.

The check/calibration procedure:

a)   Mount the unit on a tripod at one end of the 60 m range and level it. Set X and Y-axis grade counters at zero. With auto leveling transmitters, turn the transmitter control switch to the AUTO position and wait for the Auto Mode Indicator Lamp to stop flashing.

b) Station a rodman with a receiver at the other end of the range 60 m away.

c) Align the laser, using the sighting scope or groove, such that the 'X' is pointed directly at the rodman. Make sure the penta mirror is rotating and the Auto Mode Indicator Lamp has stopped flashing (if appropriate).

d) Have the rodman take a precise reading to within 2 mm and mark the reading as X1.

e) Rotate the transmitter 180 degrees and wait at least 2 minutes for it to re-level. In non-auto leveling transmitters, manually re-level the transmitter and the rodman take another accurate reading and mark it down as X2.

If the difference between X1 and X2 is less than 6 mm, no adjustment is necessary and the laser can be assumed to give the correct reading. If the difference is between 6 mm and 38 mm the transmitter then needs to be calibrated and this can be done locally in the field.

Calibration of the transmitter:

The procedure for calibrating the transmitter locally is given below:

a) From the two previous readings calculate the 'X' average = (X1 +X2)/2 and have the rodman adjust the detector on the rod to the 'X' average. (Center the detector between the two readings).

b) Locate the 'X' calibration screw and adjust it to align the beam to the 'X' average at the detector. If gentle turning of the calibration screw cannot align the beam, return the unit to an authorized service center for calibration.

c) After adjusting the beam, allow for the unit to stabilize before taking the next reading, then repeat the entire above mentioned procedure to check your work and do a fine readjust if necessary to get it just right.

d) After adjusting the 'X' axis rotate the transmitter 90 degrees to the 'Y' axis. Point the 'Y' axis directly at the rodman, using the sighting scope or groove and repeat the above steps a through c.

e) Measured the readings Y1 and Y2 and calculate the 'Y' axis average.

The same procedure may be employed by directing the beam onto a wall 60 m away. Instead of having the rodman recording on the staff, make a mark on the wall at X1 and X2 and then draw a line in the center. The beam is then adjusted until it is recorded at the centerline.

8.4.2 Measurement of strength of laser beam

a) Vibration test: The test to be conducted by mounting the emitter unit on a vibrator provided with adjustable vibration frequencies. The test should have a duration of 60 minutes at a vibration frequency during the transport of laser unit under field transport condition. The accuracy & repeatability across the X and Y axis of the rotating emitter be observed by calibration of unit before and after the operation.

b) Temperature gradient test: The test is conducted by keeping the laser emitter unit rotating at 595 rpm recommended for field operation at temperature of 55ºC ± 5ºC, continuously for 5 hours to assess the performance of laser emitter by measuring the deviation of laser beam in X and Y axis per 30 m of length.

c) Receiver & control box endurance test: It is tested in auto and manual mode for 6532 and 500 cycles, respectively for five days. The test is conducted at the working pressure recommended for field operation.

8.4.3 Land uniformity coefficient (LUC)

Levelling Index (LI) and Land uniformity coefficient (LUC) are measured using the following equations:

$$LI = \frac{\sum | DL_i - AL_i |}{N}$$

$$LUC = \left(1 - \frac{\sum | DL_i - AL_I |}{\sum DL_i} \right)$$

Where,

DLi   *- depth of cut (or filling) before levelling at point i (cm)*

ALi   *- depth of cut (or filling) after levelling at point i (cm)*

N   *- number of points of grid samples*

The minimum of LI is zero and it shows accurate levelling. LUC is between zero and one and values closer to one are accurate levelling. For taking grid readings, laser leveler should be put in manual control mode and then leveler is placed in the grid points on the ground until the height difference between the points are obtained.

The other laboratory tests and field tests (with suitable tractor) shall be carried out, recorded, analysed and reported. A sample datasheet for calibration and field test is shown in annexures 8.1 & 8.2.

Illustrative Example for Determining the theoretical time to level a field

The length of time taken to level the field can be calculated by knowing the average depth of cut from the cut/fill map, the dimensions of the field, the volume of soil that can be moved by the bucket and the tractor operating speed.

Given:

Field dimensions =100 m X 50 m; Average depth of soil to be cut = 25 cm,

Leveling bucket dimensions =2 X 1 X 1 m; Bucket fill =50%

Tractor speed (average of speed when the bucket is full and empty) = 8 km/h

Solution:

Volume of soil to be moved = (Field area/2) X avg. depth cut (m) = 100 X 50 / 2 X 0.25
= 625 m³

Volume of soil in bucket = 2 X 1 X 1 X 0.5 = 1 m³

Number of trips required = (625/1) X 2 (full and empty) = 1250 trips

Average trip length (50% of field) =100 / 2 = 50 m

Total distance traveled (m) = 1250 X 50 = 62500 m

Time (hours) = distance (m) / (speed (km/h)/1000) = 62500 / 8 /1000 = 7.81 h

Therefore approximately 8 hours is required to level this field. This is a theoretical time and will vary according to the skill of the operator, the soil type and operating conditions.

Annexure 8.1

Data Sheet for Calibration of Laser Beam

A. Deflection of laser beam before field test (mm)

B. Deflection of laser beam after field test (mm)

C. Deflection of laser beam after vibration test (mm)

D. Deflection of laser beam after temperature gradient test (mm)

Type of Test: A / B / C / D

Sr. No.	At X-axis	At X1-axis	At Y-axis	At Y1-axis	Self-levelling accuracy / deflections			RPM
					30 m length			
					At X-axis (X-X1)/2	At Y-axis (Y-Y1)/2	X-Y	
1								
2								
...								
...								
10								

Self-levelling accuracy

(Deflection per 30 m of length), mm

At axis X   :

At axis Y   :

At axis X,Y   :

Standard error at X   :

Standard error at Y   :

Standard error at XY   :

Annexure 8.2
Data Sheet for Field Performance

Type of soil :

Location of test field :

Field size (l x w) :

Tractor model & HP :

Gear used (speed) :

Particulars	Test No.1	Test No.2	Test No.3	Test No.4
Location				
Duration of test, min				
Soil moisture, %				
Area, m²				
Working width, m				
Field capacity, ha/h				
Fuel consumption, l/h				
Fuel consumption, l/ha				
Levelling index, LI, cm				
Before operation				
After operation				
Percent change in LI				
Earthwork, m³				
Before operation				
After operation				
Earthwork done, m³				
Land uniformity coeff., LUC				
Before operation				
After operation				
Percent change in LUC				

Testing and Evaluation of Seed-Cum-Fertilizer Drill, Zero-till Drill and Happy Seeder

Drilling is a common practice for mechanized sowing of different types of seeds of agricultural crops. One of the oldest seeding method was broadcasting, which comprise of scattering the seeds randomly and covering them by soil. For larger plots, broadcasting experienced a low productivity. Also, broadcast seeding resulted in a random array of growing plants, making it difficult to control weeds by using mechanized methods other than manual weeding. The drill laid the foundation for developing modern day seeding equipment including direct drill seeders and air seeders. The seed drill allows the farmers to sow seeds at desired row spacing and depth and as per recommended seed rate for a given crop. Modern day seed drills use different type of metering mechanisms for seeds and fertilizer as well as various types of furrow openers suited to the field & ground conditions comprising well-prepared seedbeds, zero/no-till field conditions as well as fields with stubbles and crop mulch. The result is the desired seed and fertilizer metering, accurate placement leading to proper germination, healthy crop stand which contributes to higher crop yields and productivity. The drive to metering devices of a drill is generally taken from the ground driving wheels. However, electric and hydraulic motors are also used in drills and seeders these days.

The no-till/zero-till/ direct drills are being widely used for conservation agriculture. Similarly, the roto-drill and happy seeders have been adopted for sowing of wheat and other crops under stubble conditions [110, 157, 203, 207, 209]. In recent years, pneumatic seed drills/ air seeders are being used in advanced countries for sowing large fields. In these drills, air conveys the seeds through plastic tubes from the seed hopper to the coulters/furrow openers. It is an arrangement, which allows seed drills to be much wider than the seed hopper— as much as 18 m wide in some cases. The seed is metered mechanically into an air stream created by a hydraulically or mechanically powered on-board blower and conveyed initially to a distribution head, which sub-divides the seed into the pipes taking the seed to the individual coulters. For smaller seeds, air seeders are optimal for sowing [6].

The fertilizer should not come in contact with the seed to avoid chemical injury. Therefore, precision placement of seeds and fertilizer is necessary for achieving perfect standing of crop. It also reduces sowing time and thus overcomes the shortages of labour.

Functional requirements of a seed drills:

- It drops seeds and fertilizer at preset rates.
- It opens furrow to a uniform depth and spacing.
- It drops seeds and fertilizer uniformly without causing any injury to seeds.
- It maintains proper distance and depth between seeds and fertilizer to avoid damage to the seeds
- It covers the seeds and compacts the soil around them in order to conserve moisture.

9.1 Types of Seed Drills

9.1.1 Animal drawn seed drill

Several designs of 2 to 5 row animal drawn drills with variations in metering, power transmission systems, and frame and furrow openers have been developed during the last three/four decades in India. The design of a five row animal drawn seed cum fertilizer drill is shown in Fig. 9.1. Fluted rollers are commonly used for metering the seed and fertilizer, by a ground wheel drive comprising chain and sprocket mechanism. A beam is attached to the frame for pulling the drill. Seeds along with fertilizer are placed in the hopper and seed drill is moved with a desired speed in the field.

Courtesy: Somnath, India

Fig. 9.1: Animal drawn seed drill

Seed and fertilizer metering devices deliver the required quantity of seeds and fertilizer through the delivery tubes attached to the furrow openers. The seed rate is controlled by changing the exposure setting of the fluted roller in the metering mechanism. Furrow openers are attached to the frame with the help of shanks and row spacing can be adjusted according to the crop. Commonly used furrow openers are hoe-type with reversible shovels with split boot or double-pores. The animal drawn seed-cum-fertilizer drill is used for sowing wheat, gram, sorghum, soybean, pigeon pea, sunflower and similar seeds.

9.1.2 Seed-cum-fertilizer drill

Out of all the designs of the seed-cum-fertilizer drills used in India at present, the one employing external fluted rollers for seed metering is popular and widely used (Fig 9.2(a)). The major components of such a drill (Fig. 9.2(b)) are frame, seed hopper, fertilizer hopper, furrow openers, transport wheels, depth adjuster, seed metering mechanism, fertilizer metering mechanism, drive wheel, sprocket and chain mechanism, seed rate indicator, seed delivery tubes, fertilizer tubes, tynes and hitching links.

Courtesy: Bir Singh & Sons, India

a) Fluted roller seed metering,

b) Seed rate adjustment,

c) Drive wheel

Fig. 9.2(a): Seed cum Fertilizer Drill

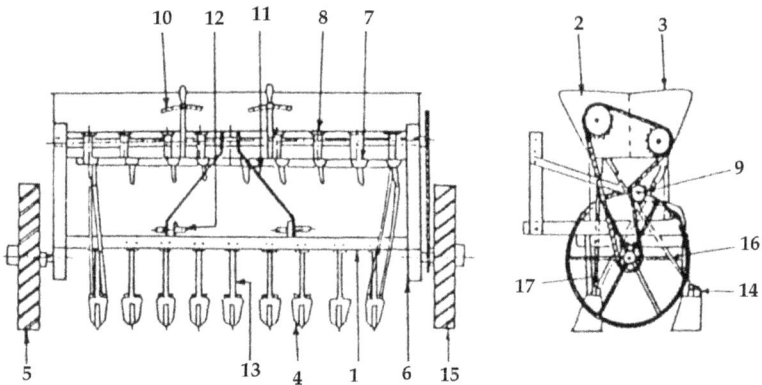

1. Frame 2. Seed hopper 3.Fertilizer hopper 4. Furrow opener 5. Transport wheel 6. Depth adjuster 7. Seed metering mechanism 8. Fertilizer metering mechanism 9. Sprocket chain tighter 10. Indicator 11. Hitching link 12. Hitch pin 13. Tine 14. Seed boot 15. Driving wheel 16. Seed tub 17. Fertilizer

Fig. 9.2 (b) Major components of seed-cum fertilizer drill

9.1.3 Zero till seed-cum fertilizer drill

It is like a seed-cum fertilizer drill but its inverted T type furrow openers are used for direct penetration into the soil. Specific type of furrow openers make the way for direct sowing of seed in the soil (Fig.9.3). All other functional requirement are met as available in seed-cum fertilizer drill.

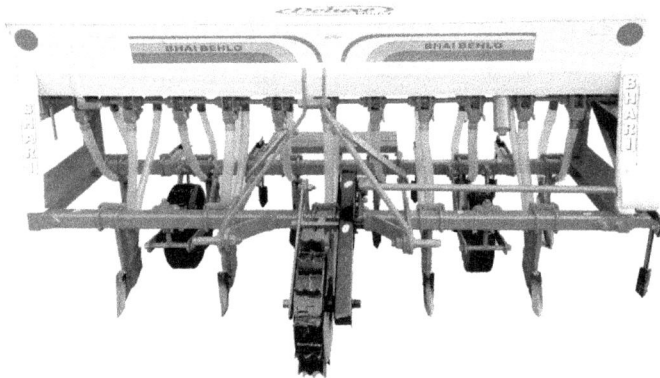

Courtesy: Bhai Behlo

Fig. 9.3 Zero till seed-cum fertilizer drill

9.1.4 Happy seeder

It is a combination of rotavator and zero till seed-cum fertilizer drill. Rotating blade of happy seeder cuts the straw and partially cuts the stubble and soil and evenly mulch on soil surface (Fig. 9.4). The Happy seeder is operated by tractor PTO shaft for rotating of blades whereas seed-cum fertilizer drill gets the power from ground wheel. This can be used for direct sowing of seed such as wheat after paddy harvesting by combine harvester.

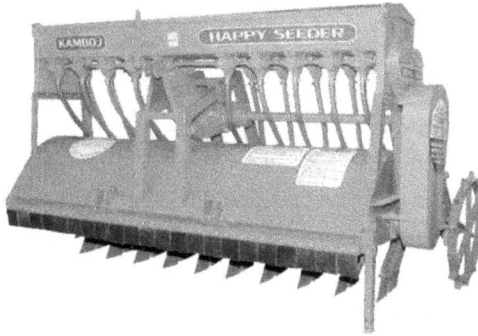

Courtesy: Kamboj

Fig. 9.4 Happy seeder

9.1.5 Pneumatic seed drill

ASABE standard S506 OCT2010 (R2014) defines air seeders as the type of machinery that uses a centralized hopper and a volumetric metering mechanism to contain and meter the seeds, respectively [11]. As the name implies, air seeders use a flow of air to distribute and deliver the seed into the soil. Despite their size, air seeders can be folded to narrower widths for convenient transport and roadwork (Fig. 9.5). The speed of the metering roller determines the seed rate and the seeding rate is automatically adjusted to changing working speeds. The metering roller selection is dependent on the grain size and seed rate. The selected metering roller volume (cm^3) should not be too large but still sufficient to spread the required quantity (kg/ha).

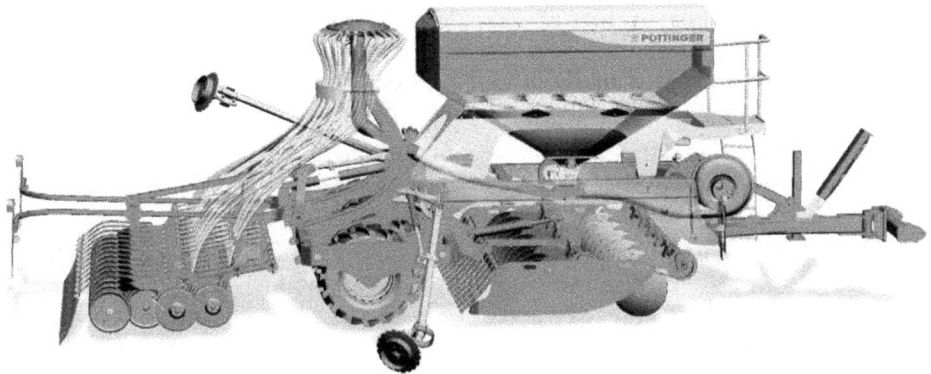

Courtesy: Pottinger, Europe

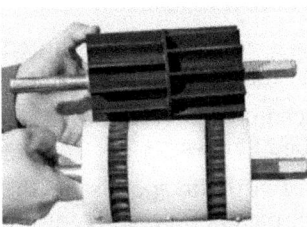

a) Metering rollers   *b) Seed rate adjustment*   *c) Disc Coulter with pressure roller*

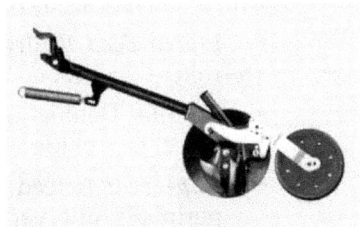

Fig. 9.5: Pneumatic Seed Drill (Air Seeder)

9.1.3 Terminology

a) **Broadcasting:** The process of scattering of agricultural inputs, such as seed, fertilizer and manure on the surface of the soil.

b) **Drilling:** The process of placing seeds in rows at uniform rate and at controlled depth with or without covering them with soil.

c) **Zero/No-till drilling:** Direct seeding in essentially in untilled seed beds.

d) **Precision Drilling:** Uniform placing of seeds in rows at a predetermined depth and seed rate.

e) **Pneumatic Seed Drill:** A seed drill in which the metering of seeds is done by air stream at set spacing in continuous parallel rows.

f) **Fertilizing:** Process of fertilizer application in or on the soil, or on the plants.

g) **Furrow Openers (Fig. 9.6 (a)):** It is a part of seed drill for opening a furrow and assisting in placing of seed.

- **Single-Disc:** A furrow opener consisting of one concave disc

- **Double-Disc:** A furrow opener consisting of two -flat discs, set at an angle to each other.

- **Shovel Type:** A furrow opener consisting of a single or double pointed shovel fastened to the lower part of the boot. Reversible shovels are also available. These are made of carbon steel and are widely used in seed drills and are best suited even in strong or root infested fields.

- **Shoe-Type:** A furrow opener sledge shaped in elevation, with a V-shaped leading edge and hollow body through which seeds are dropped. It is made of carbon steel and works well in trashy soils where the seed bed is not well prepared.

- **Stub Runner:** A furrow opener consisting of two tapered flat pieces of steel welded together, so as to form a narrow, V-shaped leading end and a hollow body into which seeds are dropped.

- **Inverted T:** It creates an inverted T type furrow groove with reduced surface exposure and thereby helps to maintain the in-groove humidity in a reasonably wet soil for better germination and emergence of seedlings.

h) **Seed Metering Mechanisms (Fig. 9.6(b)):** The mechanism of a seed drill which delivers seeds from the hopper at selected seed rates.

- **Fluted-Feed Roller:** A seed metering device with adjustable fluted roller to collect and deliver seeds into a seed tube

- **Internal Double Run:** A seed metering mechanism in which the feed wheel is provided with fine and coarse ribbed flanges

- **Cup-Feed:** A feed mechanism consisting of cups or spoons on the periphery of a vertical rotating disc, which picks up seeds from the hopper and delivers them into a seed tube

- *Auger-Feed:* A feed and/or distributing mechanism consisting of an auger, which causes a substance to flow evenly.

i) *Fertilizer Metering Mechanisms (Fig. 9.7):* A mechanism for discharging the fertilizer from hopper at predetermined fertilizer rate.

- *Spur wheel:* A wheel carrying six to twelve spurs, meters the fertilizer uniformly.

- *Serrated discs:* The circular plate carrying the serrations on the periphery rotates through an adjustable notch.

- *Star wheel:* A horizontal disc carrying stars at the periphery, meters the fertilizer without any effect of gravity.

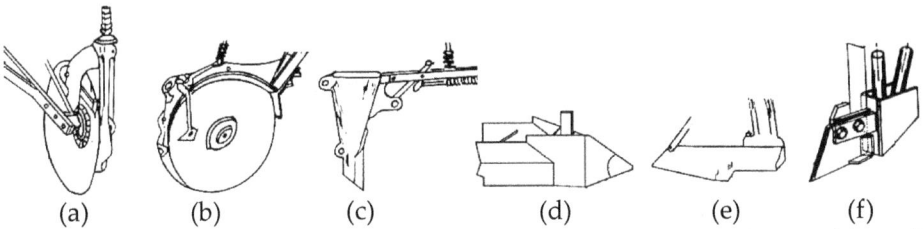

(a)   (b)   (c)   (d)   (e)   (f)

a. single disc, b. double disc, c. hoe type, d. shoe type, e. stub runner, f. inverted T

Fig. 9.6(a): Types of Furrow openers

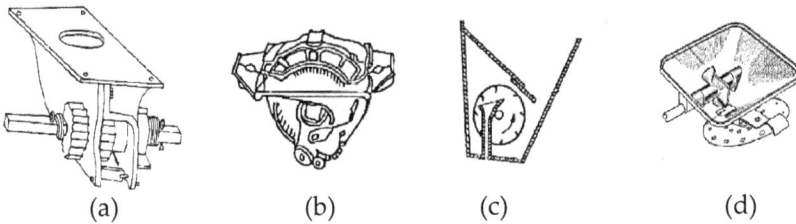

(a)   (b)   (c)   (d)

a. fluted feed roller, b. internal double run, c. cup feed, d. augur feed

Fig. 9.6(b): Types of Seed metering mechanisms

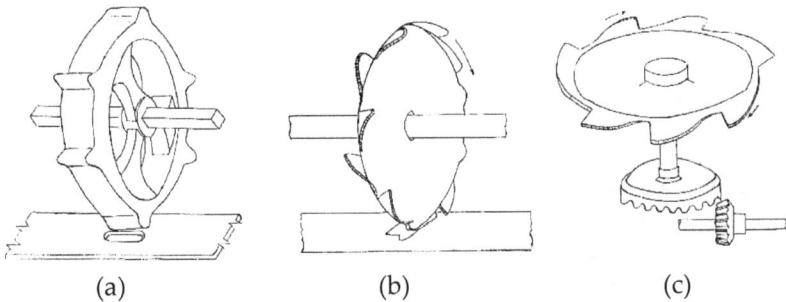

(a)   (b)   (c)

a. spur wheel, b. serrated disc, c. star wheel

Fig. 9.7: Types of Fertilizer metering mechanisms

9.3 Scope of Tests

The following aspect of the machine's performance shall be assessed.

9.3.1 Laboratory tests

a) Specification checking

b) Stationary calibration

c) Testing on repeatability of notch mark setting for calibration

d) Effect on seed discharge rate due to depth of seeds in the seed hopper

e) Effect on seed discharge rate due to different forward speeds

f) Uniformity of seed and fertilizer spacing in the row

g) Seed specifications

h) Seed damage

i) Chemical analysis and hardness test of critical components

9.3.2 Field test

a) Field calibration at different settings

b) Rate of work

c) Relative placement of seeds and fertilizer

d) Quality of work

e) Power requirement

f) Fuel consumption

g) Labour requirement and field efficiency

h) Ease of setting, operation and adjustment

i) Soundness of construction

9.3.3 Referred standards

a) IS 6316: 1993 [78]

b) IS 6813: 2000. [84]

c) ISO 6720:1989. [140]

d) ISO 7256-1:1984. [141]

e) ISO 7256-2:1984. [138]

f) RNAM Test Codes [191]

9.4 Methods of Tests

9.4.1 General conditions

i) Selection of machine

The test sample shall be selected at random from the production line by a representative of the testing station. The method of selection shall be specified in the test report. The test machine should strictly conform to the description and specifications submitted by the applicant and as generally offered for sale, except when the machine is a prototype and is not on mass production and submitted for confidential test.

ii) Running-in and preliminary adjustments

The machine should be fully assembled, examined in detail and run-in as per recommendations of the manufacturer, before the commencement of the test. At this stage, the applicant is encouraged to depute his representative for necessary adjustments and to demonstrate the operation of the machine to the testing staff. The test team should preliminarily run the machine to familiarize themselves with the operation and to check clarity of the instruction manual. All the adjustments made, should be in accordance with the instructions contained in the Operator's Manual/ literature/other written instructions on adjustments issued by the manufacturer. Adjustments made during test period which are not in conformity with the manufacturer's recommendations, shall be reported in the test report.

iii) Servicing and maintenance

The machine shall be serviced as per manufacturer's recommendations. All the tests shall be done with the seed box three-forth full. However, if necessary, this shall be repeated with half and full seed box.

9.4.2 Test conditions

9.4.2.1 Test conditions of Seeds and Fertilizer

a) **Seeds**

i) Scientific and popular name and variety

ii) Germination rate in the laboratory: Germination rate is measured by counting the number of germinated seeds in a laboratory dish in which filter paper is laid and water supplied. Usually, the laboratory dishes are put into an incubator at 25-30°C. Ratio of germinated seeds to the total seeds after three days is called germination rate.

iii) Bulk density: It is measured by filling the material in the cube or cylinder of which the volume and weight is known.

iv) Average size: It is measured to distinguish size between different types of seed and varieties.

b) **Granular fertilizer**

i) Name and kind of fertilizer

ii) Distribution of granule size: It is measured by sieving (1.0, 2.0, 2.8, 4.0 mm sieves) using a mechanical sieve shaker. Sieving time shall conform to the instructions provided by the sieve manufacturer.

iii) Bulk density

iv) Angle of repose: There is a relation between the angle of repose and the discharge rate. Therefore, it is measured and the relationship between angle of repose and discharge rate is established.

v) Moisture content

9.4.2.2 Test conditions of machine and test field

a) Source of power: manual, animal drawn or tractor drawn

b) Adjustment of parts: metering shaft speed, delivery opening, row spacing etc.

c) Test condition of field: area & shape of test plot and type of soil

d) Method of land preparation and size distribution of soil clods at surface layer (under 1 cm, 1-2 cm, 2-3 cm, 3-4 cm, over 4 cm.)

e) Soil Moisture Content and Bulk Density

f) Atmospheric conditions

9.4.2.3 *Test Condition of operation*

a) Setting of seed and fertilizer rate

b) Setting depth of drilling

9.4.3. Procedure for laboratory testing

9.4.3.1 *Specification checking*

The manufacturer shall supply all specifications about the machine in the prescribed format. These shall be verified by the Testing Station and any discrepancy, if noticed shall be mentioned in the test report.

9.4.3.2 *Stationary calibration*

For the laboratory calibration, three different type of seeds shall be selected from the applicant's recommendations. If no recommendation has been made by the applicant, the common seeds used in the area should be selected. The machine is tested on the special seed drill testing rig which can be fabricated at the testing station. The drive to the feed shafts is given in the same ratio as in the actual machine operation by an electric motor. Automatic calibration system developed may also be used for calibration of seed drills and a schematic layout of the test rig is given in Fig. 9.8, 9.9 and 9.10.

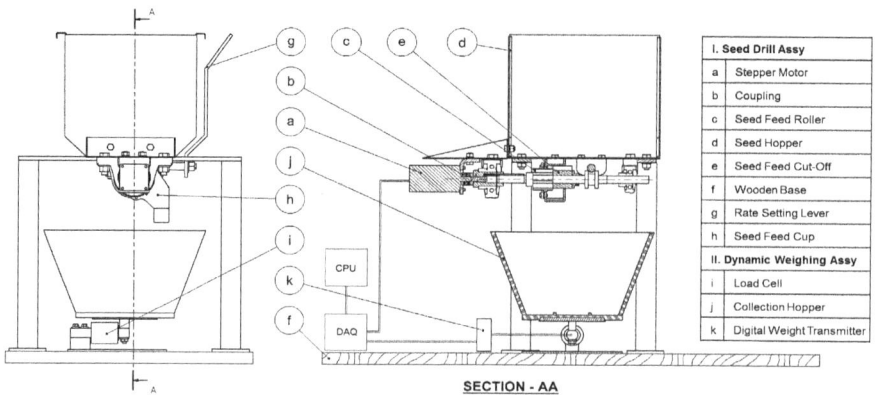

I. Seed Drill Assy	
a	Stepper Motor
b	Coupling
c	Seed Feed Roller
d	Seed Hopper
e	Seed Feed Cut-Off
f	Wooden Base
g	Rate Setting Lever
h	Seed Feed Cup
II. Dynamic Weighing Assy	
i	Load Cell
j	Collection Hopper
k	Digital Weight Transmitter

SECTION - AA

Fig. 9.8: Schematic diagram of automated calibration test rig for seed drills

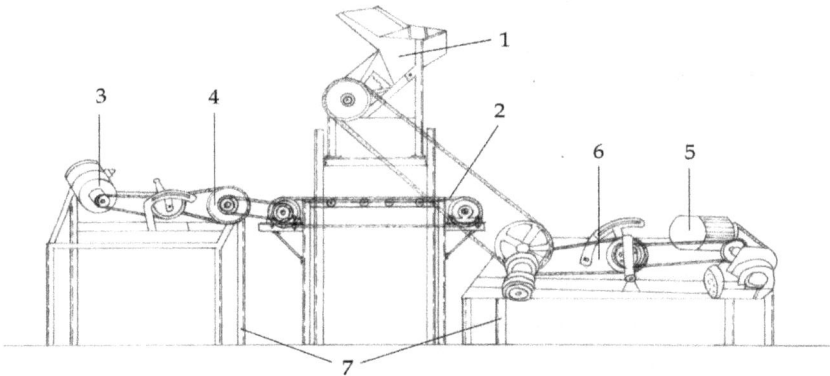

1. Seed drill 2. Greased belt 3. Motor for transmission for grease belt 5. Motor for operating seed drill
6. Variable drive for seed drill

Fig. 9.9: Schematic diagram of seed drill test rig

Calibration of seed drill

Laboratory calibration is required to check correctness of seed and fertilizer discharge rates. The procedure is as follows:

Material required for calibration

a) About 40 kilogram of seeds or the seeds required for sowing one fourth of hectare. The variety of the seed should be same as to be sown.

b) Weights and balance to weigh the seeds.

c) Measuring tape of 25 meter.

d) Cotton cloth bags or plastic bags to collect the seeds and fertilizers coming out from each of the seed/fertilizer tube.

e) Two jacks or wooden blocks for jacking up the machine above the ground, so that drive wheels could be rotated freely.

f) Marker for putting impression on drive wheel of seed drill.

9.4.3.3 Procedure for calibration

i) Measure the distance between two adjacent furrow openers: Check inter furrow distance and if necessary adjust to the recommended value. Count the number of furrow openers.

ii) Effective working width of the drill:

$W = (N \times d)/100$ meter

Where, N = number of furrow openers,

d =distance between two adjacent furrow openers (cm)

iii) Circumference of the driving wheel (L):

Circumference of the wheel (L) = $\pi D/100$ meter

Where, D = diameter of the driving wheel (cm)

iv) Area sown in one revolution of the drive wheel (A):

A = W x L (m^2)

v) Number of revolutions required to sow one hectare (R):

R = 10000/A

Number of revolutions actually required to cover one hectare considering 10% wheel slip (M): M = R (100-10) / 100 revolutions

vi) Seed rate to be sown per hectare (kg/ha) = P

vii) Seed quantity required for one revolution(s) = S:

S = Seed rate (kg/ha) / No. of revolutions required to sow one hectare

S = P x 1000 / M = G grams

viii) Seed quantity required for 50 revolutions: G x 50 = B grams (say)

ix) Similarly calculate fertilizer quantity required for 50 revolutions (say 'C' grams)

As per the above calculations, prepare the drill for calibration in the following manner:

a) Jack up the seed drill. Remove the tubes from the feed cup and in their place tie plastic or cotton bags to collect the seed and fertilizer. The bags should be numbered.

b) Marking by marker is done on the drive wheel to count the number of revolutions.

c) Rotate the drive wheel 50 times at normal field speed.

d) Weigh the seeds and fertilizer in each bag. The inter furrow variation for seeds should not exceed ± 7% and for fertilizer ± 12.5%. If the variation is higher, then the drill should be thoroughly checked and defect rectified. The calibration should be repeated. If the variation continues, the same should be reported in the test report.

e) The seeds of all the bags be collected in one big bag and weigh it. Similarly, weigh the fertilizer also. These quantities should be equal in weight as calculated above at point viii and ix. If this is less or more the metering mechanism should be adjusted accordingly and calibration process be repeated till we get desired seed and fertilizer rate. A firm mark should be put on the indicator setting accordingly.

9.4.4 Field checking

Many owners do not calibrate their seed drills, but check the operation in the field. This is done as follows:

a) Set the drill at the desired seed rate through indicator.

b) Set the drill in the field where an area of about 0.04 hectare (400 m^2) has been marked for sowing.

c)   Fill the seed box or hopper with seed up to a mark, or to the top and level off carefully.

d)   Sow exactly 0.04 hectare as marked in the field.

e)   Refill the seed box from a bag of which exact weight is known.

f)   By reweighing the grain left in the bag and subtracting from the original weight of the bag, will indicate quantity sown. If the quantity is not correct, make compensation and adjust the seed drill and check again. The field checking method is not recommended for laboratory test.

9.4.5 Testing on repeatability of notch-mark setting for calibration

The machine is calibrated (following stationary calibration method) at three random positions on the indicated scale with three replications. Each replication is attempted after shifting the calibration lever several times. The objective is to ascertain the accuracy of the notch setting in terms of repeated operations.

9.4.6 Effect on seed and fertilizer discharge rate due to level of seed and fertilizer in hopper

The effect of seed depth in hopper on seed discharge rate is determined keeping the hopper full to 3/4 full, 3/4 full to 1/2 full and 1/2 full to 1/4 full. The other variations such as position of seed drill, position of seed adjusting lever shall be kept constant. The variation in seed discharge due to box filling should not exceed 10%. Similarly effect on fertilizer rate should be checked.

9.4.7 Effect on seed discharge rate due to different forward speeds

The difference in seed discharge rate would be assessed by using recommended seeds and variation in discharge at three different speeds shall be recorded. The variation in seed discharge due to different speed should not be more than 15%.

9.4.8 Uniformity of seed spacing in the row

A sticky conveyor belt coated with thick oil or grease, is passed under the machine or the furrow openers at equivalent operating speed as in the field and the feed shaft is also driven corresponding to the field operating speed and the transmission ratio. The pattern of seed falling on the belt is recorded. The mean spacing shall be recorded and suitable index of variation from mean will be used to express the uniformity of spacing. Three replications shall be carried out and results of each run shall be recorded. This test will give the accuracy and precision at which the seed is metered and delivered into the furrows.

9.4.9 Seed specifications

The bulk density and number of seeds in one kilogram shall be determined for all types and varieties of seeds to be used during testing before actual testing of seed drill is carried out.

9.4.10 Seed damage test

The test is conducted to see the visual damage caused to the seeds passing through the metering device. Grain samples from the seed hopper as well as from the seeds passing through the metering device are taken separately and analyzed for external rind internal damage and reported.

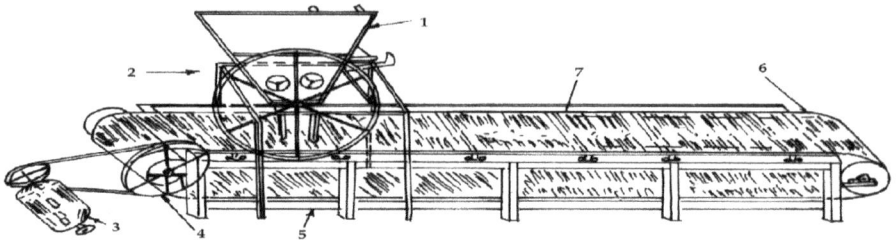

1. Seed drill 2. Tractor drive 3. Motor 4. Drive pulley for conveyer belt 5. Main frame
6. Conveyor belt (flat type) 7. Roller
Fig. 9.10: Schematic diagram for uniformity seed test rig

9.4.11 Chemical composition and hardness test of critical components

The chemical composition and hardness of furrow opening device and other critical components as decided by the testing authority shall be determined and reported. The weight of furrow openers at the start and completion of test shall be used to determine rate of wear of furrow openers.

9.5 Procedure for Field Testing

a) Field calibration at different settings

Field calibration is done with the same type of seeds which are used for laboratory calibration. A 100 meter run in well-prepared seedbed is used for field calibration. The condition of the seedbed is recorded. The machine is run for actual sowing operation except that the seeds dropped from the different rows are collected separately and weighed. The number of revolutions of the ground/driving wheel are also recorded to determine the wheel speed effect on seed rates in comparison to that obtained in the Laboratory. The calibration is done at the same notch settings as done during laboratory test and reported accordingly.

b) Rate of work

The machine has to run continuously in the field for about 25 hours under different field sizes and conditions to assess the working capacity of the machine and to record any breakdown or excessive wear. The net rate of work excludes the turning time at head land, time for filling seeds in hopper and other stoppages.

c) Relative placement of seeds and fertilizer

Operate the drill in the field under the normal seedbed conditions with average depth setting of the furrow openers. Three rows of at least 100 meter length should be covered. Then carefully remove the soil without disturbing the seeds and fertilizer at minimum 5 different places in each row. Measure the depth of the seeds below the soil surface and the vertical spacing of the fertilizer with respect to the seeds. Measure the horizontal spacing also. It will be much easier to locate the seed and fertilizer if maximum rate of application is used. The fertilizer is more easily located if white lime is mixed with it. Repeat the procedure at minimum and maximum depth setting of the furrow openers. Average value of relative placement, horizontal and vertical spacing, be reported.

d) Quality of work

The quality of work including coverage of seeds, row spacing, blockages and wheel skids would be recorded during field work.

e) Power requirement

The draft required to move the machine in a normal seedbed is observed by using an appropriate hydraulic dynamometer. The draft is measured at maximum, minimum and average depth of sowing and with 3/4th full seed hopper. Detail procedure of power measurement for trailed and mounted type seed drill is given below:

i) For trailed seed drill

The draft is defined as the horizontal component of the pull, parallel to the line of motion. Install a hydraulic dynamometer in the hitching point to measure the draft. If the line of pull through the dynamometer is not horizontal, then measure the angle between the line of pull and the horizontal line. Calculate the draft of horizontal component. Mark a 20m run space in the middle of a long row in the field with easily distinguished poles. Start the drill well in advance of the first pole and be sure it is operating smoothly when it reaches this pole. As the drill travels the marked run length, record the dynamometer reading. The more the readings taken the better the results are. Calculate the average of all the readings taken within a particular run. At the same time record the time, the machine took to cover the marked run length. From this value, calculate the speed of travel in m/s. Also, calculate the wheel slip and theoretical field capacity. The power can be calculated as follows:

$$Metric\ horsepower\ (Ps) = \frac{Draft\,(kgf)\times Speed\,(m/\,s)}{75}$$

ii) For mounted seed drill

Mark a 20m run in the field with the help of poles. Measure pull of the tractor to be used for seed drill operation by pulling it with another tractor having dynamometer in between both tractors. Similarly, pull of same tractor with seed drill in operation is noted in the same marked run. The draft of seed drill is computed by difference of these two pull readings. Time taken to cover 20 m is also recorded to calculate speed of operation. The power requirement is calculated with the help of above mentioned formula.

f) Fuel consumption

The fuel tank of tractor is filled to its full capacity before and after test. The quantity of fuel filled at the end of test divided by total hours of operation will give hourly fuel consumption. Use of digital fuel flow meters may be explored so that more accurate readings can be obtained with respect to time.

g) Labour requirement and field efficiency

The actual labour requirement for operating seed drill in the field should be assessed and recorded. The theoretical field capacity is the rate of field coverage that would be obtained if the drill was operating continuously without interrupting like turning at the ends, filling of hoppers, etc. The effective field capacity is the actual rate of work which includes the time lost in filling hoppers, turning at the end of rows, cleaning of openers, making adjustments etc. The theoretical field capacity can be determined while taking data for draft requirement. Field efficiency of the seed drilling operation is computed from the above values.

h) Ease of setting, operation and adjustments

During the field operation of the machine, all the operational and adjustment difficulties are recorded to assess the handling characteristics.

i) Soundness of construction

All major and minor breakdowns/damage of the parts occurred during entire test period are recorded and reported in the report.

Sample datasheet for laboratory and field test have been given at annexures 9.1 to 9.5.

Annexure 9.1

Data Sheet for Laboratory Test - Seed Rate

Seed Type: Turning Time: (sec)

1000 gr. w.: (g)   Revolutions of ground wheel: (sec)

Wt. of seed metered	Outlet 1				Outlet 2				Outlet 3			
	Full	¾	½	¼	Full	¾	½	¼	Full	¾	½	¼
Hopper Level												
Maximum Setting												
Minimum Setting												
Sample from above	1				2				3			
Sample weight (g)												
Damaged seeds (g)												
Seeds rate*	Maximum, kg/ha								Minimum, kg/ha			

Annexure 9.2
Data Sheet for Laboratory Test - Seed Distribution

Seed Type:	Seed Rate*:		kg/ha	100 grain weight:		g
	Row Spacing:		cm	No. of Seeds/m**:		
Seeder Speed:	Time for travelling 4m:			Sec		
Measured seed distribution per meter row length+						

Distance between 2 seeds (mm)	No. of distances (in 1m row)	No. of distances	No. of distances
2			
10			
15			
20			
..			
95			
100			

* denote average value
** use recommended values
+ in case of hill placing, indicate average number of seeds per hill brackets behind the number of distances in a certain class.

Annexure 9.3
Data Sheet for Field Testing of Seed-cum-Fertilizer Drill

Date	Operator	Type of Seed	Seed Moisture	Place of Test	Labour requirement
Variety of seed		Germination rate of seed before sowing (%)		Time of start	Name of supervisor
No load revolutions		Type of fertilizer		Time of end	

S. No.		1	2	3	4	5
Time for 20 m (sec)						
Width of sowing for 3 rows (m)						
Wheel revolution in 20m	LH drive wheel side					
	RH free wheel					
Depth of sowing	Seed (cm)					
	Fertilizer (cm)					

Vertical spacing b/w seed to fertilizer (cm)

Horizontal spacing b/w seed to fertilizer (cm)

Seed to seed distance in row (cm)

Row Spacing

Seed rate (kg/ha)

Weight of seed in 20m

Weight of Fertilizer (20m)

Fertilizer rate (kg/ha)

Draft (kg)

Annexure 9.4

Summary results for Field Test of Seed cum Fertilizer Drill

Test Team	Type of prime mover	Type of Seed	Seed moisture (%) Germination rate of seed before sowing (%) Soil moisture (%)
Place of test	Labour requirement	Variety of seed	
No load revolutions		Type of fertilizer	

S. No.		1	2	3	4	5
Date						
Time for 20 m (sec)						
Duration of test						
Width of sowing for 3 rows (m)						
Wheel revolutions in 20m	LH drive wheel side					
	RH free wheel					
Wheel slip (%)						
Depth of sowing	Seed (cm)					
	Fertilizer (cm)					
Vertical spacing b/w seed to fertilizer (cm)						
Horizontal spacing b/w seed to fertilizer (cm)						
Seed to seed distance in row (cm)						
Row Spacing						
Seed rate (Kg/ha)/ Weight of seed in 10m						
Weight of Fertilizer (10m)						
Fertilizer rate (kg/ha)						
Draft (kg)						

Area covered (ha/h)	
Fuel consumed (l/h)(l/ha)	
Atmospheric condition/ Ambient temp (°C)	

Annexure 9.5
Calculation Sheet: Seeding Efficiencies

Laboratory and Field	Unit	Symbol	Values	
			Lab	Field
Seed spacing*	cm	SS		
Seed spacing standard deviation	cm	SSD		
Seed spacing evenness = (SS-SSD)/SS	--	E_u		
Seeding depth*	cm	d^1		
Seeding depth standard deviation	cm			
Seed depth evenness = (SS-SSD)/SS	--	E_d		
From the Laboratory				
1000 grain weight	g	1000gw		
Seed dimension	mm	w^1		
Germination rate	%	b^1		
Weight of seed sample having passed through the metering mechanism	g	E_b		
Weight of broken seeds in this sample	g			
Seed breakage efficiency = $(w^1 - b^1)/w^1$	--			

denotes average value

Chapter 10

Testing and Evaluation of Planters and Transplanters

Crop planting operations include placing of seeds or tubers (such as potatoes) in the soil at predetermined depth and spacing for plant to plant and row to row distance or setting seedlings / plants in soil (such as paddy seedlings). Machines that place the seed in the soil and cover it in the same operation create definite rows, which are wide enough to permit operation of machines between them for inter-row cultivation or other cultural operations. This is called row-crop planting.

A seed planter or transplanter is required to perform all the following mechanical functions:

- Meter the seeds / seedlings

- Open the seed furrow to the proper depth and spacing (may not be necessary for transplanting)

- Deposit the seeds / seedlings in the furrow as per the cultivation requirements

- Cover the seeds and compact the soil around the seed to the proper degree for the type of crop.

Planting may be done on the flat surface of the field, in furrows, or in beds, as illustrated in Fig. 10.1. Furrow planting is practiced in semi-arid conditions because this system places the seed into moist soil and protects the young plants from wind and blowing soil. Bed planting is often practiced in high-rainfall areas to improve surface drainage. Flat planting is practiced in fields where natural moisture conditions are favorable.

Static calibration of the planters and transplanters is essential before planting in each season to ensure the required plant population, i.e. with minimum misses and high quality of feed index. Semi-automatic and automatic transplanters are commonly used in transplanting of seedlings. Testing and evaluation procedures of some of the most commonly used planters such as potato planter, sugarcane planter and transplanters like rice transplanter and vegetable seedling transplanters are discussed in this chapter.

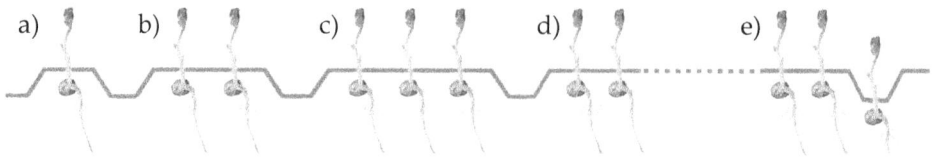

a) Ridge planting, b) Double row planting, c) Triple row planting, d) Planting on level field, e) Furrow planting
Fig. 10.1: Different types of planting practices

General scope of test for all planters given in Table 10.1 shows an overview of the tests to be conducted for different types of planters, and specific test methods with respect to different crop specific planters are given in their respective headings.

Table 10.1 Scope of test for different types of planters and transplanters

S. No.	Test Parameters	Seed planters	Potato planter	Sugarcane planter	Rice transplanter	Vegetable Seedling transplanter
(1)	(2)	(3)	(4)	(5)	(6)	(7)
I.	**General Conditions**					
1.	Selection of machine	•	•	•	•	•
2.	Running-in and preliminary adjustments	•	•	•	•	•
3.	Servicing and maintenance	•	•	•	•	•
II.	**Laboratory Tests**					
1.	Specification checking	•	•	•	•	•
2.	Seed specifications: variety, size, bulk density, moisture content etc.	•				
3.	Seed germination, %	•				
4.	Index of tuber shape		•			
5.	Age of seedlings, days				•	•
6.	Density of seedlings				•	
7.	Seedling/Plant height				•	•
8.	Rupture strength of seedling mat				•	
9.	Leaf stage				•	•
10.	Plant establishment density				•	
11.	Stationary calibration	•	•	•		
12.	Chemical analysis and hardness test of critical components	•	•	•	•	•

III.	Field Tests					
(1)	**(2)**	**(3)**	**(4)**	**(5)**	**(6)**	**(7)**
1.	Field condition: area & shape of test plot, type of soil	•	•	•	•	•
2.	Method of seedbed preparation and size distribution of soil clods at surface layer	•	•	•		•
3.	Soil moisture content	•	•	•	•	•
4.	Bulk density	•	•	•	•	•
5.	Hard pan depth				•	
6.	Soil hardness 'cone depth'				•	
7.	Water depth				•	
8.	Atmospheric conditions	•	•	•	•	•
9.	Field calibration at different settings	•	•	•	•	•
10.	Number of rows and Row spacing, cm	•	•	•	•	•
11.	Plant/seed spacing, cm	•	•	•	•	•
12.	Depth of planting, cm	•	•	•		•
13.	Cell filling errors		•			
14.	Stalk cutting, %			•		
15.	Damaged stalk eyes, %			•		
16.	Furrow depth and amount of soil over the seed cane / billets			•		
17.	Seedling biometry			•		
18.	Test of damage to sprouts on pre-sprouted tubers		•			
19.	Damaged seedlings				•	•
20.	Floating seedlings / seedlings in lying down position				•	•
21.	Buried seedlings				•	•
22.	Missing seedlings				•	•
23.	Seedling mortality				•	•
24.	Performance indices: *miss, multiple, and quality of feed*	•				
25.	Power requirement, kW	•	•	•	•	•
26.	Fuel consumption, l/h and l/ha	•	•	•	•	•
27.	Field capacity, ha/h	•	•	•	•	•
28.	Wheel slip, %	•	•	•	•	•
29.	Field efficiency, %	•	•	•	•	•
30.	Labour requirement, man-hours	•	•	•	•	•
31.	Ease of setting, operation and adjustment	•	•	•	•	•
32.	Soundness of construction	•	•	•	•	•

10.1 Seed Planters

Mechanization in row crops and vegetables helps timely completion of field activities, adds to the efficiency of the farmers in performing field operations and economizes cost of cultivation. Use of animal or tractor drawn seed drill has enabled farmers to cover large areas in a short period quite economically. However, the seed rate in sowing with the seed drill is quite high and thinning sometimes becomes essential to maintain the optimum plant stand to ensure desired quantity of sunlight, water and nutrients to each plant. This could be achieved by planting of different sizes of seeds with the help of appropriate planters. It has been reported that planters provide desired plant population with uniform plant spacing and depth of operation, which results in uniform crop stand and hence, reduced cost of cultivation as well as savings of seed and fertilizer [149, 179]. Different types of seed planters are described below.

a) Inclined plate planter

An inclined plate planter consists of a main frame with tool bar, seed boxes, furrow openers and ground drive-wheel system. The planter is provided with a number of seed boxes of modular design with independent inclined plate type seed metering mechanism. The shoe type furrow openers are mounted on tool bar of main frame through clamps. The seed box along with the seed metering system bolted above the furrow opener and seed box-furrow opener assemblies are adjustable for row-to-row spacing and work as a modular unit for sowing of each row. The drive to seed metering mechanism is transmitted from ground drive-wheel through chain and sprockets. These type of planters are suitable for planting groundnut, gram, soyabean, etc. Row-to-row distance can be adjusted and planting of different seeds in different rows is also possible.

b) Bed planter

The tractor drawn bed planter consists of a raised bed forming mechanism and a planter to place two rows of seeds on a bed in one pass. Seeds planted on the raised bed facilitate better root growth and yield enhancement. The furrows formed are used for irrigation purpose and removal of excess water. The planter comprises of inclined disc type seed metering mechanism, ridgers, ground wheel, chain and sprocket drive for transmitting power from ground wheel to the seed metering and placement device. A separate fertilizer box with fertilizer metering mechanism driven by the ground wheel and seed to seed distance is also provided. It is suitable for sowing of maize, peas, pulses, groundnut etc.

c) Precision belt planter

Circular holes are punched in a belt to accommodate the seed size and holes are spaced along the belt at specified intervals. The seed spacing is changed by varying the transmission ratio of the ground driven wheel and for different seed sizes, the seed belt needs to be changed. Usually coated seeds are used in belt planter to improve the uniformity and it is difficult to singulate irregular shaped seeds. Spacing is not as uniform as with a other type of planters and this planter can be used with seeds of tomato to watermelon.

d) Precision pneumatic planter

In pneumatic seed planters, the vacuum created by a suction blower / aspirator holds the seeds in the seed cells / holes on to the rotating seed disk. The vacuum is cut off at a certain point above the seed delivery to allow the seeds to fall into the furrow due to gravity. A mechanical system is provided to remove the extra seeds to increase the singulation. Seedway guide is used to guide the dropped seeds to the seed tube or directly to furrow and this also acts as a seed brush off device. PTO or hydraulic driven aspirator is used to create vacuum for each metering unit. Pneumatic planters give better planting performance than mechanical seeders in terms of within row uniformity, singulation and spacing as this type of metering system is less dependent on seed shape, uniformity of seed size and operating speed. Pneumatic precision planter for vegetables, developed at CMERI-CoEFM, Ludhiana is shown in Fig.10.2.

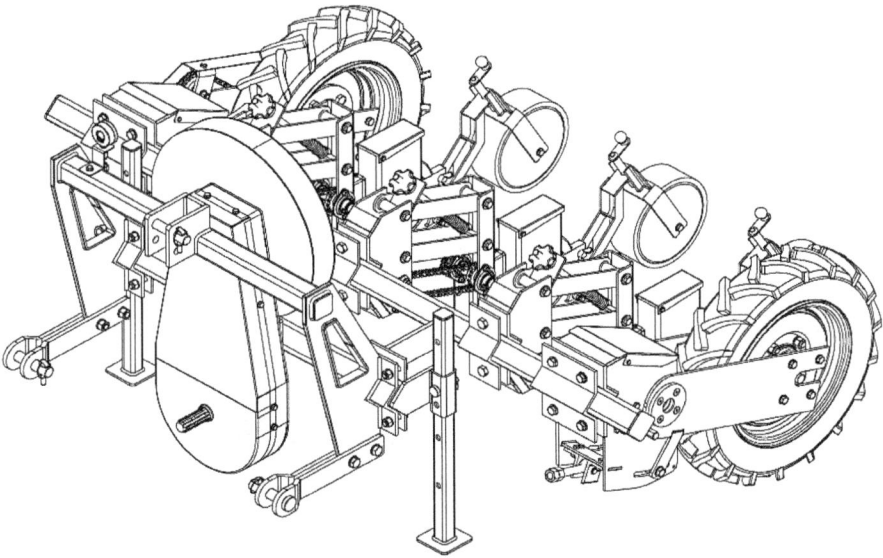

Fig. 10.2: Pneumatic precision planter

10.1.1 Performance testing

The measurements for evaluation of planters are: a) seed singulation and seed spacing, b) depth of seed placement, and c) plant stand. The measurement of this data is used for calculating the following planter performance indices [134].

a) Mean seed spacing

Mean seed spacing (S) is the mean of total number of spacing measured.

$$S = \sum_{i=1}^{N} \frac{X_i}{N}$$

Where,

N   - the total number of spacing measured

X_i - the distance between seed/plant and the next seed/plant.

b) Miss index

The missing percentage is represented by an index called the Miss Index (MI) This is the percentage of spacing greater than 1.5 times the set spacing (X) [145, 24].

$$M = \frac{n_1}{N}$$

Where,

n_1 = number of spacing > 1.5 X

N = total number of spacing measured

c) Multiple index

The multiple, more than one seed, percentage is represented by an index called Multiple Index (DI) which is the percentage of spacing that are less than or equal to half of the set spacing (X).

$$DI = \frac{n_2}{N}$$

Where,

n_2 = number of spacing < 0.5 X

d) Quality of feed index

The quality of feed index is an alternate way to present the performance as a result of combined effect of misses and multiples. The quality of feed index (A) is the percentage of spacing that are more than half but not more than 1.5 times the set spacing.

Quality of feed index (A) = 100 - (miss index + multiple index)

e) Coefficient of variation (CV)

Coefficient of variation (CV) is a mathematical equation used to measure the precision of seed placement at any population or speed. More specifically, CV is the standard deviation of seed spacing divided by average seed spacing. It is expressed as a decimal, with lower numbers indicating better in-row seed spacing, and it will vary with different crops, seed disks, planting conditions, and ground speeds. Planter seed spacing performance should always be verified by checking in-ground population and spacing.

$$CV = \frac{\text{Standard deviation of seed spacing (variation between seeds)}}{\text{average distance between seeds}}$$

Fig. 10.3 illustrates the distribution of plant spacing in the seed row at different coefficient of variation (CV). If the CV is 0 %, all plants have the exact same spacing between them, 0 – 20% is excellent, 20 – 30% is very good, 30 – 35% is good, 35 – 40 is acceptable, and 40 – 50% is classified as poor. If one percent of the seeds does not germinate, it is not possible to reach a CV 10%, and if it is 5 percent then CV will be more than 20 % .

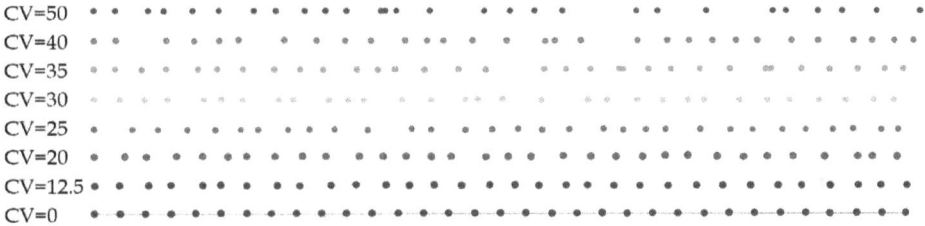

[source: www.vaderstad.com]

Fig. 10.3: Illustration of Coefficient of Variation in seed planting

10.2 Potato Planters

Uniform placement of tubers at desired spacing and proper depth and other aspects of performance of the machine is very important, which shall be obtained from testing of potato planters. During the testing of potato planter, fertilizer distribution systems should be removed from the machine, if present. Semi-automatic and automatic planters are used by farmers [103, 252]. Also, most common types of automatic planters are belt type and elevator type, shown in Fig. 10.4.

Courtesy: Droli Industries, Moga, India   Courtesy: Punjab Engineers, Meerut, India

Fig. 10.4: Belt Type and Elevator type automatic potato planters

10.2.1: Terminology

a. *Tuber distance*   : Distance between adjacent tubers in a row, measured from tuber centre to tuber centre and expressed in centimeters.

b.	*Rated planting distance*	:	Planting distance, expressed in centimeters, claimed by the manufacturer in the operating instructions.
c.	*Actual planting distance*	:	Mean value, expressed in centimeters, of atleast 100 tuber distances, discarding any misses, doubles, etc.
d.	*Row spacing*	:	Centre to centre distance, expressed in centimeters, between the adjacent rows.
e.	*Number of rows*	:	Number of rows formed at the same time by the planter.
f.	*Tuber density*	:	Number of tuber positions per hectare, expressed in ha⁻¹ and calculated by the formula:

$$\text{Tuber density} = \frac{10^6}{\text{Actual planting distance (cm) ṅ Row spacing (cm)}}$$

g.	*Tuber mass*	:	Mass, expressed in grams, of potato tuber. The average mass of each tuber in a potato batch is determined by weighing atleast 30 tubers
h.	*Tuber quantity / plant rate*	:	Total mass of potatoes planted per ha expressed in tonnes/ha and calculated by the formula:

$$\text{Tuber quantity} = 100 \; \text{ṅ} \; \frac{\text{Tuber mass (g)}}{\text{Actual planting distance (cm) ṅ Row spacing (cm)}}$$

i.	*Planting frequency*	:	Average number of tubers planted per minute and per ridge expressed in min⁻¹.
j.	*Coefficient of variation (CV)*	:	Variation of the actual distance with in a row, given as percentage of the actual planting distance.
k.	*Planting errors*	:	Deviations from the desired equal tuber distribution in a row. Planting errors are the number of misses and multiples expressed as a percentage of the actual planting distance and furthermore by the coefficient of variation.
l.	*Cell filling errors*	:	In the case of planters with cup elevators, the number of misses or multiples, expressed as a percentage per hundred cups or other metering units.
m.	*Depth of planting*	:	Distance, expressed in centimeters, measured between the bottom of the furrow and the original field surface.

10.2.2.Methods of test

a) Index of tuber shape (f)

$$f = \frac{l^2}{w \times t} \times 100$$

Where,

l – greatest length of the tuber

w – greatest width of the tuber

t - greatest thickness of the tuber

Tuber shape	Index
Round	100 to 160
Oval	160 to 240
Long	240 to 340
Very long	Above 340

Dimensions of atleast 30 tubers should be taken for calculating index of tuber shape.

b) Square measure

At least 30 tubers are passed through a set of seven square mesh type sieves. The mesh widths ranges from 25 to 55 mm in 5mm steps. The square measure is indicated by stating the maximum size of the sieve through which none of the sample will pass and the minimum size through which all of the sample will pass, thus for example 35/45.

c) Test for determine the variation of row spacing

Variation between actual and nominal row spacing shall be measured on both horizontal field surfaces, and on fields with a sideways slope of 20%.

d) Test of assessment of uniformity of tuber distribution

Test of tuber distribution in a row

The assessment to be made on rows planted separately with round, oval and long potatoes with a square measure of 35/45 and 45/55 mm. For each row, 100 measurements are necessary, to be repeated atleast four times. The coefficient of variation (CV) and planting error shall be determined.

Test of determination of cell filling errors in the case of cup elevator planters or metering units

To determine cell filling errors, samples of non-sprouted, good seed potatoes shall be prepared as follows:

Square measure	Tuber length, mm		
Test sample	I	II	III
35/40	Up to 45	Up to 56	Up to 67
40/45	Up to 51	Up to 63	Up to 78
45/50	Up to 57	Up to 73	Up to 87
50/55	Up to 64	Up to 79	Up to 97

Commercial plant tubers of several size grades and varieties are mixed thoroughly, and size graded through square sieves in 5mm steps. Each grade is then sub-divided according to tuber length, and test samples I, II and III prepared by mixing size grades /tuber lengths as in the table. Test sample I predominantly contains round tubers, test sample II oval and test sample III long tubers.

The planter mounted on a test bench in a horizontal position shall be driven by means on an engine/motor with infinitely variable speed control. The hopper shall be fed with atleast 50 kg per row of a test sample and the elevating errors shall be determined at planting frequencies of 120, 180, 240, 300 min^{-1}, etc. As the efficiency of some cup-fed planters falls off with decreasing quantities of tubers in the hopper, the test should continue until the hopper is at most one quarter full.

e) Test of damage to sprouts on pre-sprouted tubers

Sprout damage or breakage depends on the type, number, elasticity, length and arrangement of sprouts on the tubers.

Degree of sprouting can be defined as:

Single pre-sprouting	- sprouts 3 to 5 mm long
Medium pre-sprouting	- sprouts 5 to 15 mm long
Strong pre-sprouting	- sprouts 15 to 25 mm long

Assessment shall be carried out on a stationary test bench at several planting frequencies. The amount of breakage to sprouts caused by a planter shall be measured using samples of tubers having green sprouts 10 to 15 mm long.

Summary sheet for testing of potato planter is shown in annexure 10.1.

10.3 Sugarcane Planters

Sugarcane (*Saccharum officinarum* L.) is an important cash crop and cultivated between 32°N to 32°S latitude covering more than 90 countries of the world. Sugarcane contributes about 64.6%of the total world sugar production. Sugarcane is a vegetatively planted crop. Stalks or stalk sections, called billets, are planted, and the stalk buds germinate and grow to produce the next crop. Traditionally, sugarcane has been planted as whole stalks. A high planting rate consisting of an average of three to four stalks running continuously in the planting furrow is used. This expensive planting method is used to insure that an adequate stand is maintained despite the occurrence of stressful environmental conditions and diseases. Planted whole stalks can sustain partial damage from stalk rot and still produce an adequate number of plants [205, 206].

The developed countries changed the harvesting method from whole-stalk to billet harvesting during the mid-1990s. This initiated the use of billets rather than whole stalks for planting. However, the shorter seed cane billet is more susceptible to severe stalk rot damage. In fact, research has shown that planted billets are more susceptible than whole stalks to damage from any problem encountered during and after planting and also, over a period of time, whole-stalk planting will produce higher yields with fewer problems than billet planting.

10.3.1 Sugarcane cutter planter

Sugarcane can be planted by improved method of planting like, deep furrow, trench methods, ring pit method and paired row method instead of furrow system. According to methods of planting, four types of the sugarcane cutter planters having ridger, slit, furrower and disc furrow openers are used [183, 185].

a) Ridger type sugarcane cutter planter

This planter basically consists of a rectangular frame on which ridgers are mounted with a provision to adjust row to row spacing from 750 to 900 mm (Fig. 10.5). Sugarcane seed hoppers, are mounted on the main frame along with insecticide tank and fungicide tank. It also has fertilizer boxes with star wheel type agitators along with circular plate having three sizes circular openings to meter the quantity of fertilizer. The tractor PTO through a chain and sprocket drive, operates the fertilizer metering system. The sett cutting mechanism consists of cutting unit, one for each row. Cutting unit consists of two revolving knives powered by tractor PTO through universal joint and a set of bevel pinion and gear.

b) Slit type sugarcane cutter planter

Slit type furrower are used for making narrow furrows for cane planting. These are made from flat pieces of carbon steel welded together to form a cutting edge and bolted to the two points of rectangular cross section steel shank.

c) Furrower type sugarcane cutter planter

This design employs a furrower for furrow opening. It is heavier than other planters (Approx. 550 kg) although the working and power transmission is same as ridger type planter.

d) Disc type sugarcane cutter planter

This planter consists of two discs, which are configured and installed so as to produce V-shaped furrows. The discs are tilted vertically at an angle of 15° and having a disc angle of 20°. Spacing between discs can be adjusted to obtain 750 mm to 900 mm row to row spacing. The cane cutting mechanism consists of two counter rotating blades. The blades are sharpened downward to prevent upward thrust on sugarcane stalks, which are held by labourers during planting operation. There are two seats for two persons to be engaged for feeding the sugarcane.

10.3.2 Self-propelled billet planter

Longer billets than those cut for commercial harvest are used for planting with a billet planter (Fig. 10.6). An average billet length of 50 – 60 cm will provide 3 - 4 buds per billet and a length that will meter well in mechanical planters. Physical damage to billets sustained during the mechanical harvest and planting processes creates wounds that serve as entry points for stalk-rot pathogens. Good seedbed preparation with adequate soil moisture and uniform coverage with not more than 3 inches of soil are important. Good weed control during the fallow period and pre-emergence will reduce the need to make applications of herbicides after plants have emerged. A higher planting rate is typically used for billets to reduce the risk of serious stand problems, but this increases the cost. Billet planter receives, transports and plants the seed cane. The planter tills the soil with furrow opener and also covers the seed cane with loose soil.

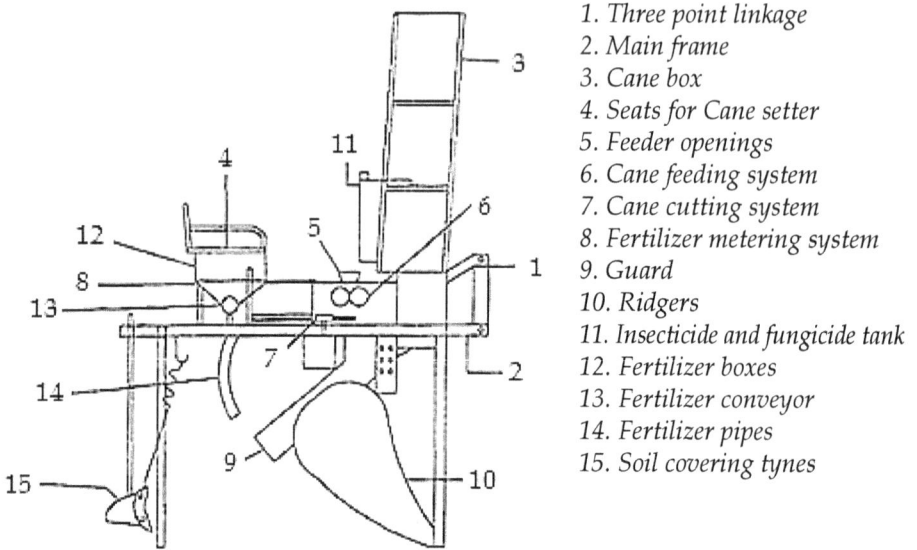

1. Three point linkage
2. Main frame
3. Cane box
4. Seats for Cane setter
5. Feeder openings
6. Cane feeding system
7. Cane cutting system
8. Fertilizer metering system
9. Guard
10. Ridgers
11. Insecticide and fungicide tank
12. Fertilizer boxes
13. Fertilizer conveyor
14. Fertilizer pipes
15. Soil covering tynes

Courtesy: VSI, India

Fig. 10.5: Schematic view of Sugarcane Cutter Planter

10.3.3 Terminology

a.	Implement width	:	Horizontal distance perpendicular to the direction of travel between the outermost edges of the implement
b.	Operating width	:	Horizontal distance perpendicular to the direction of travel within which an implement performs its intended function
c.	Percent cutting	:	Ratio of the number of stalks cut to the total number of stalks in the reservoir expressed in percentage
d.	Percent damaged stalk eyes	:	Ratio of the number of billets with damaged stalk eyes to the total number of billets dropped expressed in percentage
e.	Plant distance	:	Distance between the two sugarcane billets planted in a row
f.	Transport height	:	Overall height of the implement measured from the topmost point to its lowest point
g.	Transport length	:	Overall length of the implement measured from terminal point of the implement to the mounting point
h.	Wheel slip	:	Reduction on the travelled distance by the tractor due to the attached implement

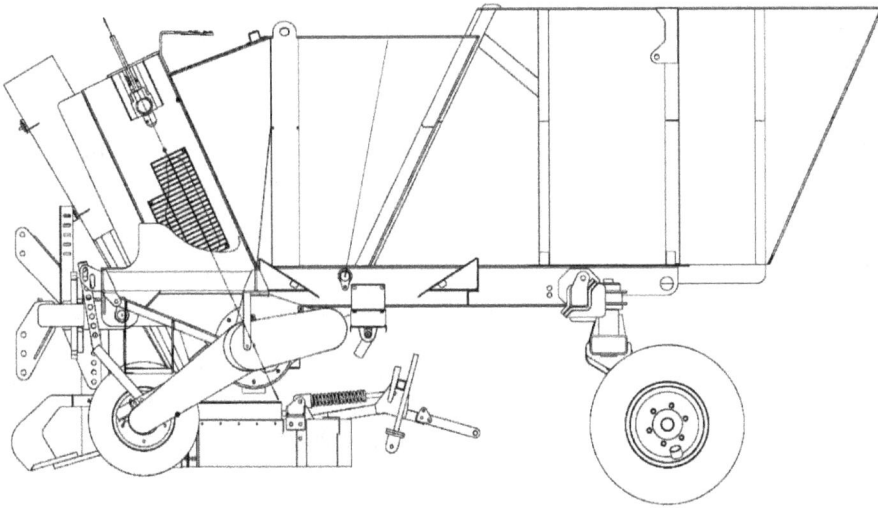

Fig. 10.6: Schematic view of tractor operated Billet Planter

10.3.4 Methods of test

a) Seedling biometry

The stalks manually harvested passed by the determinations of length (m), diameter (cm), number of internodes/stalk, number of available germs/stalk and number of not available germs/stalk (in function of bugs attack and mechanical damage). It will be sampled for repetitions of 4 meters each (with 2 furrows) spaced 20 meters apart for each furrow of each treatment. In each of these samplings, after 30 and 60 days after planting it will be counted for number of plants. Datasheet for seedling biometry results are given in Annexure 10.2.

b) Drawbar force

One can adopt the method, pulling the tractor and implement with another tractor, for measuring the drawbar force and the same may be taken using electronic instruments. Readings need to be taken in a 100 m and 200 m length with sufficient replications. Therefore, the drawbar force required (FT_p) only for the planter is:

$$FT_p = FT_c - FT_t$$

Where,

FT_c - Drawbar force required for the Tractor and Planter

FT_t - Drawbar force required only for the Tractor

The engine horse powers required for the tractors to haul the planters may be specified according to ASAE STANDARD EP391 (1983).

c) Effective capacity determination

In each test, time taken for actual planting and time taken for maneuvering and other purposed shall be noted for calculating effective field capacity.

d) Fuel effective consumption

The fuel effective consumption (l/h) may be determined with using two electronic fuel meters or a differential fuel meter.

e) Furrow depth and amount of soil over the seed cane / billets

20 readings may be taken from different rows randomly selected from the field.

10.3.5. Field performance test

- Planting speed
- Total test time
- Productive time
- Fuel consumption
- Drawbar power
- Field efficiency

- Effective field capacity
- Planting Uniformity
- Damaged stalk eyes
- Wheel slip
- Checking after planting for any defect and breakdown

Data sheet for testing of sugarcane planter is shown in annexure 10.2

10.4 Rice Transplanter

The importance of rice cultivation in India has considerably increased with the introduction of high yielding varieties which have made rice more paying than other kharif crops. Paddy occupies maximum area under cultivation in India. The operations like transplanting, intercultural, harvesting and threshing etc. are very tiresome and labour consuming. The scarcity of labour poses a big problem to complete the operations particularly transplanting in time. The transplanting period of high yielding varieties is very short and limited. The delayed transplanting cause progressive decrease in the yield and in Northern region very late transplanting may result in complete failure of crop due to cold weather at the stage of maturity. Therefore, rice transplanter is an important machine which can help the farmers in completing the work in time and achieve desired quality of work.

a) Classification of rice transplanters

i. Based upon power source
1. Manually operated
2. Power tiller operated
3. Self-propelled type:
   i) Walking type
   ii) Riding type

ii. Based upon type of seedlings
1. Root-washed seedlings
2. Mat type seedlings

iii. Based upon travelling type
1. Walking type
2. Riding type

iv. Based upon transplanting mechanism
1. Crank mechanism
2. Rotary mechanism

b) Mat type nursery for rice transplanter

Traditionally grown nursery in India is suitable for root washed type transplanters. However, transplanter using mat type nursery have been widely adopted in Japan, where this operation is completely mechanized. For raising mat type seedlings, trays or frames placed on plastic sheet are required. Soil is filled in these frames/ trays which have generally 1-8 cm height. In Japan, mat type nursery is grown on a commercial level. In India, successful efforts have been made to raise mat type nursery in field. Labour requirements at the time of sowing are comparatively higher while for nursery uprooting is lesser in mat type nursery. Mat type nursery is, generally transplanted in the age group of 25-40 days. At this age, the leaf stage is minimum 2 and the minimum plant height is 12.5 cm. Nursery mats, after uprooting are directly placed on the machine for transplanting.

c) Construction of rice transplanter

Rice transplanter consists of a prime mover, wheels, float, transmission system, planting mechanism, seedling feeding mechanism and operational control device etc. The detail of construction for walking type and riding type transplanter is shown in Fig. 10.7 & 10.8. The traveling speed is usually 0.3 - 0.7 m/s and in case of high performance transplanter speed could be 1 m/s.

The vertical distance between soil surface and hard pan is not constant in the field. Thus depth of planting depends upon change in the relative position of the float and planting finger. Accordingly height relation between wheels and floats influence planting accuracy. Generally, transplanters have automatic planting depth control to maintain uniformity.

Power transmission system is meant for transmitting power from prime mover to planting device and traveling device. Changing ratio of planting speed and traveling speed can make adjustment of hill spacing. The system consists of a set of gears, belts, roller chain, main clutch, safety clutch etc.

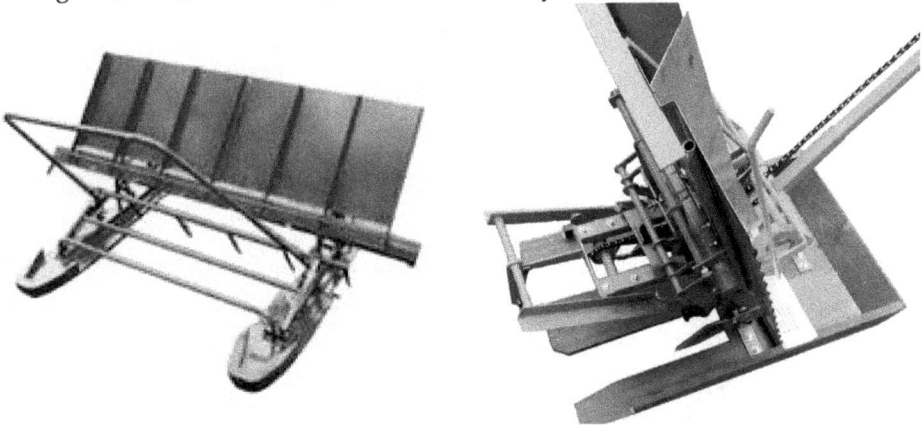

Fig. 10.7: Manually operated rice transplanters

Planting mechanism is either crank type or rotary type. Generally walking type rice transplanter are equipped with crank type mechanism and riding type with crank type or rotary type mechanism. Crank type mechanism consists of four bar linkages (crank, rocker, coupler and fixed link), planting arm, planting finger and planting fork. This mechanism makes a unique motion on a locus of the point of planting finger according to the rotating crank. Rotary type mechanism consists of eccentric planetary gears, planting arm, planting fingers and planting forks.

This mechanism enables the high speed planting work comparing with crank type.

Seedling feeding mechanism consists of cross-feeding (moves mat seedlings sideways by the reciprocation of seedling trays) and longitudinal feeding mechanism (slides down the mat seedlings on the seedling trays).

10.4.1 Terminology

a) *Root-washed seedling:* These seed lings are those which are uprooted from the traditional nursery field. After uprooting, the soil attached to the roots is washed away. Then, seedlings are separated from each other and the roots are cut to a length of 2-3 cm, so that higher accuracy of planting operation can be achieved.

b) *Mat-type seedlings:* These seedlings are raised in seedling boxes or trays. Seedlings grown in trays get their roots entangled forming a root-mat. The root-mat, including the seedling, is taken out and placed on the transplanter.

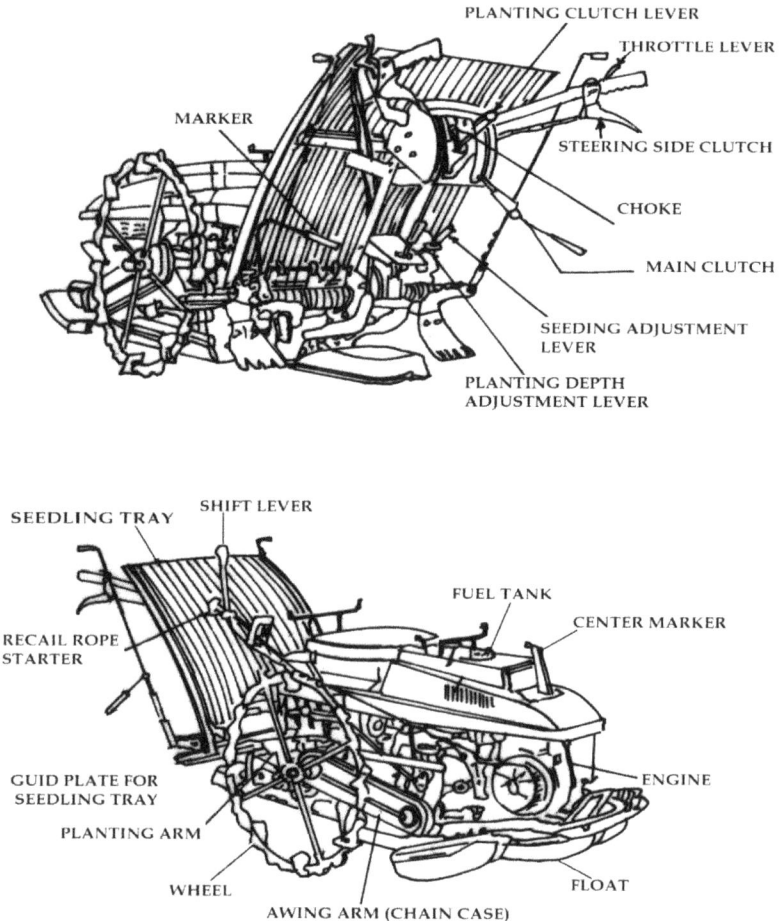

Fig. 10.8 (a): Walking type rice transplanter

c) **Leaf stage of seedlings:** Leaf stage indicates the character and length of the seedlings. Normally, seedlings are transplanted by machine at a stage having not less than 2 leaves and not more than 6 leaves.

d) **Density of seedlings:** In case of mat type nursery, the number of seedlings obtained in a section of 2 cm by 2 cm shall be converted into number of seedlings per unit area (cm^2). It influences the accuracy of planting operation and number of seedlings per hill.

e) **Rate of work:** The area covered in the field per unit time.

f) **Field efficiency:** During the field operation, time will be lost in turning at the headlands, corners and may be due to other defects. This influences the efficiency and decrease the rate of theoretical field capacity. Moreover, field efficiency will vary according to the size and shape of the field, the type and size of machine, the skill of the operator and other similar factors.

g) **Slippage:** The slippage of driving wheels is calculated by the following formula:

$$Slippage\ (\%) = \frac{(N_2 - N_1)}{N_2} \times 100$$

Where,

N_1 - The no. of revolutions of driving wheel for a certain distance in the puddled field.

N_2 - The no. of revolutions of driving wheel for a certain distance on the hard surface.

h) **Damaged seedlings:** These can be divided into two categories. Damage is caused by cutting or bending of the seedlings and by internal damage at the growing point of the seedling due to crushing by planting fork.

i) **Floating seedlings:** During transplanting, the roots get into soil and seedlings remain settled for some time. But sometimes due to higher water level and disturbance by transplanter, seedlings get up-rooted and float on the water. Such seedlings are called floated seedlings resulting in missing hill.

j) **Buried hills:** Hills which are buried under the soil after transplanting due to movement of soil in the already transplanted rows, caused by machine travel are called buried hills.

k) **Missed transplanting:** It denotes the rate of missing hill, expressed in percent.

l) **Total missing hills:** Total missing hills (%) is the sum of percentage of buried hills, floated hills and missing hills caused by planting mechanism and unevenness of seedling density in mat.

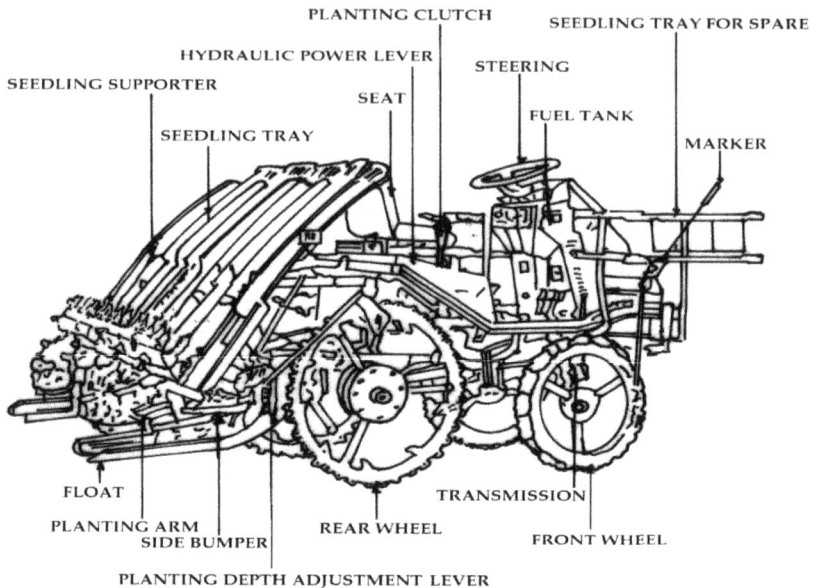

Fig. 10.8 (b): Riding type rice transplanter

10.4.2 Scope of test

The test procedure is formulated to assess the performance of transplanter, mainly based upon field test which is represented by work accuracy and work efficiency. Work accuracy is the planting accuracy which means the reliability of transplanting operation and work efficiency means working capacity. Test field and seedling conditions are very important on field performance, as these will effect work performance. For example, if short seedlings are used in a test field having very soft soil surface, then work accuracy will be poor due to high percentage of buried hills. Therefore, testing of rice transplanter should be carried out in normal conditions and that too should be defined in the test report.

10.4.2.1 Laboratory test

a) Checking of specifications and other data furnished the manufacturer

b) Engine performance test

c) Noise measurement

d) Mechanical vibration measurement

e) Turning ability test

10.4.2.2 Field test

a) Rate of work

b) Quality of work

c) Fuel consumption

e) Ease of handling and operation

e) Breakdown and repair

f) Assessment of wear of critical components

10.4.3 General conditions

a.	Condition of rice transplanter	:	The test sample shall be selected randomly from the production line and shall be well run in as per recommendations.
b.	Preliminary trial	:	The transplanter shall be tried and adjusted in the puddled field before field performance test.
c.	Fuels and lubricants	:	Fuel and lubricants used for test shall conform to the specifications prevailing in the country and be easily available in the market.
d.	Seedlings	:	Seedlings used shall be selected by the testing authority. Two kind of seedlings preferably 2-4 leaves shall be used during testing.
e.	Field dimensions	:	The test plot should have length of 50 m and width of 25m.
f.	Measuring instruments	:	All measuring instruments shall be inspected and calibrated before use.

10.4.4 Test conditions

As machine performance and operating accuracy will vary considerably according to seedling conditions, character of soil, water depth at the time of planting and adjustment of working parts of the machine. The test conditions have to be defined as below:

10.4.4.1 Field conditions

a) Area and shape of test field

b) Type and character of soil

c) Last crop grown in the field

d) Application of organic matter, if any

e) Method of land preparation

f) Period after puddling

g) Hardpan depth: The depth from soil surface to hardpan, measured by the length of a thin stick penetrated into soil.

h) Soil hardness in 'cone depth'

i) Water depth: The depth of water over the soil surface.

10.4.4.2 Seedling conditions

a) Variety of rice

b) Type of nursery

c) Soil type of seedbed or nursery field

d) Date of sowing

e) Quantity of seed

f) Germination rate

g) Nursery duration

h) Plant height: The length of a plant from the root base to the top of leaf.

i) Rupture strength of seedling mat

j) Leaf stage: The number of leaves except a coleoptile.

k) Plant establishment density in mat: The number of plant per unit area (cm^2).

10.4.4.3 Setting conditions

a) Number of workers

b) Traveling speed

c) Number of seedlings per hill

d) Row and hill spacing

e) Position of each adjusting parts

f) Any other item

10.4.5 Methods of test

10.4.5.1 Laboratory test

a) Specification checking

The main objective of specification checking is to verify and confirm the specification as claimed by the manufacturer.

b) Engine performance test

i) Two hour maximum power test

ii) Power at rated engine speed

iii) Maximum torque developed

iv) Part load test

1. At torque corresponding to maximum power at rated engine speed

2. At 85% of the torque obtained in (1)

3. At 75% of the torque obtained in (2)

4. At 50% of the torque obtained in (2)

5. At 25% of the torque obtained in (2)

6. Unloaded

c) Noise measurement

i) Maximum noise level at bystandar's position at travel speed for field operation.

ii) Noise at operator's ear level at travel speed for field operation

d) Vibration measurement

The test shall be conducted on standard concrete test track at all the specified points as horizontal displacement (HD) and vertical displacement (VD) in microns.

e) Turning ability test

Minimum turning space and minimum turning diameter on LHS & RHS shall be recorded, calculated and reported.

Prior to the tests, specifications of the machine, together with information on the manufacture's recommended performance and capacity have to be confirmed and examined. Manufacturer has to submit all the technical information like operator manual, list of spare parts. Some of the items to be examined are:

- The relation between the operating locus of planting fork and the planting posture of the seedling
- The mechanism and shape of the planting form and applicable size of seedling
- The relation of the guide chute to water and soil surface
- Adjustment of operating parts and their reliability
- The power transmission system

10.4.5.2 Field performance test

Performance tests will be carried out at five selected locations with each trial plot size should not be less than 0.1 hectare and a rectangle with the sides in the ratio of 2:1 as far as possible. The items to be measured and observed are:

a) Rate of work and labour requirement:
   - Actual forward speed
   - Operating width
   - Actual operating time
   - Time spend for supply of seedlings
   - Time spend for adjustment of machine
   - Time spent for machine trouble shooting
   - Turning radius
   - Fuel consumption
   - Time spent for turning at headland
   - Area covered per unit time
   - Time required to cover one hectare
   - Required number of workers and man-hours
   - Labour required for headland (by manual planting)

b) Quality of work
   - Spacing between two rows

- Average spacing between hills
- Average depth of transplanting
- Standing angle of planted seedlings
- Average number of seedlings per hill
- Average total number of hill/m^2
- Percentage of missed hill
- Percentage of floating seedling hills
- Percentage of buried seedling hills
- Percentage of damaged seedling hills. Total percentage of missed hills and inadequate planting
- Frequency of missed two or more successive hills
- Slippage and sinkage of machine
- Uniformity of travelling of machine
- Ease of handling and operation
- Precision and ease of adjustment of machine

10.4.5.3 *Ease of handling and operation*

The objective of this test is to ascertain the easiness of handling and adaptability for the rice transplanter. The testing authority shall investigate the adjustment of each mounting parts and measure vibration, noise. Moreover, if necessary measure and investigate the work performance under the different field conditions.

10.4.5.4 *Water proof test*

This test is conducted to ascertain dust and water proof function of mainly wheel axles and planting portion and to ascertain any abnormality or trouble in any of the parts. This test should be conducted in a water-tank to simulate the field conditions. The transplanter should be operated under the operating conditions as far as possible w.r.t. important parameters like depth of planting finger. The period of testing time should be 15 hours.

10.4.5.5 *Investigation after disassembling*

After all test are completed, the testing authority shall disassemble and check the rice transplanter. The objective of this test shall be to check the abnormality of critical components/assemblies and reported accordingly e.g. water entrance, any leakage etc.

10.4.6 Acceptable /recommended performance limits of rice transplanter

a.	No. of plants per hill	:	3 to 5 plants
b.	Planting depth	:	2 to 3 cm
c.	Row spacing	:	20 cm or as per recommendation

d.	Hill spacing	:	10 – 15 cm
e.	Missing hill (%)	:	less than 10

Data sheet for testing rice transplanter is shown in annexure 10.3

10.4.7 Methods of measuring soil hardness and rupture strength of seedling mat

a) Soil hardness

The soil hardness is measured with the depth of penetration of a drop type penetrometer so called 'cone depth'. The cone penetrometer (drop type) shown in Fig. 10.9 has an apex angle of 45 degrees and wight about 135 grams and is made of brass. The penetrometer is dropped from the height of 1 meter from the soil surface to the cone tip. When the cone has penetrated in the soil, 'çone depth' is measured as the distance between the cone tip and soil surface in centimeters.

b) Rupture strength of seedling mat

In case of mat (soil bearing) seedling, the force to separate seedlings from seedling mat is called rupture strength. The seedling mat is cut crosswise to facilitate setting of an alligator clip. The alligator clip is set to grasp five (5) seedlings, attached to the base of the shoot. The other end of the clip is tied by a string to connect it to a spring balance of 2 kg capacity. The spring balance is pulled horizontally until the seedlings are separated from the seedling mat. The maximum force (in grams) registered in the spring balance shall be the 'rupture strength'.

10.5 Vegetable Seedling Transplanters

Vegetable cultivation is the least mechanized sector in Indian agriculture and except seed bed preparation, all other farm operations are carried out manually. The common practice of vegetable cultivation is transplanting of seedlings, raised in a nursery. The removal, transportation and

Fig. 10.9: Drop type cone penetrometer

transplanting of seedlings in the field is done manually and 15 - 100 man-days/ ha is required to complete the operation depending upon crop and variety. Semi-automatic (Fig. 10.10) and automatic seedling transplanters are used in crop cultivation in America and Europe. Mechanized transplanting helps in achieving higher level of mechanization by the use of equipment for subsequent operations like interculture, earthing, fertilization, chemical application and even harvesting. The testing and evaluation procedure discussed here is based on evaluation conducted during research by various scientists and as such no official standard are available for vegetable transplanters [159].

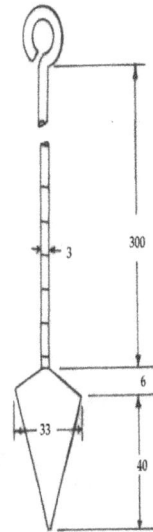

Major functions of a vegetable seedling transplanters:

- Opening a furrow to a desired and uniform depth (with/without parallelogram linkage)

- Accurate metering of the seedlings to obtain the correct spacing in rows and with a provision of different spacing for various crops.

- Placing the seedlings vertically in a furrow,

- Covering seedlings with soil sufficient enough to make the seedling stand erect,

- Firming the soil around seedlings.

The prerequisite for ensuring good mechanical transplanting and growth of vegetable crops is the quality of seed bed, which should be well pulverized to a depth more than the transplanting depth. Also, it will help in covering the seedlings from the sides and firming the soil around it. Two to three operations of cultivator or harrow and/or two passes of rotavator will give a seedbed with required tith. Soil moisture content should be optimum so that there is a better flow of soil during furrow opening, placing the seedling, covering soil from sides and firming it.

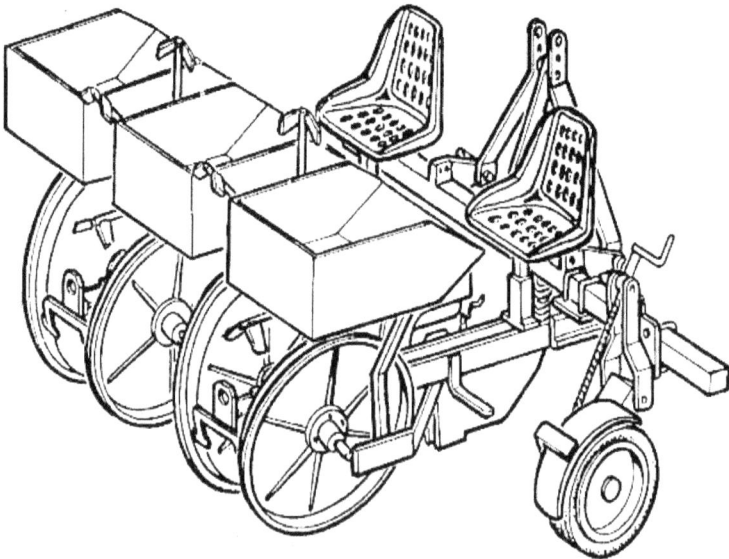

Fig. 10.10. Schematic Diagram of two row vegetable transplanter [159]

The following adjustments should be made before actual testing:

- Adjust the three pint linkage while placing the machine on a firm, level ground to ensure that the machine is level in the longitudinal direction and it will help in maintaining the depth of transplanting.

- Adjust position of furrow opener (if provision is given in the machine) so that proper depth is obtained.

- Adjust position of covering wheels both vertically and horizontally, which helps in controlling the flow and amount of soil from the sides to the seedling, and the level of soil firmness near the seedlings, respectively. This adjustments are made depending upon the type and variety of crop, and the prevailing soil conditions. Trial runs should be conducted in a row or two after making both these adjustments and it should be ensured that there is sufficient, not excessive, soil flow for proper covering.

10.5.1 Scope of test

a) Soil conditions

i) Soil moisture content, % db

ii) Bulk density of soil, kg/m³

iii) Soil mean weight diameter, cm

The mean weight diameter of soil particles is an important soil parameter to be measured in the field as it indicates the amount of pulverization of soil, which is of utmost importance for proper functioning of covering device of a vegetable transplanter.

b) Seedling conditions

i) Age of seedling, days

ii) Type of seedling

iii) Seed rate during sowing of seeds in the nursery

iv) Length of seedling, cm

v) Dimensions of soil block (in case of cup type nursery)

c) Performance testing

The vegetable transplanter needs to be tested in a field having an area of at least 0.2 ha and the following observations are taken:

i) Row spacing, cm

ii) Average plant spacing, cm

iii) Effective width of coverage, m

iv) Forward speed of the transplanter, m/s

v) Wheel slippage, %

vi) Time losses for turning, filling of trays, adjustments etc., min

vii) Field capacity, ha/h

viii) Field efficiency, %

ix) Missing hills, %

x) Seedlings in lying down position, %

xi) Seedling mortality, %

10.5.2 Methods of test

For measurement of forward speed, RADAR sensor or GPS sensors along with requisite data acquisition system may be used. Hall Effect sensors with suitable sensing sprocket or plate in each of the tractor tyres shall be installed to calculate the wheel slippage during transplanting operation. Field capacity of transplanter is the actual area covered per hour and it is obtained by measuring area covered and dividing it by the time taken. For measurement of field efficiency, theoretical field capacity is calculated by multiplying forward speed by theoretical width of coverage i.e. row spacing multiplied by number of rows. The ratio of actual to theoretical field capacity, converted to percentage gives field efficiency. Effective width of coverage is obtained by measuring width covered for at least 3 passes and dividing by number of passes. Sample data sheet for performance testing of vegetable transplanters is given in Annexure 10.4.

a) Missing hills

The percentage of missing hills is determined by counting the number of hills, including improperly transplanted seedlings which are slanted or completely lying down, in which the seedlings are transplanted for a length of 10 m. The expected number of seedlings for the same distance is determined either by theoretical calculation (or from plant spacing chart in the transplanter) or by dividing the length by average plant spacing. The expected number of hills is an integer value. The percentage of missing hills is calculated as:

$$M_h = \frac{(n_2 - n_1)}{n_1} \times 100$$

$$n_2 = \text{int}\left[\frac{d}{p}\right]$$

Where,

M_h - *missing hills, %*

n_1 - actual number of hills transplanted for a given distance

n_2 - expected number of hills for the same distance

d   - distance for which hills are counted, m

p   - average plant spacing, m

b) Seedlings in lying down position

Percentage of seedlings in lying down position is calculated by counting the seedlings lying down for a fixed distance of 10 m and total number of hills transplanted.

$$S_l = \frac{(n_4 - n_3)}{n_4} \times 100$$

Where,

S_1 - seedlings in lying down position, %

n_3 _number of seedlings in lying down position for 10m distance

n_4 - total number of seedlings transplanted in 10m distance

Seedling mortality

Seedling mortality may be due to, a) improper transplanting, b) improper irrigation, and c) diseases, wilting, waterlogging etc. Improper transplanting includes lying down and slanted seedlings. The roots of some of the transplanted seedlings may not have firm contact with the soil. In the finger type metering mechanism for bare root nursery and if the seedlings are not placed properly, with only part of the roots are protruding, they may stay temporarily erect, but later may fall down and die. In all the above cases, when irrigation is applied, these seedlings may fall down and die. Improper irrigation also causes mortality of seedlings due to the fact that local flooding in some areas and insufficient irrigation in other areas may cause plant mortality.

In seedling transplanters, plant mortality are very important parameter, which decides the final plant population / stand, readings are taken within 10 - 15 days of transplanting. For measuring plant mortality, reading may be taken from at least 5 locations of 10 m length in several randomly selected rows. Immediately after transplanting, the number of transplanted seedlings, excluding the lying down seedlings, are counted. After 10-15 days, the number of surviving seedlings are counted at the same locations. The seedling mortality is calculated as:

$$S_n = \frac{(n_5 - n_6)}{n_5} \times 100$$

Where,

S_n - seedling mortality, %

n_5 - number of transplanted seedlings in 10m excluding lying down seedlings

n_6 - number of surviving seedlings in the same 10m length after 10-15 days

Annexure 10.1
Summary Sheet for Testing of Potato Planters

1. **Technical Data**

1.1. Specifications

   a) Manufacturer

   b) Make, type

   c) Main dimensions: length, height, working width, transport width in meters

   d) Unladen weight in kg

   e) Laden weight in kg

   f) Capacity of the hopper, in kg, level and heaped up

   g) Loading height of hopper in cm

   h) Number of ridges

   i) Adjustment range for planting distance and the number of steps

   j) Adjustment range for row spacing

   k) Adjustment range for furrow opener

   l) Adjustment range for depth of setting and of the working width of ridging/covering devices

   m) Number of greasing points

1.2 Description

   a) Planting units

   b) Control and correction of missing tubers

   c) Frames and wheels

   d) Coupling method

   e) Type of drive

   f) Covering device

2. **Results**

   a) Missing tubers

   b) Multiple tubers

   c) Coefficient of variation of tuber distance within row

   d) Planting frequency

   e) Forward speed in m/s

   f) Depth of planting in cm

   g) Variation between nominal and actual row spacing

   h) Damage to sprouts

   i) Effect on performance of longitudinal and lateral slopes

3 **Performance**

   a) Surface area planter per hour, spot and rates of work

   b) Time to fill hopper in minutes

   c) Routine servicing time in minutes

   d) Turning time in minutes

   e) Time needed to adapt for use on public roads, in minutes

   f) Draught and power requirements in kW

   g) Lifting capacity for:

   h) Unladen weight of the machine

   i) Laden weight of the machine

   j) Force required to lift soil engaging components clear of the ground

Annexure 10.2
Data Sheet for Stalk Seedling Biometry [Sugarcane Planter]

S. No.	Description	Average (SD)
1.	Stalks length, m	
2.	Average stalks diameter, cm	
3.	Stalks average weight, kg	
4.	Internodes number per stalk	
5.	Number of viable gems (eyes) per stalk	
6.	Number of non-viable gems(eyes) per stalk	
7.	Total number of gems (eyes) per stalk	
8.	Average number of billets per stalk	
9.	Average length of billets, cm	
10.	Number of sampled billets (total of 30)	
11.	Weight of sample billets (total of 30), kg	
12.	Number of viable gems (eyes) per billet	
13.	Number of non-viable gems(eyes) per billet	
14.	Total number of gems (eyes) per billet	
15.	Percentage of viable gems (eyes) per stalk	
16.	Percentage of viable gems (eyes) per billet	
17.	Billet average weight, kg	

Annexure 10.3

Data sheet for Testing of Rice Transplanter

Date: Variety of Rice:

Name of Supervisor: No. of Leaves per
                                                            plant:

Name of   Place of Test:
Manufacturer:

Operator: Time of Test

Labour Requirement: From:
                                                            To:

Length of Nursery   Type of Nursery
(m):

S. No.	Time for 20 m (s) LH	Wheel rev. for (20m)		Transplanting Dimensions			
		RH	Depth	Width of 5 rows (cm)	Spacing b/w hills (mm)	No. of Plants per sq. meter	No. of Plants per hills

Time taken at head land (s)	Time taken for other delays (s) Ambient Fuel	Temperature °C			Atm. Pressure (mm Hg)
		Engine Oil	Coolant	Trans-mission	

Total:

Average:

Stoppages:

Time (mm):   Cause: Remarks:

Summary for test results of rice transplanter

1. Duration of Test (h)
2. Total Time Stopped (min)
3. Net Duration of Test (h)
4. Avg. forward Speed (kmph)
5. Slip: a - RH (%), b - LH (%)
6. Area Covered, ha
7. Fuel Consumption, lph
8. Draft Requirement (kgf)
9. Avg. Depth of Transplanting (mm)
10. Avg. Width of Transplanting (cm)
11. Avg. Spacing between hills (mm)
12. Avg. No. of Plants per Sq. Meter
13. Plant missing (%)
14. Avg. No. of Plants per hill
15. Rate of Work a - ha/h, b - h/ha
16. Avg. Length of Nursery
17. Avg. No. of Leaves per Plant

Annexure 10.4
Data Sheet for Testing of Vegetable Transplanters [159]

1.	Name of machine	:
2.	Brief specifications	:
3.	Size of field, length x width, m	:
4.	Details of seedbed preparation	:
5.	Test Conditions	
5.1	Soil conditions	
	Soil moisture content, % db	:
	Bulk density of soil, kg/m^3	:
	Mean weight diameter of soil	:
5.2	Seedling conditions	
	Name of crop	:
	Variety	:
	Type of nursery	:
	Density of sowing in nursery, g/m^2	:
	Dimensions of cell/cups of plastic trays, length x width, mm or diameter, mm	:
	Age of seedlings, days	:
	Length of seedling, mm	:
6.	Performance Results	
	Length of run, m	:
	Time taken to travel 15 m distance, min	:
	Forward speed, m/s	:
	Row spacing, cm	:
	Effective width of coverage, m	:
	Average plant spacing, cm	:
	Total time taken, h	:
	Width covered, m	:
	Field capacity, ha/h	:
	Time lost, min	
	Turning, min	:
	Filling seedlings in trays, min	:
	Adjustments, min	:
	Stoppages due to miscl. reasons, mm	:
	Field efficiency, %	:
7.	Missing hills, %	:

Sr. No.	seedlings transplanted in10 m length (including lying down seedlings)	Missing hills, %
1.		
2.		
..		
..		
Average		

8. Seedling in lying down position :

Sr. No.	Total number of seedlings transplanted in10 m length (incl. lying down seedlings)	Total number of seedlings in lying down position	Seedlings in lying down position, %
1.			
2.			
..			
..			
Average			

9. Seedling mortality at 10-15 days from date of transplanting :

Sr. No.	Total number of seedlings transplanted in10 m length (excl. lying down seedlings)	No. of seedlings in 10 m length after 10-15 days	Plant mortality, %
1.			
2.			
..			
..			
Average			

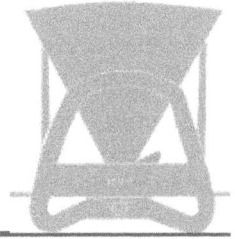

Chapter 11

Testing and Evaluation of Fertilizer Broadcasters

Granular fertilizer, lime, and other soil amendments are needed to be applied uniformly and accurately in the field. Performance failures of agricultural materials can be directly related to improper, uneven, and/or careless application or separation of blended material due to size and density differences. Uneven distribution can reduce crop yields and the effects are most significant in fields that are low in soil fertility, because response to applied nutrients is the maximum on these soils.

Proper selection of application equipment and their calibration and evaluation are essential for optimum results. Even the best equipment requires calibration checks. This is especially important when changing from one type of material to another, to different rates of application, and when altering speeds or other operating conditions and parameters [221].

Broadcaster/Spreaders use single or double spinner discs or a reciprocating (pendular) tube to distribute fertilizer on the field. These devices are mounted on drawbar trailing or 3-point hitch units. The spinner disc distributor requires a mechanism that accurately places fertilizer in the correct location of the rotating disc. Correct placement allows the fertilizer to discharge from the disc to achieve an even spread pattern on both sides of the spreader. A double spinner system has a flow divider that requires front to back adjustment such that fertilizer is placed in the correct location on the spinner discs.

Drill type spreaders are mounted on various types of planting equipment. Augers, stainless steel conveyor chain, belt conveyors, or a series of flutes rotated by a shaft, are used to meter fertilizer to a downspout tube. The downspout is attached to a soil opener device, which places the fertilizer in the soil at a specific location, adjacent to the seed so as to provide nutrients but not damage the seed at the vulnerable germination stage. This process of metering and placement of fertilizer will involve calibration of the spreader and careful observation of the placement of the fertilizer in the soil.

Pneumatic type spreaders can be installed on boom style machines that are 3-point hitch, self-propelled or trailing units. The boom spreader system can apply fertilizer more accurately than broadcast spreaders because of the design of the metering flute mechanism and individual nozzles that are spaced evenly on the boom. However, to be accurate, the boom type spreaders require as much or more attention to setup and proper adjustment. Booms must be parallel to the soil surface to achieve proper overlap at each nozzle on the boom.

This procedure is applicable for the evaluation of various types of manually-operated and power-driven fertilizer broadcasters.

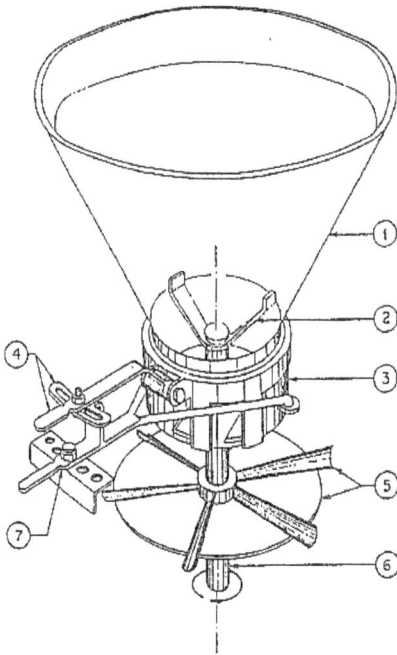

1. Hopper
2. Agitator
3. Dosage mechanism
4. Adjustment lever
5. Distribution mechanism
6. Drive shaft
7. Distribution pattern adjustment

Fig. 11.1: Centrifugal fertilizer distributor

11.1 Terminology

a) *Application, one-direction*: An application method in which successive adjacent swaths are made in the same direction of travel (racetrack or circuitous application). This method produces a right-on-left overlapping of adjacent patterns.

b) *Application, progressive:* An application method in which the spreader applies adjacent swaths in alternate directions (back and forth application). This method produces a right-on-right pattern overlap alternately with a left-on-left pattern overlap.

c) *Application rate:* Application rates are as defined in ASAE S327, is the amount of any material applied per unit area.

d) *Application, single-pass:* An application method in which the spreader applies one swath over the collection trays.

e) *Collector efficiency:* The percentage of true application rate caught in a collection device; ie, the weight of material caught in the collection device divided by the area of the collection device and expressed as a percentage of the true application rate at that point in the pattern.

f) *Swath spacing:* The lateral distance between spreader centerlines for adjacent swaths.

g) *Swath width, effective:* The swath spacing that will produce acceptable field deposition uniformity for the intended application.

11.2 Scope of Test

11.2.1 Laboratory tests

a) Checking of specifications

- Metering mechanism and method of changing feed rate
- Type of drive mechanism
- Method of distribution
- Height and depth controls
- Hopper sizes

b) Calibration test

- Metering mechanisms
- Fertilizer for tests
- Application rate

11.2.2 Field tests

Tests under field conditions will be carried out covering at least 1/10 ha on three soil conditions using various fertilizers to enable observations to be made on practical aspects of use of the machine.

a) Application rate

b) Wheel slip

c) Wheel sinkage

d) Draft

e) Fertilizer placement & distribution

- Transverse distribution
- Longitudinal distribution

f) Effect of vibration

11.2.3 Referred standards

g) IS 12337: 2009 [55]

h) ASAE S281.3 DEC96. [8]

i) ASAE S327.2 DEC95. [9]

j) ASAE S341.3 FEB04. [10]

11.3 Method of Tests

11.3.1 Optimum bout width

Using the results of the transverse distribution tests, histograms of the total spreading rate at various points of overlap across the bout can be drawn. The optimum bout width can be established from these results. Where machines place the fertilizer in rows the delivery of each spout will be recorded. For other machines, the spread material will be divided in longitudinal strips equal to the number of outlets and the amounts weighed.

11.3.2 Application rate

Application rates in kg/ha can be calculated from rates measured in the laboratory over 1/100 ha. The distance for each test run is calculated as follows:

$$\text{Length of test run (m)} = \frac{100}{\text{Nominal width of machine (m)}}$$

The broadcaster /applicator will be operated over a level floor area for the distance required at the speed recommended by the manufacturer and with the hopper half full. At least two replications will be made each at maximum, minimum and average application rate settings.

If a range of forward speeds are recommended by the manufacturer, further tests will be made to establish effects of forward speed on the application rate. Tests may also be made to establish effects of depth of fertilizer in the hopper. In each case, the total weight of the fertilizer collected during the test run is used to calculate the application rate in kg/ha.

The application rate of a broadcaster machine at a given feed setting and forward speed can only be determined when the optimum bout width has been established. Bout width is dependent on the degree of overlap required for the machine to produce an even distribution and is determined by tests for transverse distribution. If the broadcaster is PTO driven, the application rate is established by running the machine for a calculated time equivalent to covering one hectare. The time is calculated as follows:-

$$\text{Time (min.)} = \frac{600}{\text{bout width (m) ń forward speed (km/h)}}$$

For hand-wheel driven machines, the distance travelled is calculated as for the full width machines. The hopper is filled and the broadcaster is run for the calculated time or distance. The weight of fertilizer to refill the hopper is measured and the rate in kg/ha is calculated. At least two replications will be made each at maximum, minimum and average feed settings.

11.3.3 Spread pattern test

Spread pattern tests indicate the degree of uniformity of distribution of material across the swath being spread. The spread pattern test shall be accomplished by operating the spreader in a line perpendicular to a line of collection trays spaced equally on the ground. An odd number of trays should be used, and the spreader should be driven astride the center pan. Material collected in each tray should be weighed or measured volumetrically.

The actual delivery rate and the spreader settings used to achieve these rates shall be reported. All spreaders to be compared shall be tested at the same rate, if possible. If field performance of an individual spreader is desired, the application rate shall be selected based upon the agronomic requirements of the test. Application rates of approximately 25%, 50%, and 75% of the maximum application rate for the test material are suggested if multiple rates are to be used.

The spreader tested shall be rated for uniformity of distribution. The coefficient of variation (CV) shall be used to determine and express the uniformity of distribution of applications. When overlapping of swaths occurs, a simulated field application of multiple adjacent swaths shall be used to compute the CV. The simulated field distribution for each swath width to be evaluated is constructed by accumulating the sample weights from the simulated overlapping swaths at each collection tray location. Individual replicates of the swath distribution pattern (not averages) shall be used. The method of spreading used shall be reported; i.e, either progressive (back and forth) or one direction (race track).

The mean value, standard deviation, and CV shall be determined. Only the central portion of the simulated or measured overlapped distribution data is needed to compute the CV. Data points equidistant from the centerline to a distance halfway to the centerline of the next pass on each side shall be used. Data from enough adjacent swaths shall be included so that the region for calculation as indicated above would be unaffected by the addition of distribution data resulting from additional overlapping swaths. The results of this test may also be presented graphically as shown in Fig. 11.2.

a) Transverse distribution
During tests for the application rate, at average feed rate setting and with half-full hopper, the delivery from each spout will be recorded. The spread material will be divided in longitudinal strips equal to the number of outlets and the amounts weighed. The results of weighings will be presented as a histogram and the percentage variation from the average of the highest and lowest outputs will be recorded.

b) Longitudinal distribution
With the average feed rate setting, test runs will be made over a distance of 5 m at the nominal speed for application rate tests. Using four individual outlets, one will be tested at a time and the fertilizer distributed over each 50 cm length will be aggregated. The results of weighing's will be presented in histograms and the percentage variation of the highest and lowest values from the average will be recorded.

11.3.4 Effect of vibration
This test will be a repetition of the longitudinal distribution test with the addition of standard bumps 4 cm in height placed at 1 meter intervals beneath alternate

wheels. Runs of 15 m will be made and the results will be compared with those of the longitudinal distribution test.

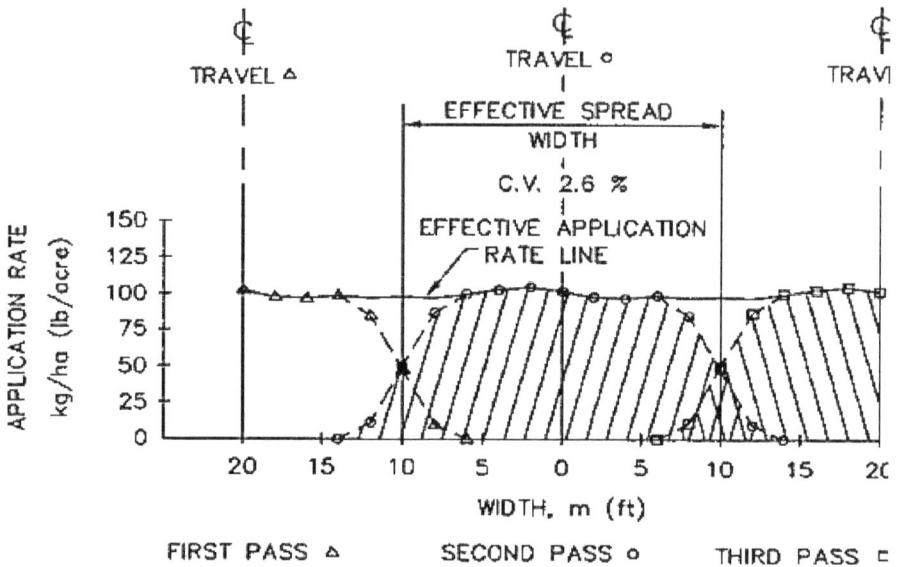

Fig. 11.2: Graphical representation of spread pattern -
one direction application method [10]

11.4 Adjustment in Spread Pattern

Spread patterns for a twin disc, solid fertilizer applicator can be classified into six different types (Fig. 11.3). The flat top, oval and pyramid patterns are most desirable because they allow more uniform overlapping of swaths [239].

The most common undesirable patterns are the M, W and offside (skewed or lopsided) patterns. The M pattern may be improved by making one or more of the following adjustments:

- Move the delivery chute toward the applicator to change the point of delivery of the material closer to the outer edge of the spinners.

- Move spinner blades in the opposite direction of the spinner rotation.

- Increase the spinner speed (recommended spinner speed: 550 - 650 rpm)

The W pattern may result from applicator conditions similar to those causing the M pattern, but have a heavy deposit at the center. The W pattern has a heavy band of material occurring in the center of the swath in addition to concentrations on both the right and left sides.

Lopsided patterns, either right or left, may result from twin spinner applicators because of uneven delivery of fertilizer material to the spinners. An improperly adjusted flow divider is usually the cause. Operations on steep slopes can also produce heavier flow to the downhill side if an effective flow divider is not included in the system.

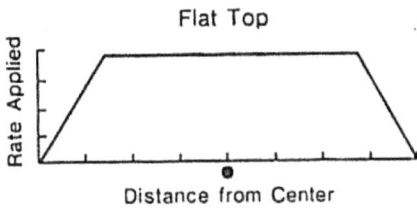

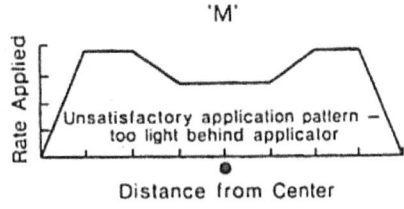

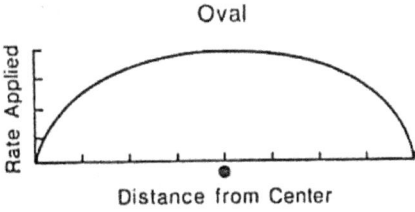

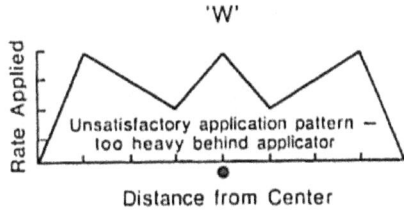

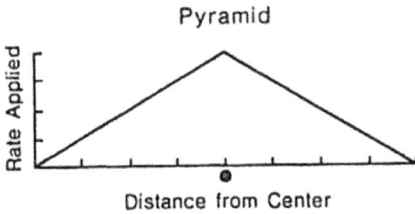

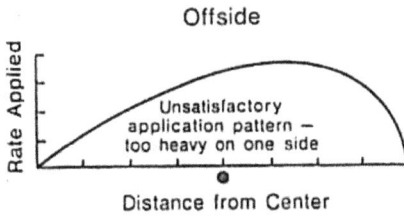

Fig. 11.3: Typical double-spinner spreader distribution patterns [239]

Calibration sheet for fertilizer broadcaster is given in Annexure 11.1.

Annexure 11.1

Calibration sheet for Fertilizer Broadcaster

Calculations to measure the existing application rate

1. Broadcast fertilizer spreader:

   A = application rate (kg/ha)

   B = weight of fertilizer (kg) collected from spreader in sixty seconds

   C = distance travelled (meter) in sixty seconds

   D = swath spacing width (meter)

2. Drill type fertilizer spreader:

   A = application rate (kg/ha)

   B = weight of fertilizer (kg) collected from down spout(s) from one row during C revolutions

   C = number of revolutions of drive wheel

   D = "actual" drive wheel circumference (meter)

   E = row spacing (meter)

Calculations to determine weight of fertilizer (B) for desired application rate

1. Broadcast fertilizer spreader:

   A = application rate (kg/ha)

   B = weight of fertilizer (kg) collected from spreader in sixty seconds

   C = distance travelled (meter) in sixty seconds

   D = swath spacing width (meter)

2. Drill type fertilizer spreader:

   A = application rate (kg/ha)

   B = weight of fertilizer (kg) collected from down spout(s) from one row during C revolutions

   C = number of revolutions of drive wheel

   D = "actual" drive wheel circumference (meter)

   E = row spacing (meter)

Chapter 12

Testing and Evaluation of Sprayers and Dusters

Plant protection from insects, pests and weeds is very important in agricultural production process. If not controlled in time, the entire crop might get affected and to suffer quantitative and qualitative loss of the produce. Hence, crop protection and care is necessary to achieve optimum crop productivity. Among the various methods of pest control, chemical is the most effective. Chemical pesticides have played a major role in the rapid advancement of agriculture. In addition to improvement in crop growth, quality and yield through proper application of specified insecticide and pesticide. The use of chemical herbicides have reduced labour intensive weeding requirements.

12.1 Types of Plant Protection Equipments

a) Manually operated hand sprayer

The manually operated hand sprayer is a small, light and compact unit. The capacity of the container is about 500 ml. It is meant for small spraying jobs in and around the house, e.g., spraying small flowerbeds and vegetable plots in kitchen gardens.

b) Hand compression sprayer

The typical hand compression sprayer comprises of a cylindrical tank for holding the spray fluid with a handle, filler hole, spray lance, nozzle and cut-off device. The capacity of the tank varies from 10 - 20 litres.

c) Knapsack sprayer

This type is commonly used in India and is lever operated plunger or diaphragm type. It has a flat bean-shaped tank. The body of the sprayer is so shaped so as to conveniently fit it on the back of the operator. The capacity of the container is from 15 - 20 litres. It is generally made of brass or plastic. In some cases it is provided with a built –in double barrel pump of piston type, or of diaphragm type, with a lever for operating it. These are manually operated or battery operated.

d) Foot sprayer

The foot sprayer consists of a plunger assembly, stand, suction hose, delivery hose, extension rod with a spray nozzle etc. One end of the suction hose is fitted with a strainer and the other with a flexible coupling. Similarly, the delivery hose has one end fitted with a cut-off valve and the other with a flexible coupling. It is operated by foot. The pump is fixed on an iron stand and a pedal attached to the plunger rod, operates the sprayer by its upward and downward movements. It does not have a built-in tank. It is used for crop and fruit trees up to 4 meters in height. It may or may not be mounted on a trolley.

e) Rocker sprayer

The rocker sprayer consists of a pump assembly, platform, operating lever, pressure chamber, suction hose with a strainer, delivery hose, extension rod with a spray nozzle, etc. The rocking movement of the handle operates the pump, which results in building up pressure in the pressure chamber. There is no built-in tank, therefore, a separate spray tank is necessary. As high pressure can be built up using this sprayer, it can be used for spraying tall field crops and trees up to 5 m in height.

f) Dusters

Equipment that are used for distributing dust formulation are called dusters. All machines used for applying dust consist essentially of a hopper (dust chamber) which usually has an agitator in it, an adjustable orifice or other metering mechanism and delivery tube. The rotary fan supplies the air stream. Generally two types of dusters are available namely the plunger type and the crank or rotary type.

g) Power sprayers

i)   *Hydraulic power sprayer:* These are based upon utilizing hydraulic energy for atomizing and spraying the liquid. They are fitted with any of the pumps, namely, piston type, plunger type, roller vane type, diaphragm type, gear type or centrifugal type.

ii) *Pneumatic power sprayer-cum-duster:* These are based upon pneumatic energy. The pneumatic pressure is thus utilized to agitate the spray or dust. Pneumatic sprayers can be of knapsack type which is mounted on the back of operator & barrow type which is mounted on wheel barrow. While using this sprayer as duster, one has to close the spray delivery tube and open the dust delivery tube. Set the air supply for dusting and then start the engine for dusting.

h) Tractor mounted PTO operated sprayer

It can be used for plant protection, weedicides spraying and liquid fertilizing. It consists of a centrifugal pump, tank and booms of flexible hose pipes on which nozzles are fixed. This boom is tied up on rigid beam by clamps. Spacing of the nozzles on the boom can be adjusted to suit the row crop spacing by fixing clamps at desired place. A pressure gauge with a pressure regulator valve is provided to control the spray pressure. The pump is driven by tractor PTO.

i) Self-propelled sprayer

These are high clearance sprayers mounted on four wheel having a suitable engine for its operation and movement. Generally, boom type spraying unit is fixed behind this machine to spray on large field. These sprayers can be used in the standing tall crop also.

j) Electrostatic sprayer

The air-assisted electrostatic sprayers produce droplets 900 times smaller than those produced by conventional or hydraulic sprayers. After atomization, the droplets are then given an electrical charge and are carried deep into the plant canopy in a turbulent air-stream. The electrostatic charge, though safe, pulls the spray towards the plant with a force 75 times greater than the force of gravity. Droplets change direction and move upwards against gravity to coat all the plant surfaces. The "wrap around" effect also causes the spray to adhere to the surface rather than being blown past the target, drifting away or falling to the ground. The result provides more than twice the deposition efficiency of traditional hydraulic sprayers [21, 182].

The most important goal in the application of agricultural pesticides is to get uniform distribution of the chemicals throughout the crop foliage / canopy. Reduced application and deposition may not give the desired coverage and control needed. On the other hand, over application is expensive as it wastes pesticide and increases the potential for groundwater contamination. Different types of sprayers are illustrated in Fig. 12.1.

12.2 Terminology

a) *Agitation:* An operation which produces and maintains uniform spray mixture in the tank, and in the case of dusts or granules to facilitate their flow from the hopper.

b) *Hydraulic agitation:* Agitation of the spray mixture by using an auxiliary pump flow or a partial flow of the main pump.

c) *Mechanical agitation:* Agitation of the spray mixture, dust or granules by means of mechanically-operated stirrer inside the tank or hopper.

d) *Pneumatic agitation:* Agitation of the spray mixture, dust or granules inside the tank or hopper using an air flow.

e) *Application rate:* The quantity of spray mixture, dust or granules distributed by an appliance per hectare.

f) *Centrifugal fan (centrifugal blower):* An appliance for producing an airflow at right angles to the blower shaft.

g) *Droplet size:* The diameter of the droplet expressed in microns.

h) *Hydraulic injector:* A device using the velocity of a jet of liquid to produce a vacuum in a suction pipe for the purpose of filling a tank.

i) *Spray angle:* The angle formed close to a hydraulic spray nozzle by edges of the spray.

j) *Spray area:* Area covered by the spraying expressed in hectare. In case of overhead spraying, this will be equal to land area; in case of strip spraying, this will depend on row width and strip length; and in case of plantation spraying, the height of the tree and strip length.

a. Manually operated   b. Hand compression   c. Knapsack sprayer   d. Foot sprayer
   hand sprayer   sprayer

e. Rocker sprayer   f. Duster   g. Power sprayer-   h. Electrostatic
                                                    cum-duster   sprayer

i. Tractor mounted boom sprayer   j. Self-propelled sprayer

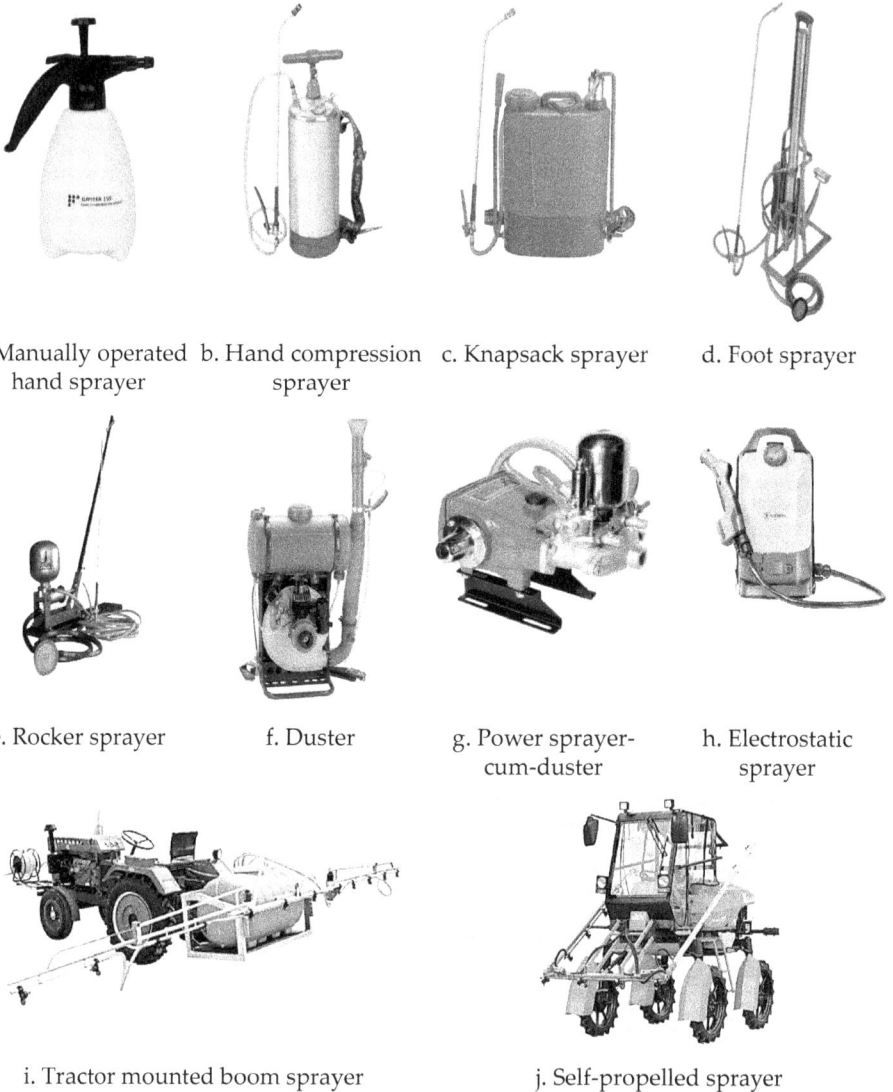

Fig. 12.1: Different types of sprayers and dusters used in crop protection

k) **Spray boom:** A device on which the nozzles are mounted and which may form or support one or more pipelines which are carrying the liquid to the nozzle.

l) **Spray nozzles:** An assembly of parts having orifice, which transforms the fluid being ejected under pressure into a spray.

Fan nozzle - A hydraulic spray nozzle that produces a narrow elliptical spray pattern.

Hollow cone nozzle - A cone nozzle in which the formation of an air core within the orifice and the swirl chamber due to a high rotational velocity results in production of a hollow cone of the liquid.

Hydraulic spray nozzle - A type of nozzle used for hydraulic spraying.

Pneumatic spray nozzle - A type of nozzle used for pneumatic spraying.

Solid cone nozzle - A cone nozzle in which extra liquid enters in the swirl chamber centrally from its base so that the air core is filled to form a solid cone of the droplets.

m) **Spray volume:** Total volume of spray mixture applied to an area.

High Volume (HV) - Spray volume -more than 560 l/ha.

Medium Volume (MV) – Spray volume more than 56 and less than 560 l/ha.

Low volume (LV) - Spray volume more than 5.6 and less than 56 l/ha.

Ultra-Low Volume (ULV) – Spray volume more than 0.56 and less than 5.6 l/ha.

Ultra- Ultra-Low Volume (U- ULV) - Spray volume less than 0.56 l/ha.

n) **Volume Median Diameter (VMD):** The droplet size that divides the spray into two equal parts by volume, one-half containing droplets smaller than this diameter, the other half containing larger droplets.

o) **Coarse spray:** The distribution of droplets with a VMD value of more than 400 μm.

p) **Fine spray:** The distribution of droplets with a VMD value in range of 100 to 400 μm.

q) **Mist:** The distribution of droplets with a VMD value in range of 50 to 100 μm.

r) **Aerosols:** The distribution of droplets with a VMD value of below 50 μm.

12.3 Scope of Test

12.3.1 General tests

a) Visual inspection

b) Checking of specifications

c) Checking of material of construction of different components.

12.3.2 Performance tests

a) Discharge rate

b) Volumetric efficiency

c) Pressure development

d) Pressure retention

e) Liquid remains

12.3.3 Component tests

a) Pressure chamber, pump cylinder and pressure tank

b) Hose and hose connection

c) Strap and its assembly

d) Gasket test

e) Spring test

f) Operating lever, handle and piston rod for knapsack sprayer

g) Handle, piston rod, foot rest and stirrup for stirrup sprayer

h) Frame, piston rod, pedal/hand lever and extension for foot and rocker sprayer

i) Frame, operating lever, connecting rod and handle for charge pump.

j) Valve assembly test

k) Tank impact test

l) Fatigue test

m) Test for spray lance

12.3.4 Performance test for cut-off device

12.3.5 Spray distribution pattern for nozzle

12.3.6 Endurance test

12.3.7 Engine test for power sprayers

12.3.8 Referred standards / test codes

a) IS 1970: 1995 (2009) [68]

b) IS 3062: 1995 (2006) [69]

c) IS 3652: 1995 (2006) [70]

d) IS 3906: 1995 (2006) [71]

e) IS 5135-2: 1994 (2009) [74]

f) IS 7593–1: 1986 (2012) [86]

g) IS 10134: 1994 [37]

h) IS 11313: 2007 (2012) [44]

i) IS 12482: 1988 (2009) [57]

12.4 Methods of Tests for Sprayers

12.4.1 General test

a) Visual & specification checking

The equipment shall be visually inspected with respect to the functional requirements given by the manufacturer. Conformity or otherwise of the requirements shall be reported. Specifications of the machine shall be checked and compared as per the manufacturers' data sheet and relevant Indian standards and reported accordingly.

b) Checking of material of construction

The material of constructions of various parts of the sprayer shall be checked as per test codes or otherwise and reported.

12.4.2 Performance test

a) Discharge rate

This test is applicable to continuous knapsack sprayers, foot sprayers, rocker sprayers and stirrup sprayers. This test is conducted on a standard test rig as shown in Fig. 12.2, having crank mechanism to operate sprayer at the speed of 16±1 cycles per minute. The discharge is measured at relevant pressure norms.

b) Volumetric efficiency

This test is applicable to knapsack sprayer, foot sprayer, rocker sprayer and stirrup sprayer. The discharge of water in 10 successive cycles is collected and measured at the same speed of 16±1 cycles per minute on a standard test rig. The test is replicated four times and average value of discharge is taken. Accordingly, the volume of water discharge in one cycle is calculated. This is called actual discharge of sprayer. To calculate theoretical discharge. The piston displacement is calculated by measuring the inner diameter of pump cylinder and the actual length of one stroke. Volumetric efficiency is calculated by dividing the actual discharge by theoretical discharge, and the same be compared with the norms and reported in the test report.

c) Pressure development

The test is applicable to compression sprayer and is designed to check the pressure development in tank. The tank of sprayer is filled with clean water upto 2/3rd of its total capacity. The discharge outlet is closed and pressure gauge is fitted on the tank to read the pressure. The handle of sprayer is operated continuously 100 times at a constant speed on the rig as shown in Fig. 12.2. The pressure so developed in the tank is recorded and this test should be repeated four times and average value is reported and compared with relevant standard as per test code.

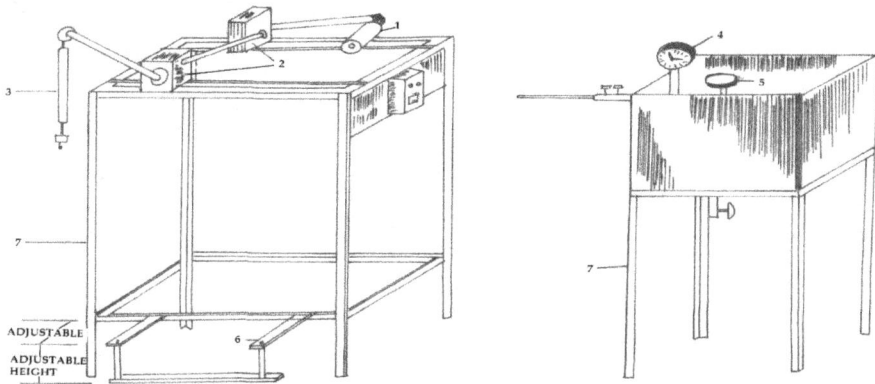

1. 5hp motor, 2. Gearbox, 3. Crank, 4. Pressure gauge, 5. Valve, 6. Platform, 7. Frame

Fig. 12.2: Test rig for discharge rate measurement, Pressure retention and endurance test

d) Pressure retention

The test is applicable to compression sprayer (pressure retaining type) in order to check the pressure retention in the tank (Fig. 12.3). The tank of sprayer is initially charged with an air pressure of 275 kPa. The desired quality of liquid is then pumped inside the tank at a pressure of 833 kPa by charge pump. This test is repeated 100 times and at the end of each test, reading of air pressure is taken and average is reported. For the purpose of routine and acceptance check the sprayer is subjected to a minimum of 5 repetitions instead of 100 times for its conformity.

e) Test for liquid remains

This test is conducted for compression sprayer (pressure retaining type). Initially, the sprayer is charged with water by applying two to three strokes of charge pump. Then conduct the operation for pressure retaining. When it is not possible to take out liquid, the air pressure shall be released from pressure release device. Then the amount of liquid left over, if any, in the tank shall be measured and reported.

12.4.3 Component tests

a) Test for leakage & deformation of pressure chamber, pump cylinder, pressure tank

i) *Pneumatic test*

A hose shall be fitted to the opening of the pressure chamber or pump cylinder or tank. In case there are more openings, all will be sealed except to which the hose is fitted. The pressure chamber or pump cylinder or tank shall be then pneumatically pressured to a minimum of 1.5 times the normal working pressure of the sprayer. The pressure shall be retained for a period of one minute. The component shall be disconnected, immersed in water and examined for any leakage and deformation. The pressure chamber or pump cylinder or tank shall deemed to have passed this test if no leakage or deformation during this test is occurred.

ii) *Hydraulic test*

The sprayer tank shall be filled with water up to its capacity. The hose connections are to be made as per pneumatic test. Then the pressure chamber or pump cylinder or tank shall be pressurized to a static hydraulic pressure of a minimum two and a half times the normal working pressure of the sprayer. The pressure shall be retained for a period of one minute. The pressure chamber or pump cylinder or tank shall be deemed to have passed this test if no leakage or deformation is found during the test.

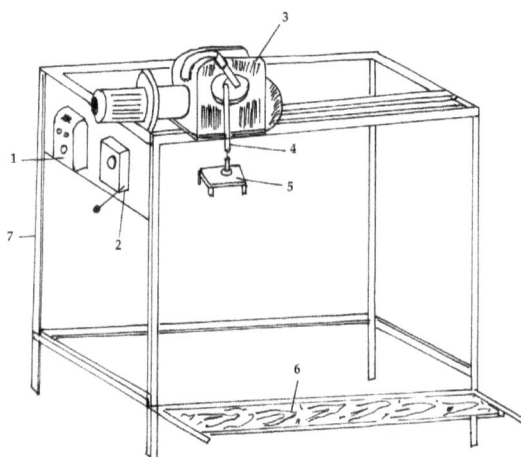

1. Main switch, 2. Starter, 3. Gearbox, 4. Crank, 5. Lock, 6. Sprayer platform, 7. Angle iron frame

Fig. 12.3 : Test rig for pressure retention in compression sprayer

b) Test for hose and hose connection

The inlet of the hose pipe shall be connected to a hydraulic pump through hose connection. The other end of hose pipe shall be connected to the cut-off device. The outlet of the cut-off device shall be closed in such a way that no discharge is allowed. A minimum hydrostatic pressure of 1.5MPa, using water as a liquid shall be developed in the hose assembly and retained for a period of one minute. The hose and hose connection shall conform this test if no leakage or crack is observed during the test is observed

c) Test for strap and its assembly

This test is applicable to knapsack sprayer in order to check the strength of strap and its assembly on the test rig. The tank is filled with clean water to its specified capacity. The sprayer is suspended form a solid support through its strap, simulating the conditions on the shoulder of the operator. Raise the tank vertically to a height of 300 mm and allow to drop freely. Repeat the operation 24 times. The assembly conform test, if no part of strap, bracket or clamp etc. is broken.

d) Gasket test

A new set of gasket of the sprayer shall be immersed in a mixture consists of 60% kerosene, 5% benzene, 20% toluene and 15% Xylene for a period of 72 hours at a temperature of 27 to 33°C. Then gasket are dried in air at the same temperature range for a period of 24 hours. Then these gaskets be fitted in their original positions on sprayer. The sprayer complete with its discharge line shall be operated at its normal working speed and conditions for 8 hours. The gasket shall conform this test if no leakage is found.

e) Spring test

The test is applicable for the springs provided in foot sprayer. The free length of spring is measured. The spring shall be attached to test rig in vertical position. It shall be compressed, up to touching position of coil at a rate of 20 stroke per minutes for a period of one hour. The free length of spring shall be measured after test. The spring shall conform this test if the difference in free length of the spring does not exceed 5%.

f) Test for operating lever, handle and piston rod for knapsack sprayer

This test is applicable to knapsack sprayer piston type in order to check their strength. The discharge outlet of sprayer is closed and handle is operated to develop the pressure in the sprayer to the minimum two and half times the normal working pressure. When the handle, operating lever and piston rod are operated at this pressure, there should not be any distortion or crack in order to confirm this test.

g) Test for handle, piston rod, foot rest and stirrup

This test is applicable in case of stirrup type sprayer to check the strength of components. The procedure to carry out is same as given in test for operating lever, handle and piston rod. This test deemed to have confirmed if no breakage or deformation is observed.

h) Test for frame, piston rod, pedal lever or handle lever and extension

This test is applicable in case of foot and rocker sprayers. The procedure for testing is same as for operating lever, handle and piston rod test. Frame, piston rod and pedal lever in case of foot sprayer whereas frame, piston rod, handle lever and extension in case of rocker sprayer shall not break, deform or crack when the force is exerted on them in order to pass this test.

i) Valve assembly test

This test is applicable for compression knapsack sprayer. The piston rod and piston assembly is taken out. The discharge outlet is connected to a hydraulic pump through hose connection. The outer openings except discharge are sealed. The tank shall be pressurized to a static hydraulic pressure of minimum two and a half times the normal working pressure of the sprayer. This pressure is retained for a period of five minutes. The valve assembly is confirmed if no drop in pressure is observed.

j) Tank impact test

This test is applicable in case of compression sprayer and is conducted on rest rig as shown is Fig. 12.4. The tank is filled with clean water upto 2/3rd of its capacity and pressurized pneumatically to the normal working pressure of the sprayer. Take out the discharge line and plug the discharge outlets. The tank shall be dropped for 25 times from a height of 600 mm in following positions:

  i)   7 times with its long axis horizontal

  ii) 6 times with its long axis vertical

  iii) 6 times with the long axis inclined at 75 degree to the horizontal

  iv) 6 times as in (iii) with the position of impact diametrically opposite.

The platform on which the tank is dropped shall consist of a plane solid teak wood or similar hard wood of 60 mm thick, placed on a hard level surface. The sprayer is deemed to have passed this test, if tank shall not burst and bottom of tank or any other part of sprayer shall not extend below the bottom of skirt.

k) Fatigue test

This test is conducted for compression sprayer on test rig as shown in Fig. 12.5. The pump assembly and discharge line are taken out from the sprayer. An opening of the tank is connected to the manifold of test rig through a hose and shut-off cock. The tank is completely filled with clean water. The manifold is also filled with clean water through the filler hole. Air pressure with the help of compressor shall be developed equal to the normal working pressure of the sprayer within the range of ± 10 through a three-way valve. A timer switch connected in the circuit is used to open and close the three-way valve within a range of 3 to 5 times per minute. The tank is subjected to 1200 such pressure cycles. A counter connected in the system will indicate the number of pressure cycles. The tank shall conform this test, if no leakage, crack or deformation of the tank is occurred.

l) Test for spray lance

The inlet of spray lance shall be fitted to a hydraulic pump directly or through a delivery hose. The outlet of the lance shall be closed so that no discharge is allowed from the lance. A hydraulic pressure of 1 MPa or two and half times of the normal working pressure of the sprayer, whichever is more shall be applied to the lance for a period of 5 minutes. If during this test there is no leakage, crack of burst then the test is confirmed.

12.4.4 Performance test for cut-off device

This test shall be conducted with water liquid containing 5% suspension of DDT. The cut-off device shall be rigidly mounted on the test rig as shown in Fig. 12.6 & 12.7 for trigger type and knob type respectively. The test liquid shall be applied to the inlet of cut-off device under a static pressure of 300±30 kPa. The cut-off device shall be operated for 5000 cycles at the speed of 15 cycles per minute and further for 500 cycles at a pressure of 600±60 kPa. The cut-off device is confirmed if the leakage is not observed.

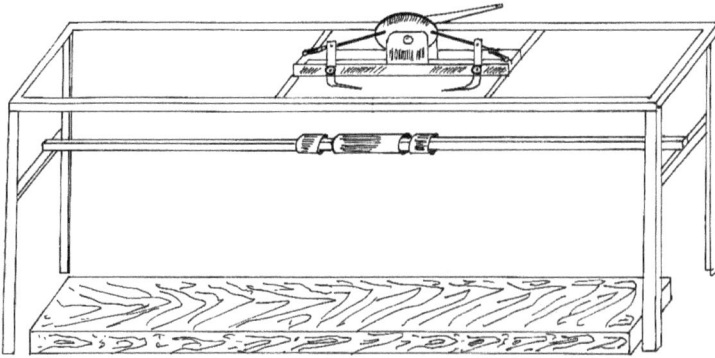

Fig. 12.4: Test rig for tank impact test

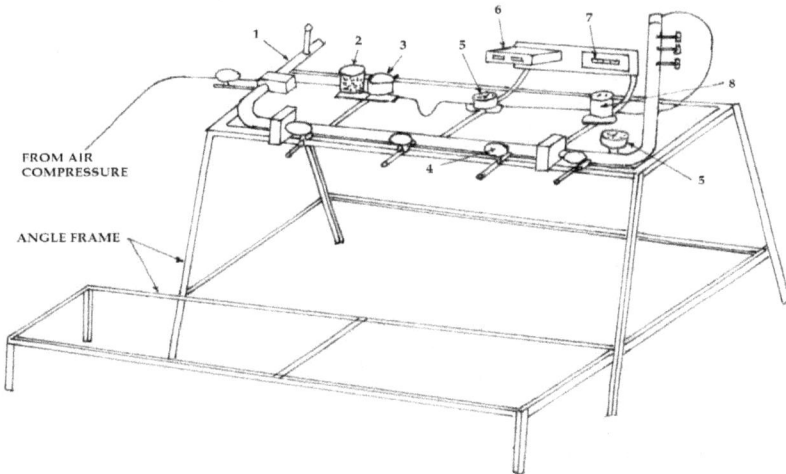

1. Brass pipe, 2. Air filter, 3. Air filter, 4. Valve, 5. Pressure regulator, 6. Timer, 7. Counter meter, 8. Solenoid valve

Fig. 12.5: Test rig for fatigue test on compression sprayer

12.4.5 Distribution pattern test for nozzle

To calculate spray angle, the nozzle is fitted at the height of 545 mm on a patternator. A long scale is placed on the patternator to have impression of discharge. Accordingly, spray angle is calculated. The value so obtained should not have variation of more than +3° than the declared value in order to conform this test.

12.5.6 Endurance test

The sprayer is operated in accordance with the normal working pressure for a minimum period of 48 hours on the test rig. This period preferably should be in continuous stretches of 6 hours. The sprayer is deemed to have passed this test if no leakage of breakdown is observed during this test.

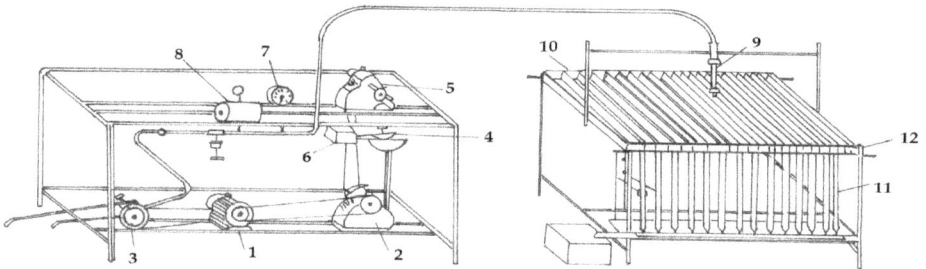

1. Motor, 2. Gear box, 3. Hyd. Pump, 4. Cut off device, 5. Gearbox, 6. Spray box, 7. Pressure gauge, 8. Pressure chamber, spray nozzle, 10. Channels, 11. Burette, 12. Up-down slope

Fig 12.6: Test rig for trigger type cut-off device and patternator for spray pattern

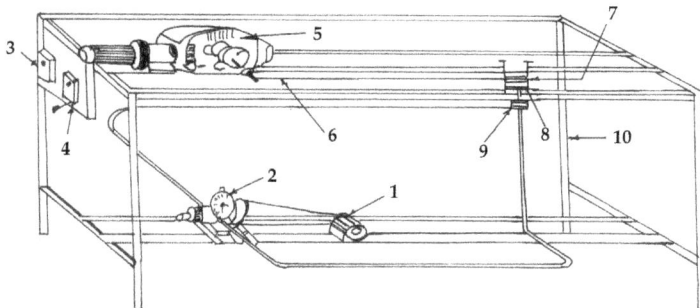

1. Motor, 2. Pump, 3. Main switch, 4. Starter, 5. Gearbox, 6. Crank, 7. Rack, 8. Pinion, 9. Cut-off device, 10 Angle iron frame

Fig 12.7: Test rig for knob type cut-off device

Recommended norms and parameters for different types of sprayer are shown in Annexure 12.1.

12.5 Scope of Test for Dusters

12.5.1 General test

a) Material of construction for different components.

b) Constructional requirements of hopper, feed control mechanism, agitator and feeder.

c) Test for agitator

d) Strap drop test

12.5.2 Performance tests

a) Air output test

b) Dust delivery test

c) Dust throw test

d) Leakage test

12.6 Testing Method for Dusters

12.6.1 General test

a) Material of construction for different components

All the metallic parts coming in contact with the pesticide should preferably be of the same material to minimize electrolyte potential deterioration. The material used for different components declared by the manufacturer be compared and reported.

b) Constructional requirement of hopper, feed control mechanism, agitator and feeder

The hopper shall have a concave shape or conical bottom so that the dust contained in it, moves towards the feeding aperture. On the top of the hopper, a filler hole of at least 130 mm diameter should be provided. The hole shall be covered with a lid. A feed control mechanism with locking device should be provided to control the flow of dust through the aperture. An agitator either integral with feeder or separately shall be incorporated within the hopper to keep the dust agitated and to avoid the clogging of the aperture of feeding the dust to the aperture smoothly.

c) Test for agitator

The hopper shall be filled up to 3/4th of its total capacity with talc powder used for insecticidal formulation. The duster shall be fixed rigidly in place and shall be operated continuously at medium discharge rate setting of feed control mechanism as specified by the manufacturer till the discharge at the outlet seizes. The duster shall be considered to conform the test if the dust remains in the hopper is not more than 0.5% of the total mass.

d) Strap drop test

The hopper shall be filled with talc powder to its total capacity. The duster shall be suspended from a solid support by its straps. It shall be lifted to a height of 300 mm and allow to drop. The test be repeated for 25 times. If no breakage or deformation occurred, the test is confirmed.

12.6 2 Performance test

a) Air output test

The air output test is conducted either by pivot tube method or by air flow method. Average value of 5 readings is calculated and reported. The fan should be able to deliver not less than 0.3 m³ of air/minute in order to pass this test.

b) Dust delivery test

The hopper shall be filled up to 3/4th of its total capacity with talc powder. The duster with its all working accessories shall be weighed. The duster shall be fixed rigidly and operated uniformly at a speed of 35 rpm for at least 2 minutes. The mass of the duster shall again be determined. The total mass of the dust discharge shall be computed and the rate of discharge/minute determined. The delivery rate at maximum discharge setting should not be less than 150 gm/minute in order to conform this test.

c) Dust throw test

The hopper shall be filled up to 3/4th of its total capacity with talc powder. Set the duster and delivery pipe at its horizontal position. Operate the duster continuously and uniformly at a speed of 35 rpm. Measure the horizontal distance from the outer most point of the delivery pipe and the outer most point where dust falls on the ground. The duster should be able to throw the dust up to a minimum distance of one meter in order to conform this test.

d) Leakage test

The hopper shall be filled up to 3/4th of its total capacity with talc powder. Set the duster at its horizontal position and plug the outlet. Operate the duster at a speed of about 30 rpm for two minutes. The duster is considered to conform this test if no leakage of dust is occurred.

Recommended norms and parameters for different types of dusters are shown in Annexure 12.2.

Annexure 12.1

Recommended Norms & Parameters for Different Types of Sprayers

Name of test	Specified requirements as per BIS for different type of sprayers				
	Hand operated continuous knapsack sprayer	Rocker sprayer	Foot sprayer	Stirrup sprayer	Hand operated compression sprayer
(1)	(2)	(3)	(4)	(5)	(6)
Performance requirement of sprayer					
Discharge rate	Minimum (min.) 500 ml/min at min. pressure of 200 kPa	1200 ml/min at pressure of 450 kPa	1200 ml/min at pressure of 450 kPa	Min. 400 and 500 ml/min at minimum pressure of 150 kPa and 450 kPa resp.	Pressure development in tank should be minimum 400 kPa
Volumetric efficiency	Min 80%	Min 80%	Min 80%	Min 80%	Min 80%
Performance requirement of nozzle	Rate of discharge shall be as per norm The rate of discharge shall be within ±5% of declared value.	Rate of discharge shall be as per norm The rate of discharge shall be within ±5% of declared value.	Rate of discharge shall be as per norm The rate of discharge shall be within ±5% of declared value.	Rate of discharge shall be as per norm The rate of discharge shall be within ±5% of declared value.	Rate of discharge shall be as per norm The rate of discharge shall be within ±5% of declared value.
Constructional Requirements					
Piston	Mini height-13mm Mini thickness-2.5mm	Mini height-16mm Mini thickness-3.5mm	Mini height-16mm Mini thickness-3.5mm	Mini height-13mm Mini thickness-25mm	Mini height 13mm Mini thickness-2.5mm
Pump Cylinder dia.	Not more than 55mm	Not more than 55mm	Not more than 55mm	Not more than 30mm	Not more than 40mm
Pressure chamber	Min. capacity should 8 times the piston displacement	Min. capacity should be 8 time the piston displacement	Min. capacity should be 8 times the piston displacement	--NA--	--NA--

(1)	(2)	(3)	(4)	(5)	(6)
Strainer	The apertures of the strainer shall be not more than 625 µm	The apertures of the strainer shall be not -more than 625 µm	The apertures of the strainer shall be not more than 625 µm	The apertures of the strainer shall be not more than 625 µm	The apertures of the strainer shall be not more than 625 µm
Springs	--NA--	--NA--	Two springs shall be provided. Variations of spring after test shall be not more than ±5%.	--NA--	--NA--
Lance	The length of lance should be between 500 to 900mm	The length of lance should be between 500 to 900mm	The length of lance should be between 500 to 900mm	The length of lance should be between 500 to 900mm	The length of lance should be between 500 to 900mm
Cut-off device	Maximum torque required for trigger actuation shall be not more than 3.5 Nm	Maximum torque required for trigger actuation shall be not more than 3.5 Nm	Maximum torque required for trigger actuation shall be not more than 3.5 Nm	Maximum torque required for trigger actuation shall be not more than 3.5 Nm	Maximum torque required for trigger actuation shall be not more than 3.5 Nm
Total mass	The mass of sprayer shall not be more than 9 kg.	The mass of sprayer shall not be more than 11.5 kg.	The mass of sprayer shall not be more than 11.5 kg.	The mass of sprayer shall not be more than 5 kg.	The mass of sprayer shall not be more than 9 kg.
Gasket Test	The gasket shall be deemed to have passed this test, if no leakage observed from points during 8 h test	The gasket shall be deemed to have passed this test, if no leakage observed from points during 8 h test	The gasket shall be deemed to have passed this test, if no leakage observed from points during 8 h test	The gasket shall be deemed to have passed this test, if no leakage observed from points during 8 h test	The gasket shall be deemed to have passed this test, if no leakage observed from points during 8 h test
Endurance Test	If no leakage or breakdown is observed during 48 h test, the sprayer shall be considered passed the test.	If no leakage or breakdown is observed during 48 h test, the sprayer shall be considered passed the test.	If no leakage or breakdown is observed during 48 h test, the sprayer shall be considered passed the test.	If no leakage or breakdown is observed during 48 h test, the sprayer shall be considered passed the test.	Tank filled with water at 2/3 of total capacity & at pressure set 400 to 600 kPA is required 100 times repetition in order to clear the test.

Annexure 12.2

Recommended Norms & Parameters for Different Types of Dusters

Name of test	Specified requirements as per BIS for different type of dusters	
	Hand rotary duster (Shoulder mounted type)	Hand rotary duster (Belly mounted type)
(1)	(2)	(3)
Performance requirement		
Air output	The fan should be able to deliver not less than 0.8 m³ of air per minute.	The fan should be able to deliver not less than 0.8 m³ of air per minute.
Dust delivery	The delivery rate at maximum discharge setting should be not less than 150 gm per minute.	The delivery rate at maximum discharge setting should be not less than 150 gm per minute.
Dust throw	The duster should be able to throw the dust up to a maximum distance of one meter.	The duster should be able to throw the dust up to a maximum distance of one meter.
Hopper capacity	The total capacity of hopper should be 0.005 to 0.0075 m³. The tolerance on the declared capacity should be ±5%.	The total capacity of hopper should be 0.004 to 0.006 m³. The tolerance on the declared capacity should be ±5%.
Test of agitation	After completion of test the left out dust powder should not be more than 0.5% mass of total dust.	After completion of test the left out dust powder should not be more than 0.5% mass of total dust.
Constructional requirements		
Hopper	A filler hole of atleast 130 mm diameter shall be provided on the top of the hopper.	A filler hole of atleast 130 mm diameter shall be provided on the top of the hopper.
Suction pipe	The suction pipe shall have an internal diameter of not more than 45 mm.	The suction pipe shall have an internal diameter of not more than 45 mm.
Mass of duster	The total mass of the duster should not exceed 8 kg.	The total mass of the duster should not exceed 8 kg.

Chapter 13

Testing and Evaluation of Irrigation Pumps

Generally, two types of pumps are being used in agriculture and required testing: a) Centrifugal pumps, and b) Submersible pumps. Centrifugal pump is one of the most important and popular water lifting devices for irrigation purpose because of low initial cost, greater flexibility in application, wide working range in size and capacity, simplicity in construction, constant steady discharge and ease of operation, maintenance and repairs etc. In the last decade in India, there has been a four-fold increase in consumption of energy in the agricultural sector, out of this, roughly 85% energy is consumed by the agricultural pumps. The system efficiency of some of the pumps used in agriculture is around 40%. This is colossal waste of energy. For example, in case of India, if 90% of existing electric pumping systems are improved/rectified, 4000 to 5000 MW of power can be saved annually. Measures to save this energy waste is essential and some headway has been made by BIS (Bureau of Indian Standards) by formulating minimum performance standards for pumping system and related equipment.

The impeller is rotated at a very high speed by the motor or engine, which is coupled to the pump. Water coming at the centre of impeller is picked up by the vanes and accelerated to a high velocity by the rotation of the impeller and thrown out by the centrifugal force. The horizontal centrifugal pump has a horizontal shaft on which vertical impeller is mounted.

The centrifugal pump, shown in Fig. 13.1, consists of following major components:

i) *Impeller:* This is the heart of the pump and serves to exert a centrifugal force upon the water contained therein.

ii) *Casing:* This component performs the function of converting the centrifugal force created by the impeller effectively into pressure.

iii) *Suction port:* This is used for sucking the water.

iv) *Discharge port:* This is used to discharge the water.

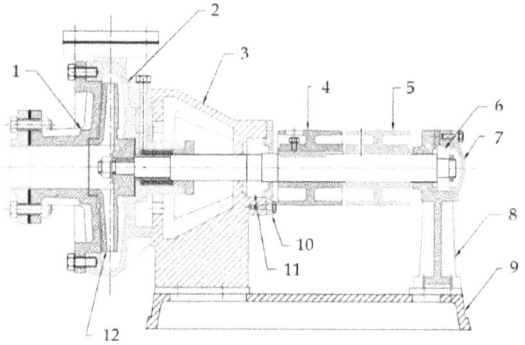

1. Section plate, 2. Pump cover, 3. Large bracket, 4. Fast pulley, 5. Loose pulley, 6. Bearing, 7. Bearing flange, 8. Small bracket, 9. Pump base, 10. Bearing flange, 11. Bearing, 12. Impeller

Courtesy: Kalsi pumps, India

Fig. 13.1: Construction of Centrifugal pump with section view

The submersible pump consists of a pump and motor assembly, head assembly, discharge column and cable to supply power to the motor. The pump and motor are submerged in the water in the well at all times. The propelling shaft, made of stainless steel is very short and impeller is mounted on it. Impeller may be closed, semi-open or open and are arranged in series. Water enters the pump through a screen located between the motor and the pump. The pump column and the submersible are supported by the discharge pipe. The submersible motor form a complete unit and are made smaller in diameter and much longer than the ordinary motor so that they can be inserted in small diameter tube wells.

Major components of a submersible pump are Fig. 13.2:

   i)   Motor Body

   ii) Rotor

   iii) Stator

   iv) Cable Sealing Arrangement

   v)   Breather Diaphragm

   vi) Impellers

   vii) Shaft

   viii) Suction Casing

   ix) Casing Wear Rings

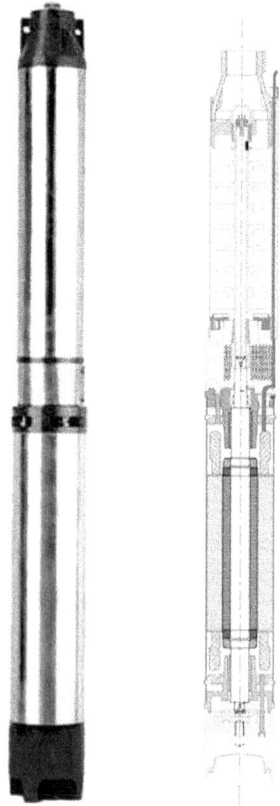

Courtesy: Lubi & Kalsi, India

Fig. 13.2: Sectional view of motor and pump of a submersible pump

13.1 Terminology

a) **Suction lift:** The vertical distance between the centre line of the pump shaft and the pumping water level in the well.

b) **Delivery head:** The vertical distance between the centre line of the pump shatt and the centre line of delivery pipe at the discharge end.

c) **Static head:** The vertical distance between pumping water level and discharge water level.

> Static head = Static suction head + Static delivery head

d) **Total head:** The actual head against which the pump has to work

> Total head = Static head + All head losses

e) **Friction head:** The head requires to overcome the resistance of water in fittings in the pipe line. It varies with, (a) rate of flow (b) pipe size (c) interior condition of pipe and (d) type of pipe. This will include the friction in strainers, elbows, bends, foot valve, reducing sockets, tees, valves etc.

f) **Net positive suction head requirement (NPSHR):** This is a function of pump design and varies from one make of pump to another, between different model of same manufacturer and speed of pump. Therefore NPSHR value must be obtained from manufacturer and for cavitation free performance. This value should be less than net positive suction head available.

g) **Velocity head:** The vertical distance through which the liquid must fall to acquire a given velocity and is calculated using the following formula:

$$h = \frac{V^2}{2g}$$

Where,

h	- velocity head
V	- velocity of water in the pipe
g	- acceleration due to gravity

h) **Pump characteristics:** The relation between speed, head, discharge and horsepower of a pump are represented by number of curves known as "Characteristics Curve".

i) **Specified speed:** The speed of geometrically similar pump when delivering a volume of one cubic meter per second against a head of one meter height.

13.2 Selection of Parameters

The performance of the total pumping system is the sum total of the efficiency of each part of the system. Besides, each part affects the efficiency/performance of the other part. It is necessary that a "system approach' is applied to the selection of a pumping system. Pump selection is therefore, a process of selecting the prime mover, pump, foot valve, suction and delivery pipes keeping in view the type of soil, crops grown, area to be irrigated and the hydrogeological conditions, so that the pump gives timely and trouble-free services at the minimum cost per litre of water delivered.

For proper selection of pump, following parameters are applied:

- Total head: It consists of, a) Static suction head, b) Draw down, c) Delivery head, and d) Frictional head
- Discharge in litre per second desired from the pump: This will be determined by, a) Size of the land to be irrigated, b) Crops to be grown, c) Type of soil, and d) Capacity of the aquifer.

Careful scrutiny of the following points is also necessary for better results from the pumps:

- Depending upon the soil conditions, water losses due to seepage and evaporation should be considered at the time of calculating the water requirement.
- Suction lift should be kept to the minimum possible and it should never be more than 6 m.
- One size higher of the suction, delivery pipe and foot valve should be used which would result to fuel saving.

13.2.1 Causes for low overall efficiency of a pumping system

- Low efficiency of the pump
- Poor suction lift characteristics (NPSH) of the pump
- High friction losses in the foot valve/reflux valve
- Low efficiency of prime mover
- High specific fuel consumption in case of engine driven pump
- Improper selection of the pump and prime-mover such that they do not operate in their best efficiency zone
- High friction losses in the pipes because of:
  a. High coefficient of friction of pipe material
  b. Under size pipes
  c. Sharp bends, unwanted lengths and heights of piping
- Air leakage in suction side
- Excessive power losses in power transmission from prime mover to pump
- Poor maintenance and lack of service facilities
- Improper installation.

13.2.2 Efficiency of pump

The efficiency has a direct bearing on power consumption. Higher the efficiency, more will be the energy conservation. An agricultural pumping system does not operate at a fixed duty point, like an industrial pump. When the pump starts, suction lift is low, but while pumping, the suction lift increases due to draw down. The total lift as well as total head increases. Again the operating points vary from season to season. Thus, agricultural pumps operate in a wide range of head and discharge. The efficiency of a pump should be reasonably good in the entire range and minimum declared efficiency should not be less than as shown in Fig. 13.3 to 13.5.

Fig. 13.3: Efficiency in percent for horizontal centrifugal pumps for agricultural purposes

Fig. 13.4 : Efficiency in percent for monoset pumps for agricultural purposes

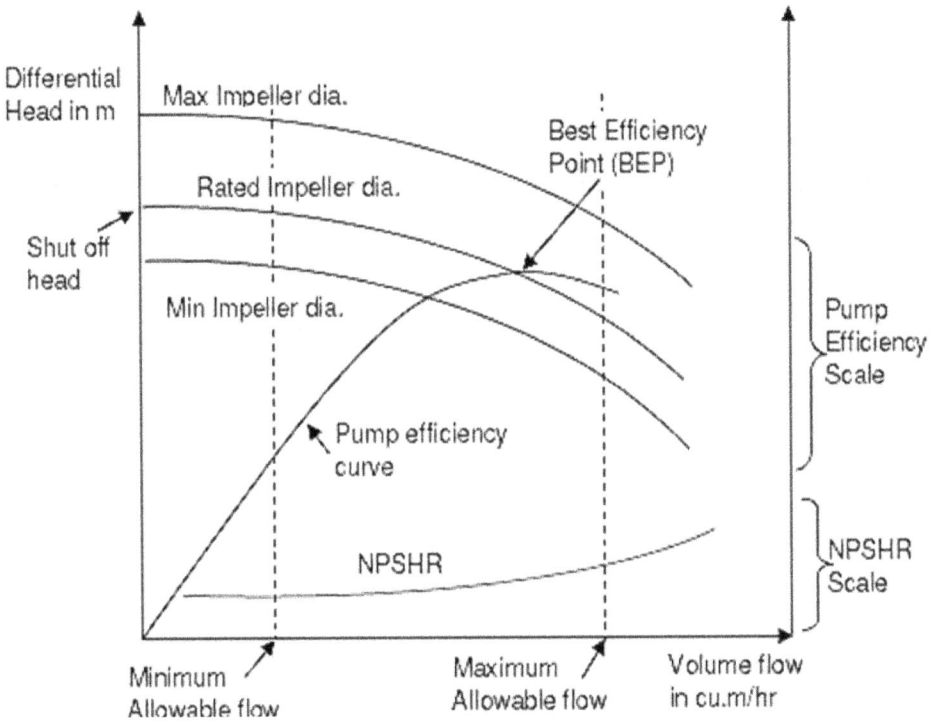

Fig. 13.5: Centrifugal pump performance curve

Foot valve/reflex valve: The foot valve or reflux valve is an integral part of the pumping system. The foot valve is preferred while pumping water from open water, whereas reflux valve is used in case of tubewells. It plays an important role in the performance of the pumping system. Higher friction losses in the foot valve/reflux valve not only increase the total head of installation, but it also increases the total suction lift, ultimately reducing the discharge and efficiency of pump. Friction losses in a foot valve/reflux valve are given as:

$$H = \frac{K \dot{n} V^2}{2g}$$

Where,

        H   - frictional losses

        K   - coefficient of friction

        V   - velocity of flow in suction pipe

        g   - acceleration due to gravity

To get the best performance from the pumping set, the valve of K should be as low as possible. Foot valves/reflux valves available in the market are having K valve ranging from 0.5 to 12. The value of K should be less than 1 so that the pumping system will give more discharge while consuming less energy.

13.2.4 Data required for selection of centrifugal pump

For selecting the right size of the centrifugal pump for a particular set of conditions, the following information are required:

a) The source of water supply and level fluctuations in summer and rainy season.

b) Vertical suction lift: the vertical distance from water level to centre of the pump.

c) Static discharge lift: the vertical distance from centre of the pump to delivery point.

d) The length of delivery line in case the water is not discharged just outside the well.

e) The diameter of the tubewell.

f) The discharge capacity required.

g) The type of power: whether electrical energy or diesel engine.

13.2.5 Quality criteria

Normally a complete pumping set is not manufactured by any manufacturer and there are separate Original Equipment Manufacturers (OEM's) for prime mover, pump base frame, coupling, and other accessories required to make a complete pumping unit. A manufacturer collects all such components and assembles the complete pumpset and provides it to the dealer.

Prime mover

Diesel engine: Diesel engine is the most important and costly unit in the whole pumping system and a quality engine should be able to give:

i) Low specific fuel consumption

ii) Low lubricating oil consumption

iii) Low exhaust smoke

iv) High reliability and should be of IS or other standard mark

Electric motor: It should be able to withstand a voltage variation from +6 to -15% of rated voltage. The motor should have class E or higher class of insulation capable of withstanding temperature up to 115°C. The efficiency should not be less than 77% for 3 hp and 80% for 5 hp and above.

Centrifugal pump

A good quality pump should have high reliability and high efficiency. Reliability is achieved by better materials, heat treatment, surface finish, proper design of bearing supports and appropriate tolerances etc. The main consideration should be the overall efficiency and minimum power consumption. A pump having minimum efficiency of 60% or above should be preferred.

Coupling of prime mover with pump	There should be minimum power transmission loss through coupling. The quality of rubber and canvass used in coupling should be of good quality. However, monoblock pump set have the minimum losses and be preferred.
Suction condition and net positive suction head requirements	Water level variation between summer and monsoon seasons forces the farmer to use higher suction lift. The pump may start working with a suction lift of 3 m and go on pumping upto 8 m in most of the cases. This would result in drastic loss of efficiency in pumps with poor NPSH characteristics. NPSH requirements of the pump should be checked before the installation to avoid cavitation. The suction line should be airtight and any leakage in the suction line would result in poor performance of the pump.
Suction and delivery hoses	Flexibility of hoses especially in suction hose is very important. It has been observed that cheaper quality hose gets cracked and results in leakage of water or clogging of inside surface due to unnatural union causing obstacle in the passage of water. A quality hose should have successfully tested for working temperatures, pressure and inside finish as per available standards.
Piping system	Generally, farmers feel that if the pump is discharging water with higher velocity, they will get more water. This results invariably in the use of smaller piping and foot valve, carrying a loss of nearly 25 to 30% due to friction. The size of pipe should be suitable for flow rate. The friction losses in pipe should not exceed 10% of total length of the pipe. Pipe material and inside surface should be as smooth as possible so that the pipe offers the least possible friction. There should be minimum possible number of bends in the piping system. Pipe fittings should match with the pipes and all the joints in the suction line should be air tight. Normally, the farmer decides the purchase of a system based on the rating of the prime mover like 7.5 hp engine or a 5 hp motor. The specifications of the pump, foot valve, reflex valve and piping's are neglected.

13.2.6 Selection of pump-set

It has been observed that the farmer's first choice is always for a quality pumpset, but he is tempted to go in for sub-standard pumpsets only on price consideration. But it is desirable to recommend quality units to him even if it costs a little higher. Usually a pump is selected from the chart of total head v/s discharge rate supplied by the manufacturer. In this chart, the efficiency at different total heads is generally not indicated. So, either manufacturer should be asked to indicate efficiency at different total heads or to supply performance characteristics curves of the pump. Draw the system head characteristics on the total head v/s discharge rate curve, of the pump and find its intersection point with the total head curve for different pumps (Fig. 13.6). Pump which gives highest efficiency and meeting the discharge requirements may be selected. A flat efficiency curve characteristic pump is always preferred. It should also be ensured that the pump is working in summer also, for both NPSH and total head requirements, as the water level goes down in summer. If a single pump is not sufficient to work in all seasons a separate pump may be selected for summer season. If the yield of the well or power consumption is a limiting factor, the period of operation may be adjusted suitably to irrigate the required area.

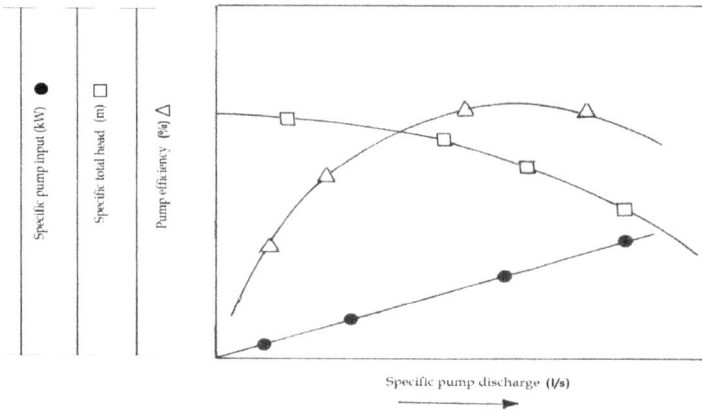

Fig. 13.6 Pump performance curve chart

13.2.7 Referred standards and test codes

a) IS 6595: 2002 (2017) [80]

b) IS 9694–1: 1987 (2017) [98]

c) IS 9694–2: 1980 (2017) [99]

d) IS 9694–3: 1980 (2016) [100]

e) IS 9694-4: 1980 (2016) [101]

f) IS 14582: 1998 (2008) [64]

g) IS 8034: 2002 (2017) [88]

h) IS 9283: 2013 [97]

i) IS 14536: 1998 (2009) [63]

13.2.8 Case study on selection of pumpsets

A farmer has 7.5 hectares of land. He plans to grow sugarcane in 5.5 hectares and vegetables in 2.0 hectares. The source of water is a Well and water table is 4.5 meters. The water is to be discharged at a height of 4.0 m from the ground level. There is no draw-down problem in the well. Select the most suitable pump set.

Solution:

i) Water requirement:

Total area to be irrigated	7.5 ha
Sugarcane area	5.5 ha
Vegetable area	2.0 ha

Irrigation interval for sugarcane	10 days
Depth of water required for sugarcane	10 cm
Irrigation interval for vegetable	10 days
Depth of water required for vegetable	2 cm

Area to be irrigated per day for sugarcane, 5.5/10 = 0.55 ha/day

Water required for sugarcane, 0.55x10 = 5.5 ha-cm/day

Area to be irrigated per day for vegetable, 2/10 = 0.2 ha/day

Water requirement for vegetable, 0.2x2 = 0.4 ha-cm/day

Total water requirement for sugarcane and vegetable, 5.5 +0.4= 5.9 ha-cm/day

Corresponding discharge for 5.90 ha-cm/day (Refer Annexure 13.1) = 24 lps

ii) Suction and delivery line:

It is assumed that pump is of 100 x 100 mm size and flange is used to make it suitable for 125x 125 mm suction and delivery. (Refer to Annexure 13.2)

Static suction head	4.5 m
Length of suction pipe	5.5 m (upto bend)
One medium type bend (Ø125 mm, 90°)	1.0 m
Length of suction pipe from bend to pump	2.6 m

It is assumed that diameter of pipes and footvalve is 125 mm

Flange size (suction side)	100 × 125
Delivery head	4.0 m
Length of delivery pipe	5.5 m
One medium type bend (Ø125 mm, 90°)	1.0 m
Flange size (delivery side)	100 x 125

Suction and delivery pipes are C.I. pipes

iii) Head loss in suction side:

Length of suction pipe	8.1 m
Head loss in foot valve (125 mm), (Annexure 13.3) eqv. to length of pipe	10 m
Head loss in bend	3.6 m
Head loss in flange	1.0 m
Equivalent length of pipe for suction head loss in suction pipe, foot valve, bend and flange.	8.1 + 10 + 3.6 + 1 =22.7 m
Head loss at 24 lps (1440 lpm) [Annexure 13.4]	0.463 m/10 m
Actual head loss	(0.463x22.7)/10 =1.01 m

Total suction head	4.5 + 1.01 =5.51 m

iv) Head loss in delivery side:

Length of delivery pipe	5.5 m
Head loss in bend	3.66 m
Head loss in flange (100x125)	1.0 m
Total head loss equivalent to pipe length	5.5 + 3.66 + 1 = 10.16 m
Head loss in meter for 24 lps (1440 lpm)	0.463 m per 10m pipe
Actual head loss	(0.463x10.16)/10 = 0.47 m
Total head loss delivery pipe	4.0+0.47 =4.47 m
v) Total head = suction head + delivery head	5.51+4.47
	9.98 m or 10m (say)

A pump has to be selected for total head of 10m and 24 lps discharge keeping in view the best efficiency.

13.2.9 Guarantee and purpose of testing

The tests are intended to ascertain the performance of pump and to compare this with the Manufacturer's Guarantee. The following parameters are generally guaranteed:

    a)   Discharge rate

    b)   Total head

    c)   Pump input and pump efficiency

    d)   Net Positive Suction Head (NPSH)

It is necessary to specify the pump speed or the electrical supply frequency and voltage for the motor pump unit. Therefore, the guaranteed operational data shall form the basis of testing.

13.3 Testing Procedure

Measurements shall be taken on not less than five different discharge values starting from full flow rate to nil discharge, and atleast one of them shall be measured at a head lower than the specified head. A suggested performa for observation and calculation is shown in Annexure 13.5.

13.3.1 Testing parameters

    i)   Total head   iv) Specific Pump output

    ii) Discharge rate   v)   Pump efficiency

    iii) Specific pump input

    The testing apparatus shall have adequate provisions for the testing of pump performance. Installation of test sample is shown in Fig. 13.7.

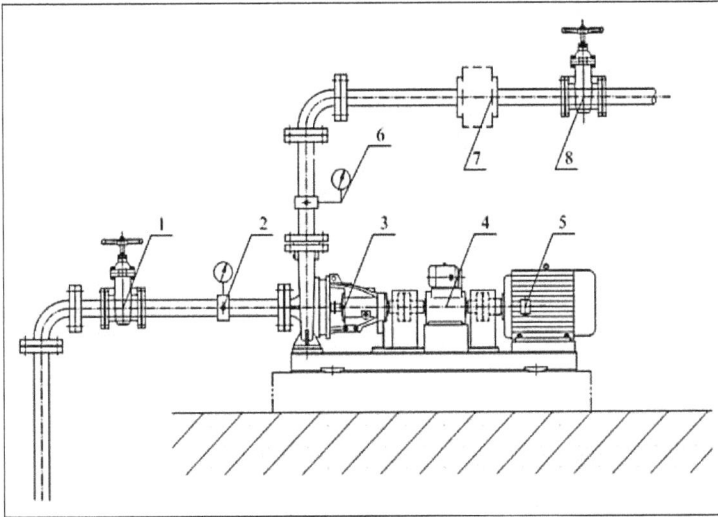

1 - inlet valve; 2 – vacuum gauge; 3 – centrifugal pump; 4 – torque meter; 5 – motor; 6 – pressure gauge; 7 – flow meter; 8 – outlet valve

Fig. 13.7: Schematic diagram of centrifugal pump performance test apparatus

13.3.2 Measuring instruments

i) Total pump head

Before using, calibration of the gauge shall be made with a standard weight type pressure or standard liquid column type mercury pressure manometer. The measuring instruments used for the measurements of pressure shall be as follows.

- Bourdon-tube gauge and vacuum gauge
- Liquid column gauge
- U-tube mercury gauge
- Pressure transducer

ii) Discharge rate

The flow meter shall be calibrated before use, and the calibration shall be performed in accordance with weight method, by using a sensor or volume method by using a mean of container (Fig. 13.8) The measuring instruments used for the measurements of discharge rate, shall be as follows:

- Weirs:
    - Right-angle triangular weirs
    - Rectangular weirs
    - Full-width weirs
- Float type area flow meter
- Electromagnetic flow meter
- Turbine type flow meter

iii) Shaft power requirement

The shaft power requirement shall be obtained by measurement of the input of the driving motor of known characteristics through an accurate testing or by using dynamometer.

iv) Measurement of speed of rotation

The speed of rotation shall be measured directly as far as possible with the help of tachometer.

v) Accuracy of measurement

The measuring instruments should not exceed the systematic error limit as specified below as well as the overall error limit.

Parameters	Permissible error of measuring instruments (%)	Permissible overall error (%)
Rate of flow, total head	±2.5	±3.5
Pump power input (electric)	±2.0	±3.5
Speed of rotation	±1.4	±2.0
Motor efficiency	±2.0	--
Pump efficiency	--	±5.0

vi) Calculation of performance parameters

a. Discharge with 'V' notch

$$\text{Discharge, lps} = \left(\frac{\text{Head over notch, cm}}{20600} \right)^{2.48}$$

b. Velocity head (suction or delivery side)

$$\text{Pipe area} = \frac{\pi d^2}{4}$$

$$\text{Velocity, } v = \frac{\text{Discharge}}{\text{Area}}$$

$$\text{Velocity head} = \frac{v^2}{2g}$$

c. Total Head = Suction head + Delivery head + Manometer distance + Velocity head (delivery side) – Velocity head (suction side)

d. $\text{Speed factor (N)} = \dfrac{\text{Actual measured speed}}{\text{Specified declared speed}}$

e. Specific discharge = $\dfrac{\text{Actual measured discharge}}{N}$

f. Specific head = $\dfrac{\text{Actual measured head}}{N^2}$

g. Specific motor input = $\dfrac{\text{Actual measired horsepower}}{N^3}$

h. Specific pump input = (Specific motor input x motor efficiency) – transmission losses

(Transmission loss should be taken as 6% for flat belt and 3% for V-belt)

i. Specific pump output or specific water hp = $\dfrac{\text{Specific discharge (1/s) ṅ Specific head (m)}}{76}$

j. Pump efficiency = $\dfrac{\text{Specific pump output}}{\text{Specific pump input}} \times 100$

13.3.3 Case study for analysis of testing data

A pump of size 100 x 100 mm was tested with motor coupled by flat belt. Different readings were taken by throttling delivery valve. Discharge was measured with 'V' notch. Specified speed of pump is 1450 rpm. Guarantee point declared by manufacturer are Head = 21 m, Discharge = 28 lps, Efficiency = 64%. Calculate the performance of pump.

Observed data for one set is as follows:

Actual speed of pump	= 1440 rpm
Suction head (vacuum gauge)	= 4.5 m
Delivery head (pressure gauge)	= 14.0 m
Manometer distance, (distance between two gauges)	= 0.725 m
Hook gauge initial reading at 'V' notch	= 354.0 cm
Hook gauge final reading at 'V' notch	= 128.5 cm
Voltage during testing	= 400 volt
Current during testing	= 26.2 amp
Motor input reading	= 11.3 kW
Motor efficiency after calibration	= 85.2%

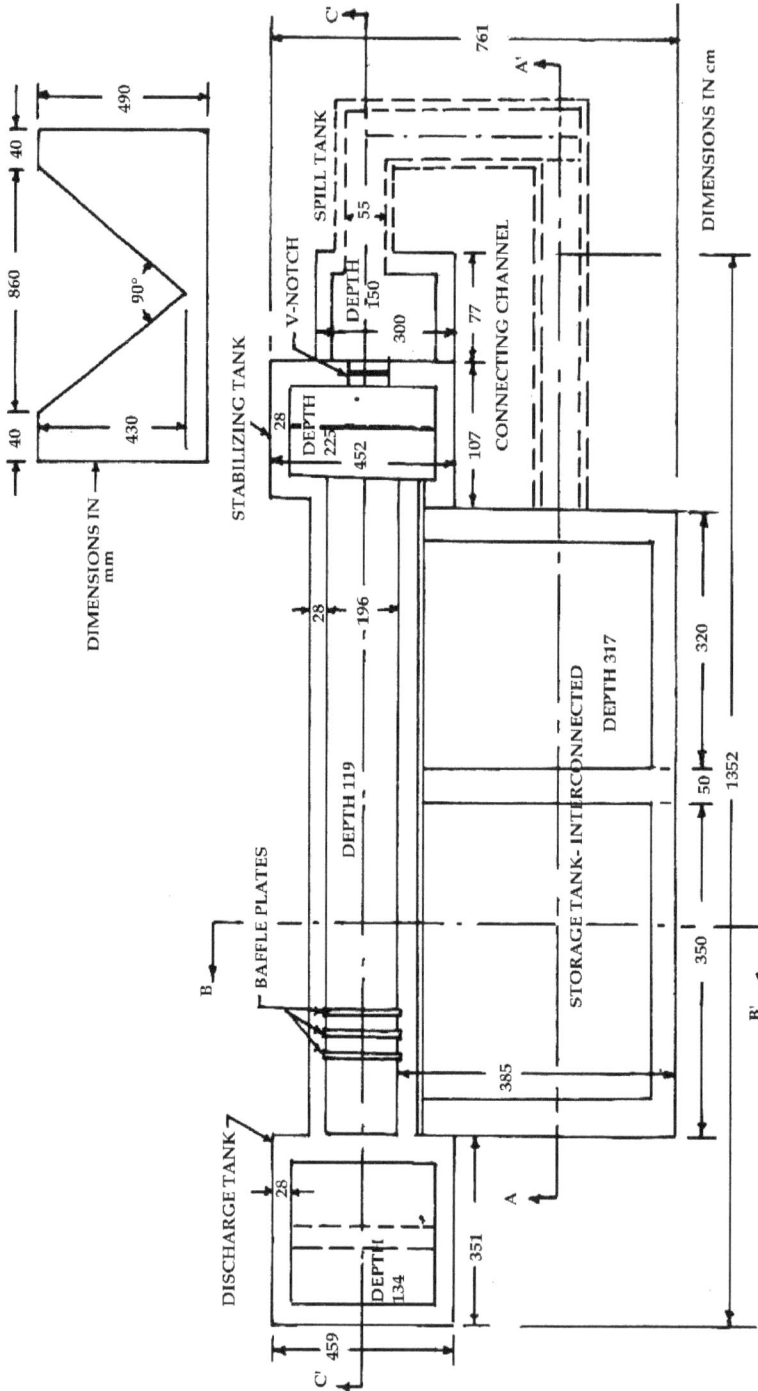

Fig. 13.8: Suggested Centrifugal Pump Discharge Measurement Structure

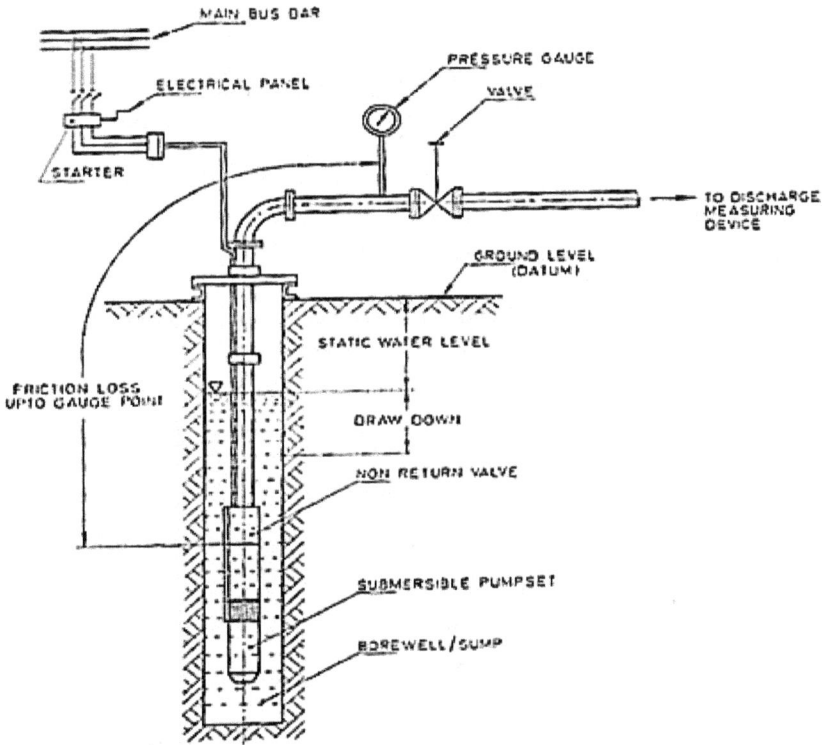

Fig. 13.9: Typical testing arrangement of submersible pump

Solution:

Head over 'V' notch = 354 - 128.5 = 225.5 cm

Discharge= $\left(\dfrac{225.5}{20600}\right)^{2.48}$ = 33.26 lps = 0.03326 m³/sec

Suction pipe (cross sec. area) = 3.14 ×100² / (4×1000) = 0.00785 m²

Velocity (V) = 0.03326 / 0.00785 = 4.23 m/s

Velocity head (suction) = (4.23)² / (2 × 9.81) = 0.914 m

Velocity head (delivery) = 0.914 (since suction & delivery pipe diameter are same)

Total head = 4.5 + 4.0 + 0.725 + 0.914 - 0.914 = 19.22 m

Speed factor (N) = 1440 / 1450 = 0.993

Specific discharge = 33.26 / 0.993 = 33.49

Specific head = 19.22 / (0.993)² = 19.49 m

Specific pump input = 11.3 / (0.993)³ × 0.852 = 9.83 kW

Specific pump input (after deducting 6% belt losses) = 9.83 (100-6) / 100 = 9.24

Specific pump output or Sp. water hp =33.49 × 19.49 / 76 = 8.59 hp

Specific pump output (kW) = 8.59 × 735.5 / 1000 = 6.31 kW

Pump efficiency = (6.31 / 9.24) 100 = 68.29%

Similarly, other sets of reading are calculated and characteristics curves are drawn as shown in Fig. 13.6.

13.3.4 Verification of guaranteed duty points

A) Head and discharge

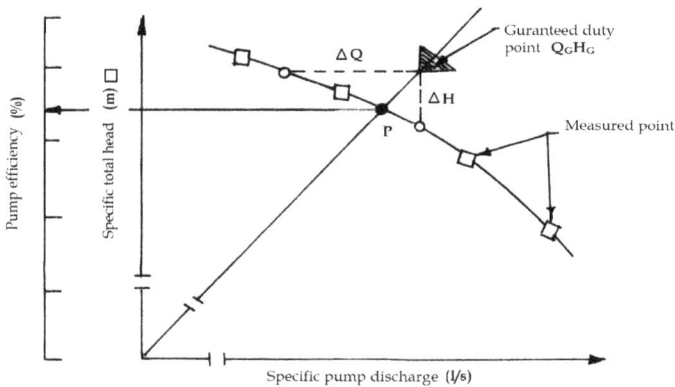

Fig. 13.10: Verification of Guarantee Points

The guarantee duty point declared by the manufacturer, i.e. discharge (QG) and head (HG) are plotted as straight line on the H-Q curve of the pump undergone testing as shown in Fig. 13.10. The point is extended vertically and horizontally so that the lines intersect on the drawn curve. Then measure the distance ΔQ and ΔH from the test curve. Tolerances $\pm X_Q$ and $\pm X_H$ respectively shall be applied to the guaranteed duty point QG and HG. In general, the values, $X_Q = 0.07$ and $X_H = 0.04$, may be used.

The following test formula is applied for verification of a guarantee:

$$\left(\frac{HG \times X_H}{\Delta H} \right)^2 + \left(\frac{QG \times X_Q}{\Delta Q} \right)^2 \geq 1$$

If the calculated value is greater than or equal to 1, the guarantee condition will be considered as qualified and if the value is less than 1, the guarantee condition not fulfilled.

B) Efficiency

The efficiency shall be derived from the drawn Q-H curve where it is intersected by the straight line passing through the declared duty point $Q_G H_G$ at point P as shown in Fig. 13.10. The efficiency at the point of intersection shall be at least 95% of the specified value as the maximum permissible overall error being ± 5%. But for combined motor pump units the maximum permissible overall error will be ± 4.5%.

Example for verification of guaranteed duty point:

Declared duty point:

Head (HG) = 21 m;   Discharge (QG) = 28 lps; Efficiency = 64%

Calculation:

$\Delta H = 0.6$ and $\Delta Q = 0.9$ (measured from test graph)

$[(21 \times 0.04) / 0.6]^2 + [(28 \times 0.07) / 0.9]^2 \geq 16.7 \geq 1$

Hence, guarantee point confirms the declared efficiency of 64%

After permissible error = (64 × 95) / 100 = 61.8%

Derived from Q-H curve = 66% [which is greater than declared. [Hence confirms].

Annexure 13.1

Area (ha) Irrigated in 8 hours for different pumping capacities and irrigation depths

Hectares Irrigated in 8 hours (depth in cm)							
lps	2.5 cm	5.0 cm	7.5 cm	10.0 cm	12.5 cm	15.0 cm	ha-cm
2	0.2	0.1	0.07	0.05	0.04	0.033	0.05
4	0.4	0.2	0.13	0.10	0.08	0.07	1.00
6	0.6	0.3	0.20	0.15	0.12	0.10	1.50
8	0.8	0.4	0.27	0.20	0.16	0.13	2.00
10	1.0	0.5	0.33	0.25	0.20	0.17	2.50
12	1.2	0.6	0.40	0.30	0.24	0.20	3.00
16	1.6	0.8	0.53	0.40	0.32	0.27	4.00
20	2.0	1.0	0.67	0.50	0.40	0.33	5.00
24	2.4	1.2	0.80	0.60	0.48	0.40	6.00
32	3.2	1.6	1.07	0.80	0.64	0.53	8.00
40	4.0	2.0	1.33	1.00	0.80	0.67	10.00
60	6.0	3.0	2.00	1.50	1.20	1.00	15.00
800	8.0	4.0	2.67	2.00	1.60	1.33	20.00
100	10.0	5.0	3.33	2.50	2.00	1.67	25.00
120	12.0	6.0	4.00	3.00	2.40	2.00	30.00

Note: The hectare-centimeters will have to be calculated from the tabulated data and the requirement at site. A suitable pumping set must then be selected from selection charts of centrifugal pump catalogue for desired capacity and head. Source: [13]

Annexure 13.2

Recommended Size of Suction and Delivery Pipe at Various Flow Rate

Flow rate (lit/sec)	Nominal size of suction pipe (mm)	Nominal size of Delivery pipe (mm)	Flow rate (lit/s)	Nominal size of suction pipe (mm)	Nominal size of Delivery pipe (mm)
0.50	20	20	12.50	100	100
1.00	30	25	16.00	125	100
1.25	40	32	20.00	125	125
1.60	40	40	25.00	150	125
2.00	40	40	30	150	150
2.50	50	40	40	200	150
3.20	65	50	50	200	200
4.00	65	50	60	250	200
5.00	65	65	80	250	250
8.00	80	80	100	300	300
10.00	100	80	125	350	350

Note: Values have been calculated on the basis of flow velocity of 1.5 m/s and 2.0 m/s in the suction and delivery pipes respectively.
Source: [233]

Annexure 13.3

Length of straight pipe in meters giving equivalent resistance of flow in pipe fittings valve etc.

Size (mm)	Standard Elbow	Medium Elbow	Long Radius Elbow	45° Elbow	TEE	Sluice Valve full open	Globe Valve full open	Angle Valve open	Foot/ Reflex Valve
25	0.82	0.70	0.52	0.40	1.77	0.18	8.24	4.57	2.04
40	0.31	1.10	0.85	0.61	2.74	0.29	13.40	6.71	3.05
50	1.67	1.40	1.07	0.76	3.35	0.37	17.40	8.54	3.96
65	1.98	1.65	1.28	0.92	4.26	0.42	20.10	10.00	5.18
80	2.47	2.00	1.55	1.15	5.18	0.52	25.90	12.80	6.10
100	3.35	2.77	2.13	1.53	6.71	0.70	33.50	17.70	8.23
125	4.26	3.66	2.78	1.86	8.24	0.88	42.60	21.20	10.00
150	4.87	4.26	3.35	2.35	10.00	1.07	48.70	25.30	12.20
200	6.40	5.48	4.26	3.05	13.10	1.37	67.10	33.50	16.20
250	7.62	6.71	5.18	3.95	17.10	1.74	88.50	42.60	20.40
300	9.75	7.92	6.10	4.57	20.10	2.04	100.50	51.80	24.40

Source: [13]

Annexure 13.4

Friction head loss in CI and MS water pipes (Size of pipes in millimeters)

Water flow rate in LPM	25	40	50	65	80	100	125	150	200	250	300
40	1.25	*(Head loss in meter per 10 meter length of pipe)*									
80	4.69	0.57									
100	7.00	0.809									
150		1.85	0.662								
200		3.13	1.099								
250		4.75	1.67	0.54							
300		6.80	2.26	0.78							
400			3.98	1.31	0.54						
500			5.97	2.00	0.837						
600				2.84	1.163	0.292					
800				5.00	1.983	0.476					
1000					3.017	0.721					
1400						1.393	0.463	0.182			
1800						2.206	0.74	0.305			
2000						2.677	0.887	0.364			
2400							1.25	0.517	0.138		
2800							1.673	0.647	0.165		
3000							1.903	0.753	0.197	0.066	
3500							2.55	1.165	0.262	0.089	
4000							3.35	1.333	0.328	0.113	0.046
4500								1.660	0.415	0.142	0.055
5000								2.050	0.505	0.174	0.067
5500								2.50	0.582	0.200	0.082
6000									0.707	0.233	0.095
7000									0.937	0.312	0.125
8000									0.187	0.407	0.164
9000									0.487	0.497	0.205
10000										0.607	0.250
11000										0.695	0.295
12000										0.807	0.345

Source: [13]

Annexure 13.5

Data sheet for Centrifugal Pump Testing

Date: Motor make: Specified speed (rpm)

Name of Supervisor: Motor rating Phase   Head (m)

Name of Manufacturer: Motor speed (rpm)   Power (kW)

Nature of Test: Capacity measured by   Efficiency (%)

Place of Test: Suction lift head measured by   Discharge (l/sec)

Pump type: Power measured by   Atmospheric pressure

Suction size: Temp. of test liquid

Delivery size:

S. No.	Speed of pump (rpm)	Suction head (m)	Delivery head (m)	Monometer distance (m) / Hook gauge initial reading (mm)	Velocity head suction (m) / Hook gauge initial reading (mm)	Velocity head delivery (m) / Hook gauge final reading (mm)	Total head (m) / Head C' over notch (mm)	Flow Discharge (l/sec)			
1	2	3	4	5	6	7	8	9	10	11	12

| *Testing and Evaluation of Agricultural Machinery*

(1)	(2)	(3)	(4)	(5)	(6)	(7)	(8)	(9)	(10)	(11)	(12)
	Power						Pump input (kW) Motor input x Motor efficiency				
Voltage (v)	Current (a)	P.P	Energy meter reading (kW.h)		Motor input (kW)	Motor efficiency (%)			Water (hp)	Pump efficiency (%)	Remarks
13	14	15	16		17	18	19		20	21	22

Chapter 14

Testing and Evaluation of Combine Harvester

Combine harvester is a versatile machine designed for harvesting, threshing, separating, cleaning, collecting, and unloading clean grain into the grain tank of the machine and subsequently releasing it into the tractor trailer or wagon. Different functions of a combine harvester are:

a) cutting the crop and feeding to the threshing cylinder

b) threshing the grain from the ear heads

c) separating the grain from the straw

d) cleaning the grain

e) handling the grain after threshing

These functions are performed automatically as the material is moved through different systems of combine harvester. The schematic diagram of a self-propelled combine harvester and tractor drawn combine harvester is shown in Fig. 14.1 and Fig. 14.2, respectively. Combine harvester density in various countries are given in Table 14.1.

About 125 combine harvester manufacturers are manufacturing combine harvesters, around 4000 - 6000 units per year, in India. Half of which are tractor operated in which tractor is mounted above the combine as a source of power. The use of straw management system (SMS) is made mandatory in the combine harvesters to reduce the environment pollution problems.

14.1 Types of Grain Combine Harvester

a) Self- propelled combine harvester

A combine on which an engine of suitable power rating (60 to 150 hp) is mounted to serve as a source of power. It can be wheeled type or track type.

- **Wheel type combine:** A Combine in which the pneumatic wheels are used

- **Track type combine:** A combine fitted with full or half tracks instead of pneumatic wheels are used. Small track type combines are suited for lowland clayey soil in which wheeled combines have a higher chance of sinkage.

Table 14.1 Combine harvester density in various countries

Sr. No.	Country	No. of combine harvester per 1000 ha
1.	India	0.12
2.	Japan	816.7
3.	Germany	11.43
4.	United Kingdom	8.32
5.	France	4.93
6.	Italy	6.24
7.	Argentina	1.48
8.	Brazil	0.92
9.	China	1.39
10.	Egypt	0.833
11.	Pakistan	0.081

b) Tractor operated combine

A combine harvester, which requires a tractor of suitable power to serve as a source of power for its working. Presently, a tractor without wheels is mounted on the combine harvester chassis and used as source of power. When, the harvesting season is over, the tractor is removed from the combine harvester chassis and is used for other field operations to avoid multiple investment on farm tractors.

14.1.1 Components of self propelled combine harvester

a)	Knife	m)	Ear return
b)	Reel	n)	Bottom sieve
c)	Crop divider	o)	Blower fan
d)	Conveyor	p)	Slide
e)	Feeding/Retractable fingers	q)	Collector bottom
f)	Feeding conveyor	r)	Grain auger
g)	Threshing cylinder	s)	Grain elevator
h)	Straw guide drum	t)	Tank filling auger
i)	Straw walker	u)	Grain tank
j)	Catch	v)	Tank auger
k)	Flap sieve	w)	Ear auger
l)	Rake	x)	Concave

The standing crop is guided to the cutter bar platform by the reel. Dividers divide the crop which has to come to the machine and the rest standing in the field to be slightly pushed apart so that it is not damaged by the machine. The crop is harvested by the cutter bar and the cut crop falls into the pan of the cutter bar. It has a rotating auger which brings the cut crop to the centre of auger. The centre of auger is fitted with retractable fingers which conveys the crop to feeder channel. The feeding conveyor chain fitted with slats, running in feeder channel conveys the crop to the threshing unit. The crop is threshed by threshing drum and concave. To suit various crops and their threshing characteristics, drum speed and the gap between concave and threshing drum can easily be adjusted.

A stone trap is provided in front of threshing drum and concave so as to catch the stone coming along with the crop. Stone being heavy in weight drops in the stone trap and it remains there to be taken out during the servicing of the machine.

The threshed straw is brought to the straw walker by guide drum. The mixture of grain and short straw (chaff) returns to step bed through bottom pan of straw walker. A baffle plate is placed above straw walker to control the straw to the straw walker by allowing the straw to fall at the end of the straw walker thereby making it to travel complete length of straw walker so that left over grains are dropped out to the pan of the straw walker. The straw is thrown out from discharge hood. The mixture of grain and chaff dropped from the grates of the concave and also through the straw walker fall on the step bed. Due to vibrating action of step bed, grains are separated from the chaff and short straw and carried to the chaffer sieve and cleaning sieves. When the grains, chaff and short straw are being dropped on the chaffer sieve, air is directed from the blower which blows the chaff and short straw out of the combine. The unthreshed ears drop through chaffer rack to the ear return pan. Under the adjustable sieve, another fixed hole sieve is placed. The inclination of this sieve is adjustable. It is also replaceable and can be changed to suit the type of crop.

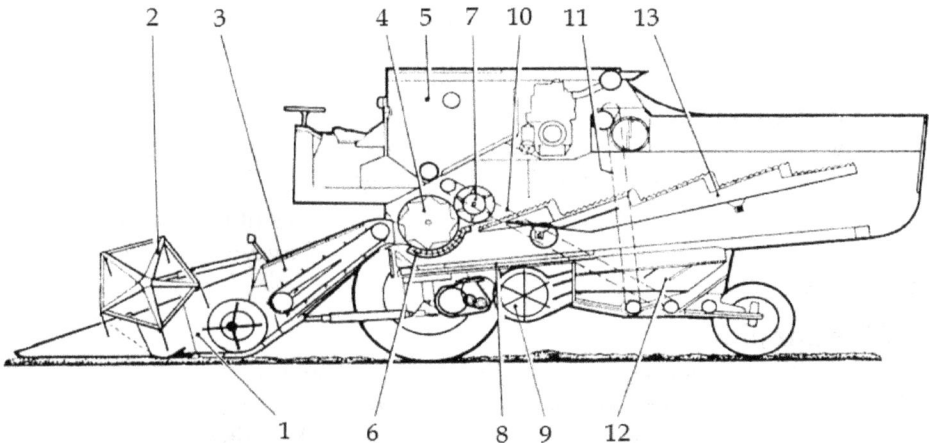

1. Header, 2. Reel, 3. Crop elevator, 4. Threshing Cylinder, 5. Grain tank, 6. Concave, 7. Beater, 8. Grain pan, 9. Cleaning fan, 10. Return elevator, 11. Grain elevator, 12. Shaker shoe, 13. Straw walkers

Fig. 14.1: Schematic diagram of a self-propelled combine harvester

Fig. 14.2: Tractor operated Combine harvester showing the tractor mounted on the combine chassis

Air from variable speed blower is arranged in such a way that it passes under chaffer rack as well as both the sieves. The air current blows out light weight dust particles, short straw, green trash and other impurities out of the combine. A slide is mounted at the end of cleaning box to trap the grain if any being carried away by the air current. The slide is adjustable in height and can be locked in any position. The clean grains are dropped from the fixed hole sieve to the inclined bed and carried to the grain elevator with the help of screw conveyor. At the top end of the grain elevator another screw conveyor is fixed to convey the clean grain to grain tank. The grain can be discharged from grain tank to the transporting trailer with the help of discharge conveyor and discharge auger.

The unthreshed ears which had fallen at the rear part of the chaffer rack find their way on the inclined bed and carried by screw conveyor into the threshing drum for re- threshing.

14.2 Terminology

a) *Cutter bar effective width:* The distance between the points at which the tips of the knife sections meet the last effective shearing edges of the guards (fingers) or shoes at the extremities of the cutter bar expressed in mm.

b) *Cutter bar working width:* The distance expressed in mm between two vertical planes passing through the points of the outermost dividers and parallel to the centre line of the cutter bar. If adjustable dividers are used, the maximum and minimum dimensions shall be stated.

c) **Threshing cylinder:** A balanced rotating assembly comprising rasp bars, beater bars or spikes on its periphery and their support for threshing the crop.

d) **Concave:** A concave shaped metal grating partly surrounding the semi circle of threshing cylinder against which the cylinder rubs the grain from the ears and through which the grains fall on the sieve.

e) **Cylinder and concave clearance:** The gap between the tip of the cylinder to the inner surface of the concave expressed in mm. The minimum and maximum clearance in a particular setting and adjustment range for both the front and the rear side of the concave shall be stated.

f) **Collectable loss:** The unthreshed grain from main grain outlet, threshed, unthreshed and damaged grains from the secondary cleaning or grading unit.

g) **Non-collectable losses:** Header loss, straw walker loss, sieve loss, secondary blower loss and grain breakage loss in main grain outlet.

h) **Header loss:** The loss of grains and ear heads being shed and left over on the ground as a result of operations of cutter bar and header unit.

i) **Pre-harvest loss:** The loss of grain or ear heads from the standing crop prior to the operation of combine harvester in the field.

j) **Processing losses:** The total grain breakage in grain tank, straw losses, sieve losses and unthreshed from main outlet.

k) **Rack loss:** The threshed grains passing out in the straw.

l) **Shoe loss:** The threshed grains blown or carried out with the chaff.

m) **Threshing efficiency:** Threshed grains from all the outlets of the combine with respect to total grain output from all sources and expressed in percentage by mass.

n) **Cleaning efficiency:** Clean grains percent in the main grain outlet with respect to the total grain obtained in the main grain outlet and expressed in percentage.

o) **Grain output:** The mass of the grain mixture delivered by the combine per unit time.

p) **Straw output:** The mass of the straw and chaff not including grain losses delivered by the combine per unit time.

14.3 Scope of Test

14.3.1 Laboratory tests

a) Specifications checking
b) Pre-test checking
c) Material analysis
d) Engine performance
e) Noise level
f) Vibration level
g) Visibility from operator
h) Header lifting

i) Turning ability

j) Air cleaner pullover

k) Location of centre of gravity

l) Brake performance

m) Endurance

n) Power drop

o) Wear and tear

14.3.2 Field performance test

a) Rate of work

b) Quality of work

    i) Grain losses

    ii) Rubbish content

    iii) Grain damage

    iv) Threshing efficiency

    v) Cleaning efficiency

c) Output of straw and grain

d) Fuel consumption

e) Night trial observations

f) Ease of operation and handling

g) Operational comfort and safety

h) Soundness of construction

i) Labour requirements

j) Nature of break downs and repair

k) Wear of critical components

14.3.3 Referred standards / test codes

a) IS 8122-1: 1994 [89]

b) IS 8122-2: 2000 [90]

c) IS 9877-2: 1981 [104]

d) IS 15806: 2008 [67]

e) ISO 8210: 1989 [142]

f) ISO 5702:1983 [139]

14.4 Method of Testing

14.4.1 Specification checking

The specifications of the combine harvester given by the manufacturer shall be checked, compared and reported by the testing authority. While checking various dimensions, following items need to be considered:

a) Name and address of the manufacturer

b) Country of origin

c) Make

d) Model

e) SI. No.

f) Prime mover and its detail

g) Chassis detail

h) Wheels with their details

i) Reel assembly with details

j) Cutter bar assembly details

k) Combining system details

l) Threshing Drum assembly details

m) Straw walker details

n) Cleaning and chaffer sieves details

o) Grain conveying mechanism details

p) Speeds of various assemblies

q) Lubrication of the combine harvester and lubrication schedule

r) Fuels and lubricants to be used

s) Light arrangements

t) Operational comfort and safety

14.4.2 Pre-test checks

The combine is parked on a plain surface and examined for pre-test checks with particular attention to:

a) Bearings

b) Drives and other moving parts

c) Correctness of various controls and system

d) Proper adjustments of various assemblies

e) Tightness of bolts and nuts

f) All fuel and lubricants

14.4.3 Material analysis

The hardness and chemical analysis of critical components like knife section, raspbar, peg tooth, ledger plate and knife guards shall be made and reported.

14.4.4 Engine performance test

This test is conducted only in case of self- propelled combine harvester. In case of tractor fitted combine harvester, the test report of tractor will be referred. Performance characteristics curves of engine as shown in chapter 5 at Fig. 5.2 and 5.3.

14.4.4.1 General requirements

Run-in of the engine

a) The engine of the combine harvester will be run-in in accordance with the procedure laid down in the instruction manual. During this period, drives and bearings will be checked tor overheating, alignment, excessive vibration etc.

In the absence of specific instructions about run-in, the testing station shall carry out run in as under:

i) Five hours at quarter load at speed specified by the manufacturer for continuous operation

ii) Five hours at half load at speed specified by the manufacturer for continuous operation

iii) Five hours at three quarters of load at speed specified by the manufacturer for continuous operation

iv) Five hours at full load at speed specified by the manufacturer for continuous operation.

b) The torque and power values in the test report should be obtained from the dynamometer without correction for losses in engine auxiliaries.

c) In all tests, universal shaft connecting to the dynamometer should have the minimum angularity.

d) During maximum power test, ambient temperature should be $27\pm7°C$ and atmospheric pressure should not be less than 96.6 kPa. Atmospheric pressure and ambient temperature should be noted during all the test readings for power and the average is reported.

e) For all tests, all accessories shall be in position on the engine. All attachments necessary for running of the engine shall be attached in the same position.

f) The various tests shall be carried out continuously, the governor control lever be placed in the position recommended by the engine manufacturer for obtaining continuous maximum power.

g) In addition to the performance measurements, the following should also be noted:

- The temperature of fuel on fuel measuring apparatus, the temperature of engine lubricating oil near the drain plug and the exhaust gas temperature near the final junction of exhaust manifolds shall be measured.

- The coolant temperature shall be measured at the outlet of coolant flow line between cylinder block or cylinder head and thermostat. In case of air cooled engine, the temperature of air emitting from cylinder block shall be measured at 2 points at a maximum distance of 4 cm from cylinder fins.

- The air temperature measurements shall be made approximately 2 m in front of the air inlet and approximately 1.5 m above the ground to ascertain the effect of exhaust gases on the air intake.

- For engines fitted with a blowing device, air temperatures shall be taken at approximately 2 m behind the engine and approximately 1.5 m above the ground and at the engine air intake.

- Smoke density - Smoke density exceeding 5 Bosch number is unacceptable.

- Pressure of the engine lubricant.

- Relative humidity of air.

- The specific fuel consumption values in the test report should be given as grams of fuel per horsepower per hour. The specific fuel consumption at all test shall be noted but the specific fuel consumption at corrected maximum horsepower shall be considered as base value.

h) To obtain hourly fuel consumption the volume and work performed per unit volume of fuel conversion of weight to units of volume should be made using the density value at 20°C.

The comments regarding specific fuel consumption for test on combine engines may be assessed on the following:

Rating	Specific fuel consumption (gm/hp/h)
Normal	170 to 200
Slightly high	200 to 225
High	Above 225

i) When consumption is measured by volume, the specific fuel consumption should be calculated using the density corresponding to the appropriate fuel temperature corrected to 20°C.

j) Oil consumption: The observation for this test should be taken from 5 hours test (high ambient) and report of oil consumption be made as gm/hp h.

14.4.4.2 Procedure of engine tests

a) Varying speed test

The varying speed test shall be conducted under natural ambient conditions (temperature range: 27±7°C). It shall be repeated under high ambient condition (temperature range: 43±2°C) to assess engines performance under tropical climatic conditions.

b) Test at maximum power

The engine shall be operated for a period of two hours after it has warmed up, for maximum power to become stabilized. The test shall be carried out under natural ambient conditions. A minimum of six readings at 20 minutes interval of time shall be made during two hours test period. Temperature, pressure and other observations will be recorded simultaneously.

The maximum power quoted in the report should be the average of the readings made during the two hour period. If the power variation exceeds ±2 percent from the average, the test should be repeated. If the variation continues to exceed ±2 percent it should be mentioned in the report.

c) Tests at varying load

In the zone controlled by the governor, the torque, speed and hourly fuel consumption shall be noted as a function of power. In addition the no-load engine speed shall be recorded. The data required to complete the varying load test shall be recorded in the following loads and sequence of measurements. Each load shall be maintained for a duration of 20 minutes.

i)   A load corresponding to maximum power

ii) 85 percent of load obtained at maximum power

iii) 75% of the load defined in (ii)

iv) 50% of the load defined in (ii)

v)   25% of the load defined in (ii)

vi) On minimum load

d) Five hours high ambient rating test

The engine shall be run at 90% of maximum output continuously for 4 hours under high ambient temperature conditions (43±2°C). The engine shall be run at a load corresponding to maximum power for a period of 1 hour running after 4 hours of continuous high ambient test.

The readings of speed, power, torque, temperature, pressure and fuel consumption shall be taken after every half hour during the 4 hours period and after every 15 minutes during the 5th hour run. Coolant and lubricating oil consumption shall be noted and reported in the test report.

14.4.4.3 Presentation of curves

The test report should include the following curves made for full range of engine speeds available under natural ambient and high ambient conditions:

a)   Power as a function of engine speed.

b)   Equivalent crank shaft torque as a function of engine speed.

c)   Hourly fuel consumption as a function of engine speed.

d)   Specific fuel consumption as a function of power.

e)   Specific fuel consumption as a function of speed.

14.4.5 Header lifting test

The combine harvester is parked on a level ground and is put on rated engine speed for fieldwork (under operating conditions). The assemblies/sub-assemblies other than hydraulic system shall remain disengaged. The cycle of lifting and lowering shall be kept continuous and is tested for 1000 times. Before test, oil temperature shall be 65±5°C.The leakage of hydraulic fluid from any part of hydraulic system is noticed & reported.

14.5.6 Turning ability test

Minimum turning diameter and minimum turning space is measured after brake is applied in LHS & RHS on independent brake pedal. Turning space is defined as the diameter of the circle formed by the outer most point of the combine when it takes shortest turn. Turning circle or turning radius is defined as the distance from the turning centre to the centre of ground contact of the wheel forming the largest circle when the combine takes shortest turn. Turning space and turning circle readings shall be taken for left and right turn when the combine is running at speed less than 4 km/h.

14.4.7 Location of centre of gravity
The combine harvester fitted with all standard accessories & all reservoirs full, grain tank full with wheat and operator replaced with 75 kg mass on the seat and header assembly in raised position, the location of centre of gravity is calculated.

14.4.8 Visibility from operator seat
The visibility test is conducted to assess the visibility from the normal sitting position of the operator. The cutter bar is kept at 150 mm above the ground level during the test. The height of vision during test shall be maintained at 760 mm on a vertical plane from the centre of operator seat. The result of visibility test shall be represented graphically indicating the area visible and that obstructed by the combine harvester.

14.4.9 Brake performance test
The test is conducted on standard concrete test track at a maximum travel speed of 25 km /h. The tyre pressure shall be that as recommended by the manufacturer for the filed work [50]. The following test shall be conducted:

a) **Cold Test**
   i) Force applied on brake
   ii) Mean deceleration rate
   iii) Stopping distance

b) **Hot test (Brake fade test)**
   i) Force applied on brake
   ii) Mean deceleration rate
   iii) Stopping distance

c) **Parking brake test:** The force, necessary to apply at the control of the parking braking device, to hold the combine harvester stationary, when facing up and down on 12 percent gradient in a condition recommended for road transport, shall be measured. The maximum actuating force shall not be more than 400 N for hand operated and 600 N for foot operated parking brake device.

The combine shall be declutched and the brakes applied successively increasing force on the pedal, until that giving the shortest stopping distance is found. The details of the instrument used for brake tests shall be indicated in the test report. The following shall be recorded.

- Deceleration as measured by a 'maximum decelerometer'
- Stopping distance
- Force exerted on the brake pedal, and,
- Braking efficiency

14.4.10 Mechanical vibration test

The amplitude of mechanical vibration of various sub-assemblies and components of the combine shall be measured by using precision vibration meter. The specifications of the instrument used shall be included in the report. The observations shall be recorded by parking the combine on a level concrete surface and operating the combine at the speed recommended for field work. The cutter bar shall be kept at 15 cm above ground level. The readings of vibration may be taken on important assemblies/sub-assemblies/ components of the combine. The maximum displacement of vibration shall be measured in horizontal and vertical position and represented in microns for horizontal and vertical displacement.

14.4.11 Air cleaner oil pull over test

The mass of oil in the air cleaner assembly (with recommended grade of oil) shall be 5% in excess than the marked level. Then engine is run for 15 m followed by sudden acceleration and deceleration made after every 30 secs for 15 mts. The oil pull over in grams and percent of oil pull over shall be calculated for 5 different position of combine i.e. horizontal, tilted 10° forward 10° backward, 10° left side and 10° right side in direction of travel of the combine harvester.

14.4.12 Noise level test

Sound level meter shall be used for noise level test on a concrete standard test track. The microphone shall be installed 1.2 m above the ground level and 7.5 m from the line of travel. The cutter bar height shall be kept at 15 cm above the ground level. Measurement shall be made with no-load, in sufficiently silent and open space of 50 m radius. Also, measurements shall be made in good weather conditions with little or no wind.

The test will be conducted in all gears and travelling speed as recommended for field operation. Noise level of combine harvester at operator ear level in all gears at maximum travel speed will also be recorded & reported.

Test method

The combine shall be operated at the recommended speed for field work. All the mechanism in the combine shall be in working position. The cutter bar height shall be kept at 15 cm above the ground level. Noise measurements at a height of 1.2±0.1 m from ground level shall be made at the following four points when the microphone is facing the combine.

a) In front of combine 3 m from the fore-most point of the combine.
b) In the rear of the combine 3 m from the rear most point of the combine.
c) Both the sides of the combine 2 meters away from the wheels or outer end.
d) Measurement shall be considered valid if the difference between the

2 consecutive measurements from the same side of the combine is not greater than 2 dB. The value shall be that corresponding to the highest sound level.

Noise at the driver's ear level

a) Measurement of sound frequency spectrum shall be carried out by using a frequency analyzer fitted with octave filters.

b) The microphone shall be positioned atleast 4 cm and not more than 7 cm to the side of the driver's fore-head.

c) The report shall state whether the test was carried out with or without a cab on the combine.

d) The highest noise level and any unusual characteristics giving noise to the operator shall be stated in the report.

14.4.13 Field test

The manufacturer's representative shall demonstrate the operation of the combine harvester to the test team in actual field conditions. An experienced operator should drive the machine. Before starting the test, the machine should be adjusted following the instruction manual in order to obtain the highest possible performance. During test run, no alteration in adjustment or speed is allowed. However, the combine may be readjusted between successive test runs in order to have maximum possible performance.

a) Field and crop conditions

The combine shall be operated for minimum of 50 hours preferably for the following crops:

Wheat: 25 h ; Paddy: 25 h

The test crop shall be ripe (ready for harvesting), and standing angle should be more than 60 degrees. Other crop conditions like variety, height of plant, length of ear head, tilling, no. of grains in ear head, plant density and moisture content of grain and straw etc. shall be recorded for each test. Field conditions, like topography, type of soil, weed intensity, size and shape of field etc. should also be recorded together with important observations on crop conditions and plant parameters.

b) Material required for measurement and sampling during field test

i) Canvass sheets or cloth sheets for collection of straw walker and sieve sample during 20 m test run.

ii) Platform balance or tripod balance for weighing straw and chaff samples.

iii) Sighting poles 4 nos.

iv) Whistle or flag for giving signal at the beginning and end of measurement test run.

v) Stop watch for recording running time of combine harvester during test run.

vi) Bags for collection of grain and mixture sample.

vii) Measuring tape and rulers for measurement of test run, test plot, crop and plant parameters.

viii)Pan balance for weighing sample for moisture analysis.

ix) Containers for grain sample and moisture sample.

c) Procedure for test

i) When the combine harvester is ready for test, it should be topped up and be operated in the plot at uniform speed as shown in Fig. 14.3 and 14.4.

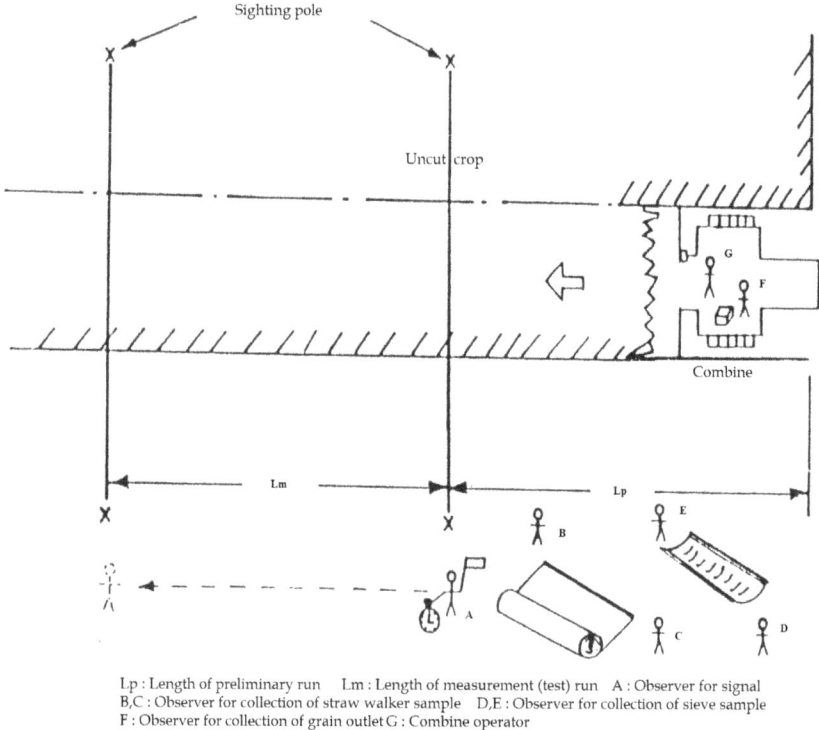

Lp : Length of preliminary run   Lm : Length of measurement (test) run   A : Observer for signal
B,C : Observer for collection of straw walker sample   D,E : Observer for collection of sieve sample
F : Observer for collection of grain outlet G : Combine operator

Fig. 14.3 Arrangement for field testing of combine harvester

ii) A test run of 20 m is selected from the test plot and marked with sighting poles.

iii) Pre-harvest losses at three different places randomly selected having an area of 1 m x half the cutter bar width be determined in kg/ha. Use a rectangular frame of appropriate size for marking the area.

iv) Two roll of canvass or cloth having 30 m length and one and half times of width for straw and sieve out-let are rolled over on the specially attached rollers behind the combine to collect straw and sieve sample.

v) Arrangement for collection of sample from different outlets shall be made.

vi) The combine, when enters the test run, signal shall be given to record the time taken to cover the 20 m test run and to collect different samples.

vii) The header loss shall be determined from that portion from where pre-harvest losses have been recorded and during test, it was protected by cloth sheet.

viii) All the data shall be recorded as shown in Annexure 14.1.

d) Observations to be recorded during and after test

   i)   Area covered

   ii)   Time of operation

   iii)   Operating speed

   iv)   Time for any stoppage

   v)   Time loss in turning should be recorded for at least one hour of operations

   vi)   Average working width

   vii) Time required to discharge the grain from main grain outlet

   viii) No load and on load engine speed

   ix)   Maximum temperature of engine oil. Coolant, transmission oil, hydraulic oil and ambient temperature.

   x)   Fuel consumed (l/h and l/ha)

   xi)   Lubricating oil consumed

   xii) Coolant (water) consumed

   xiii) Average forward speed (km/h)

e) Sampling technique

Three samples of 100 gm each from the main grain outlet shall be taken and analyzed for threshed, unthreshed, broken and rubbish content. Similarly, complete sample for the test run from straw walker and sieve shall be analyzed.

14.4.14 Wear test of critical component

The engine (in case of self-propelled combine) parts, clutch, transmission system, starter and alternator etc. shall be dismantled and assessed for rate of wear (percentage basis). The assessment will be made keeping in view the initial average values (when the engine is new) and also the maximum permissible wear as specified by the manufacture. The rate of wear in respect of rasp-bar, peg tooth, concave shall be computed in terms of percent mass basis on the basis of original mass before test and final mass after test.

14.5 Data Analysis

The data obtained during field test and sample analysis will be used for analysis and presented and the following results can be obtained:

14.5.1 Rate of work

It can be represented in terms of:

   a)   Area covered (ha/h)

   b)   Net grain output (kg/h, kg/ha)

   c)   Grain throughout (kg/h, kg/ha): It consists of net grain output, header loss, threshing cylinder loss, rack loss and shoe loss.

   d)   Straw output (kg/h, kg/ha)

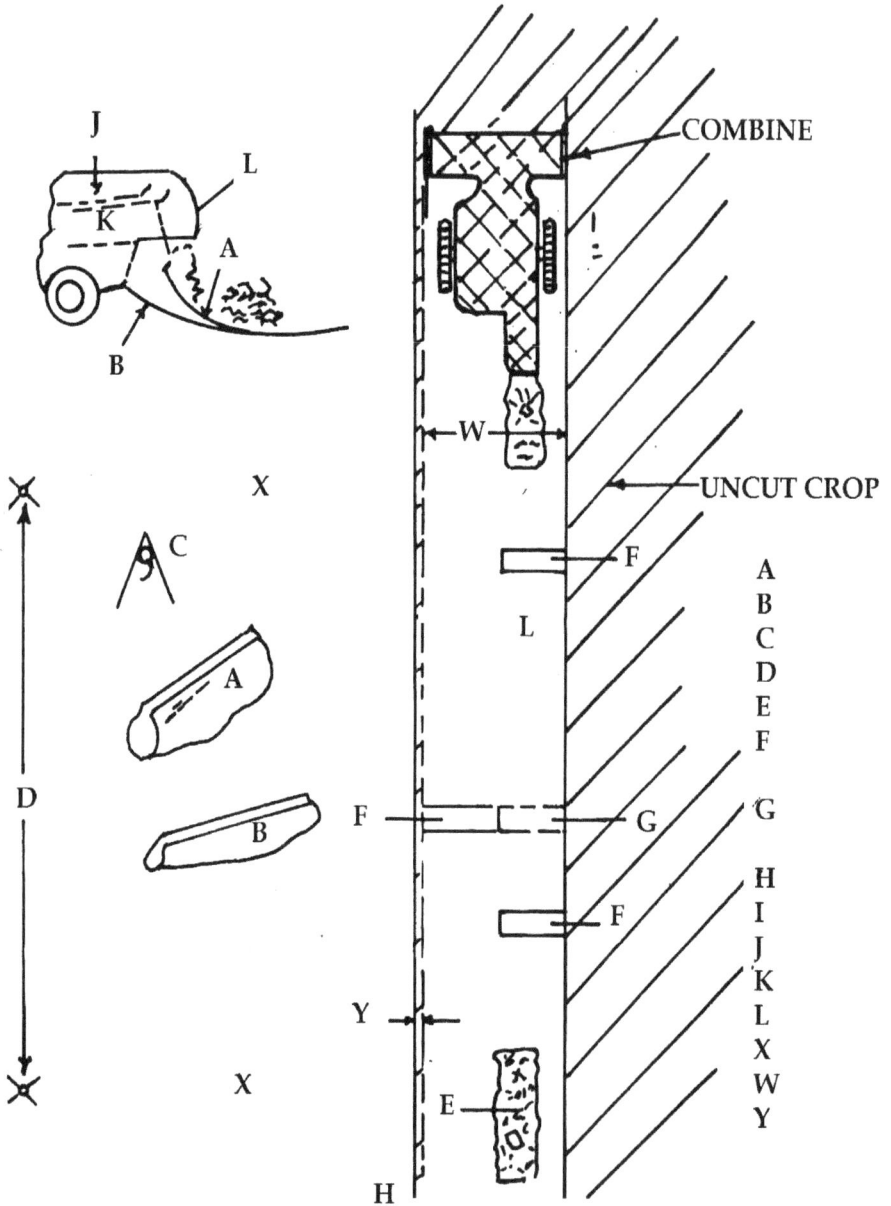

A. Calico sheet for straw walker efflux; B. Calico sheet for sieve efflux; C. Tripod and balance;
D. Distance representing 1/200ha; E. Swath; F. Wire rectangles for cutter bar losses;
G. Second rectangle to locate *half width* rectangle at left of cut width; H. Entry point;
I. Length of test run; J. Straw walkers; K. Sieve; L. Deflector; X. Stakes; W. Width of cut; Y. Uncut strip
Fig. 14.3: Continuous collection methods to determine combine losses [127]

14.5.1.1 Observations to be recorded during and after test

i) Area covered

ii) Time of operation

iii) Operating speed

iv) Time for any stoppage

v) Time loss in turning should be recorded for at least one hour of operations

vi) Average working width

vii) Time required to discharge the grain from main grain outlet

viii) No load and on load engine speed

ix) Maximum temperature of engine oil. Coolant, transmission oil, hydraulic oil and ambient temperature.

x) Fuel consumed (l/h and l/ha)

xi) Lubricating oil consumed

xii) Coolant (water) consumed

xiii) Average forward speed (km/h)

14.5.1.2 Sampling technique

Three samples of 100 gm each from the main grain outlet shall be taken and analyzed for threshed, unthreshed, broken and rubbish content. Similarly, complete sample for the test run from straw walker and sieve shall be analyzed.

14.5.2 Quality of work

a) Grain loss from a combine harvester

i) **Pre-harvest loss:** The grain found on the ground before harvesting indicates this loss. The sample should be collected from 3 places randomly selected from an area having 1 m length in the direction of travel and full or half the width of cutter bar of combine. All the loose grains, complete and incomplete ear heads fallen in the marked area have to be picked up manually without disturbing the plants before harvesting and this will give pre-harvest loss.

ii) **Cutter bar loss / Header loss:** Heads of grain shattered out of the head as the knife cuts the straw and grain dropped to the ground before reaching the feeding platform; grain shattered out when the reels strikes the standing grain; and heads of grain thrown out of the reel may cause cutter bar loss

$$\text{Header loss (\%)} = \frac{\text{Grain collected from } 1\text{n}\frac{1}{2} \text{ area after harvesting } \cdot \text{grain collected from same area before harvest}}{\text{Gross yield}} \times 100$$

It is determined on those portions of ground, which are protected from combine afflux by the use of rolls of cloth. The loose grains and complete and incomplete ear heads fallen on the marked area, where pre-harvest losses were determined, shall be collected manually. This gives the header loss. It is also called cutter bar loss.

$$\text{Cylinde loss (\%)} = \frac{\text{Untreshed grain collected from straw rack and sieve}}{\text{Gross yield}} \times 100$$

iii) Cylinder loss / Threshing unit loss: There are two types of cylinder loss: a) unthreshed grain left in the heads and carried to the rear of the combine by the rack, and b) cracked grain in the grain tank caused by higher cylinder speed or low concave cylinder clearance.

iv) Rack and shoe losses : The rack loss is the loose grain that has not been separated from the straw as it passes over the rack and is carried out of the machine with the straw. The shoe loss is the grain that is carried over the rear of the sieves with the chaff or blown out of the combine with the air by the blower.

For determining the rack and shoe loss, the straw and chaff afflux is collected separately. To collect these, two rolls of cloth 30 m in length and one-and-half fold the width of straw/ chaff outlet is suspended, especially attached fittings beneath the rear of machine. As the sheets of cloth unroll, one sheet retains the afflux from straw walker and other from sieve for 20 m run length. Unrolling operation starts 5 m in advance and terminates 5 m ahead of end point.

$$\text{Rack loss (\%)} = \frac{\text{Free grain cllected from straw rack sample}}{\text{Gross yield}} \times 100$$

$$\text{Sieve loss (\%)} = \frac{\text{Free grain collected from sieve sample}}{\text{Gross yield}} \times 100$$

b) Grain crackage

It is determined from the samples (atleast 3) taken from the grain tank. Only visible damaged grains are separated and expressed in percentage of sample taken.

$$\text{Grain crackage (\%)} = \frac{\text{Damaged grain wholly or partially collected from sample}}{\text{Gross yield}} \times 100$$

Net yield = Grain collected in the bag from combine test area.

Gross yield = Net yield + header loss + cylinder loss + rack loss + shoe loss

Total combine loss = Cutter bar loss + cylinder loss + rack loss + shoe loss

$$\text{Performance efficiency (\%)} = \frac{\text{Net yield}}{\text{Gross yield}} \times 100$$

$$\text{Unthreshed (\%)} = \frac{\text{unthreshed grain in tank + cylinder loss}}{\text{Gross yield}} \times 100$$

$$\text{Cleaning efficency (\%)} = \frac{\text{Clean grain}}{\text{Total grain collected from main outlet}} \times 100$$

$$\text{Threshing efficiency (\%)} = \frac{\text{Threshed grain from all outlets}}{\text{Grain output in tank}} \times 100$$

c) Accepted range of losses

Losses vary greatly depending upon the type, variety and condition of the crop. Total loss for normal crops of wheat and barley might vary between 1 and 4% of the total yield. Under optimum harvesting condition the following per cent losses might be expected while harvesting the small grain crops like wheat and paddy:

Grain breakage in main grain tank	=	2.5% (Maximum)
Non-collectable losses	=	2.5% (Maximum)
Threshing efficiency	=	≥ 98%
Cleaning efficiency	=	≥ 96%

14.5.3 Fuel consumption

Fuel consumed during each test shall be computed by topping up at the start of test and at the finish of test and expressed in l/h and l/ha.

14.5.4 Ease of operation and handling

Observations shall be made on skill and intensity of effort required to operate various controls of the machine. Adequacy and accessibility of controls and visibility of the header and instrument panel shall also be recorded. The operator's working condition, the ease of setting adjustment, routine maintenance and other similar features shall also be recorded.

14.5.5 Checking for safety

The safety devices such as slip clutches, shear pin, signal horns, indicator lights etc. provided for various systems shall be assessed for their effectiveness. Provision of stone trap, spark arrester and automatic intermittent horn while reversing of combine shall be checked and reported accordingly.

14.5.6 Soundness of construction

Observations of those features shall be recorded, which adversely affect the operation and efficiency of combine in the field. The test will be completed with a thorough check of the test combine, with dismantling wherever necessary to note any failure or cracks, which has shown since the pre-test check, and which may affect the life of the combine. Modifications, which could bring about improvement in the quality of work, shall be noted. Any failure and cracks, which has been seen since the pre-test check or excessive wear will be reported together with details of breakdowns, defects and replacement of parts.

14.5.7 Labour requirement

The labour required for operating the combine harvester for routine servicing and adjustments shall be recorded and reported. The time required for daily maintenance should also be reported.

14.5.8 Unloading of grains

The unloading time to unload the grain tank should be recorded & reported.

14.5.9 Sample analysis

Three samples of 100 gm, each from the main grain outlet shall be taken and analyzed for threshed, unthreshed, broken and rubbish content. Similarly complete sample for the test run from straw walker and sieve shall be analyzed. The complete analysis shall be done as per Appendix 14.2.

14.5.10 Output of straw and grain

Grain output in kg/h and kg/ha and other parameters can be calculated as per Annexure 14.3.

14.5.11 Night trial observations

A night trial of minimum two hours each shall be conducted in wheat and paddy crop to assess the intensity and suitability of the lighting equipment for the night work.

14.5.12 Summary of performance parameters required for analysis during field test

a) Duration of test

b) Forward speed

c) Rate of work (ha/h)

d) Net grain output (kg/h)

e) Fuel consumption (l/h and l/ha)

f) Pre-harvest losses (kg/ha)

g) Cutter-bar losses (%)

h) Total straw walker losses (%)

i) Total sieve losses (%)

j) Total collectable losses (%)

k) Total non-collectable losses (%)

l) Grain breakage or damage (%)

m) Threshing efficiency (%)

n) Cleaning efficiency (%)

o) Moisture content of grain (%)

p) Moisture content of straw (%)

14.6 Illustrative Example for Calculation of Grain Losses and Other Parameters

During combine testing, the following data was recorded:

Avg. time of 20m length	=	22.98 sec
Avg. width of cut	=	3.98 m
Weight of grain sample	=	29000 g

Analysis of 100 g grain sample

Healthy threshed	=	92.72 g
Unthreshed	=	0.617 g
Broken	=	1.43 g
Rubbish	=	5.233 g

Calculation value of total grain sample

Healthy threshed grain	=	26888.8 g
Unthreshed grain	=	178.93 g
Broken grain	=	414.7 g
Rubbish	=	1517.57 g

Analysis of straw sample

Weight of straw sample	=	12.2 kg
Healthy threshed grain	=	24.8 g
Unthreshed grain	=	6.4 g
Broken grain	=	0.9 g
Straw	=	12167.9 g

Analysis of chaff sample

Weight of chaff sample	=	7.8 g
Healthy threshed grain	=	42.7 g
Unthreshed grain	=	2.4g
Broken grain	=	5.1 g
Chaff & Bhusa	=	7749.8 g
Weight of pre-harvest	=	3.68 g

Losses in 1 m x half of cutter bar width

Weight of cutter bar losses	=	8.4 g in 1 m x half of cutter bar width

Calculations

Area covered in 20 m run	=	3.98 x 20 = 79.6 m²
Critical rate of work at the time of sample collection	=	(Area covered x 0.36)/t
	=	(79.6 x 0.36)/22.98 = 1.25 ha/h
Total threshed grain from all sources (Healthy + broken)	=	26888.8 + 24.8 + 42.7 + 414.7 + 0.9 + 5.1
	=	27377 g
Rubbish	=	12167.9 + 7749.8 = 19917.7 g
Total grain from all sources	=	grain sample + grain from straw & chaff sample
	=	29000 + 32.1 + 50.2 = 29082.3 g

Pre-harvest losses (kg/ha) = Weight of pre-harvest grain in 1 m x half of the cutter bar width (m) x 10)/ (half the cutter bar width (m)) (3.68 x 10)/2.16 = 17.04 kg/ha

Grain output (kg/h) = (3.6 x Weight of grain sample)/ (Average time for 20 m length) (3.68 x 29000)/22.98 = 4543.08

Grain output (kg/ha) = (10 x Weight of grain sample)/(Area covered in 20 m run) (10 x 29000)/79.6 = 3643.22

Cutter bar losses (kg/ha) = (10 x Weight of cutter bar losses in 1 m x half cutter bar width)/(half of cutter bar width)
= (10 x 8.4)/2.16 = 38.89

Grain throughput (kg/h) = (Total grain x 10)/(Area covered in 20 m)+cutter bar losses
= [(29082.3 x 10)/79.6]+38.89 = 3692.45

Grain throughput (kg/h) = Grain throughput (kg/ha) x rate of work (ha/h)
= 3692.45 x 1.25 = 4615.56

Straw output (kg/ha) = (Weight of straw x 10) /(Area covered in 20 m)
= (19917.7 x 10)/79.6 = 2502.22

Straw output (kg/h) = Straw output (kg/ha) x rate of work (ha/h)
= 2502.22 x 1.25 = 3127.77

Crop throughput (t/h) = (Grain throughput (kg/h))/1000 +(Straw output (kg/h))/1000
= (4615.56/1000) + (3127.77/1000) = 7.74

Losses due to combine (%)

a) Collectable

Unthreshed from main outlet = (178.93 x 1000)/(79.6 x 3692.45) = 0.61

Broken grain from main outlet = (414.7 x 1000)/(79.6 x 3692.45)=1.41

Total collectable losses (%) = 2.02

b) Non – collectable

i) Header loss = (38.89 x 100)/3692.45=1.05

ii) Straw walker losses:

Threshed = (24.8 x 1000)/(79.6 x 3692.45) = 0.08

Unthreshed = (6.4 x 1000)/(79.6 x 3692.45) = 0.02

Broken = (0.9 x 1000)/(79.6 x 3692.45) = 0.003

Total straw walker losses (%) = 0.103

iii) Sieve losses:

Threshed	=	(42.7 x 1000)/(79.6 x 3692.45) = 0.15
Unthreshed	=	(2.4 x 1000)/(79.6 x 3692.45) = 0.01
Broken	=	(5.1 x 1000)/(79.6 x 3692.45) = 0.02
Total sieve losses (%)	=	0.18
Total non-collectable losses (%)	=	1.33

c)Threshing efficiency (%) = (Total threshed grain)/(Total grain)
= × 100
(27377/29082.3) x100 = 94.14

d) Cleaning efficiency (%) = (Healthy threshed grain in main
= outlet x 100)/(Weight of grain sample)
(26888.8/29000) x 100 = 92.72

e)Total losses (%) = 3.35

Annexure 14.1: Data Sheet for Combine Harvester

Date:

Place of Test:

Time of Start:

Supervisor:

Gear used:

Combine

Time of End:

Operator:

Tractor

Model:

HP:

Intensity:

Crop Height:

Variety:

Length of Ear:

Grain Moisture %

Straw Moisture %

Grain/Straw Ratio

S. No.	Time for 20 m (sec)	Width of Cut for 3 Rows (W)	Time Taken to Fill the Grain Tank	Hour of Day	Temperature					Atmos. Pressure (mm of Hg)	Observations
					Amb °C	Fuel °C	Eng. °C	Cool °C	Trans °C	Hydro °C	

1. Engine Speed (no load) - rpm
2. Engine Speed (on load) - rpm
3. Threshing Cylinder Speed (no load) - rpm
4. Threshing Cylinder Speed (on load) - rpm
5. Blower Speed (no load) - rpm
6. Blower Speed (on load) - rpm
7. Beater Speed (no load) - rpm
8. Beater Speed (on load) - rpm
9. Forward speed - kmph

10. Area Covered - ha
11. Fuel Consumed - l
12. Duration of Test - h
13. Total Time Stopped - h
14. Net Time - h
15. Average Time Lost at Corners – hr.
16. Average width of cut - cm
17. Fuel Consumption
    a) Per hectare – l;
    b) Per hectare - l
18. Area Covered – ha/h
19. Time required for per ha – hr.
20. Height of Stubble - cm
21. No. of Grains/Ear

Total:

Average:

Time lost (min/sec)	Stoppages	Cause

Annexure 14.2: Analysis Sheet of Combine Testing

Test conducted at: Grain Straw Ratio: Name of Combine

Time: Weight of 1000 grins: Manufactured by:

Date   Variety:

S. No.	From	Area (sq. m.)	Time (Sec)	Weight (g)	Representative sample (g)	Healthy threshed grain (g)	Broken grain (g)	Total Threshed grain (g) (7+8)	Un-threshed (g)	Rubbish (g) 6-(9+10)	Total grain (g) (9+10+11)	The critical rate of work ha/hr at the time of sample = A×0.36/t(1)×0.36 (2)
					Sample Collected	Sample Analysis						
1	2	3	4	5	6	7	8	9	10	11	12	13
A.	Grain Outlet				100							
					100							
					100							

Average:

Calculated value:

B.	Straw								9+10 for B			
C.	Chaff & Bhusa								9+10 for C			

Total:

Annexure 14.3
Summary Sheet of Combine Field Performance

Name of Crop:

Test No.:

Calculated by:

Test conducted at:

Date:

Manufactured by:

Name of Machine:

Pre-harvest losses (kg/ha)	Grain Output	Cutter bar losses	Grain through put	Straw Output	Crop through-put
	kg/h	(kg/ha)	kg/h	kg/h	t/h
1	2	3	4	5	6

Losses due to combine (percent by mass)

Collectable			Non-collectable						Cleaning Efficiency (%)	Threshing Efficiency (%)
Header			Straw Walker			Sieve				
Unthreshed from grain outlet (%)	Broken (%)	(%)	Threshed (%)	Unthreshed (%)	Broken (%)	Threshed (%)	Unthreshed (%)	Broken (%)		
7	8	9	10	11	12	13	14	15	16	17

Annexure 14.4

Recommendations - Performance & Other Characteristics of Combine

S. No.	Characteristics	Requirements	Tolerance in Percent	Remarks
(1)	(2)	(3)	(4)	(5)
I. Prime Mover Performance				
a.	Maximum power (absolute) – Average maximum power observed during two hours, Max power test under ambient conditions, kW	To be declared by the manufacturer	-5	--
b.	Maximum power observed during the test after adjusting the no load engine speed as per recommendation of the manufacturer for field work, kW		-5	--
c.	Power at rated engine speed, kW		-5	IS 8122 (2)
d.	Specific fuel consumption corresponding to average maximum power under 2h maximum power test, g/kWh	-do-	±5	--
e.	Maximum smoke density at 80% load between the speed at maximum power and 55% of speed at maximum power or 1000 rpm whichever is higher	As per CMV rules	--	--
f.	Maximum crank shaft torque observed during the after no load engine speed is adjusted as per manufacturers recommendation for field works, Nm	To be declared by the manufacturer	-8	--
g.	Back-up torque, percent	7 percent, Min	--	--
h.	Maximum operating temperatures ºC Engine oil Coolant	To be declared by the manufacturer	--	
i.	Lubricating oil consumption	1 % of SFC at max. power (high ambient)	+10	5 h rating test
II. Brake Performance:				
a.	Maximum stopping distance at a force ≤600N on brake pedal (m)	10m or as per CMVR	--	--
b.	Maximum force exerted on the brake pedal to achieve a declaration of 2.5m/s^2	≤600 N	--	--

c.	Whether parking brake is effective at a force of 600N at foot pedal(s), or 400N at hand lever	Yes/No	--	Based on the test

III. Mechanical Vibration (Amplitude of Vibration) at:

a.	Operator's platform	120 µm, Max	--	--
b.	Steering wheel	150 µm, Max	--	--
c.	Seat (with driver seated)	120 µm, Max	--	--

IV. Air Cleaner Oil Pullover:

	Maximum percentage of oil pullover	0.25 percent, Max	--	--

V. Noise Measurement:

a.	Max. ambient noise emitted by the combine dB(A)	As per CMVR	--	IS:12180 (2)
b.	Max. noise at operator's ear level dB(A)	As per CMVR	--	IS:12180 (1)

VI. Discard Limit of:

a.	Cylinder bore diameter	To be specified by the manufacturer	--	--
b.	Piston diameter			
c.	Ring end gap			
d.	Ring grove clearance			
e.	Dia. & axial clearance of main bearing			
f.	Dia. & axial clearance of big end bearing			
g.	Thickness of brake lining			
h.	Thickness of clutch plate			

VII. Field Performance

a.	Suitability for the crops	Wheat, Paddy (essential)	--	--
b.	Grain breakage in the grain tank	≤ 2.5 %	--	--
c.	Non-collectable losses	≤ 2.5 % for wheat, paddy, gram and ≤ 4 % for soya bean		
d.	Threshing efficiency	≥ 98 % for wheat, paddy	--	--
e.	Cleaning efficiency	≥ 96 % for wheat, paddy	--	--

VIII. Safety Requirements:

a.	Guards of all moving parts/drives, hot parts	Essential	--	--
b.	Lighting arrangement	As per CMVR	--	--
c.	Grain tank over	Essential	--	--
d.	Spark arrester in engine exhaust	Essential	--	--
e.	Stone trap before concave	Essential	--	--
f.	Rearview mirror	Essential	--	--
g.	Slip clutch at drives: Cutting platform auger; Under shout conveyor; Grain & tailing elevator	Essential	--	--
h.	Anti-slip surface at operators platform and ladder and proper gripping for control levers	Essential	--	--
i.	Working clearance around the controls	70mm, Min	--	--
j.	Labelling of controls, gauges	Essential	--	IS:6283 (1/2)

IX. Materials of Construction:

a.	Guards, Knife blades and knife back	--	--	IS:6024, IS:6025, IS:10378
b.	Labelling of combine harvester	Essential	--	IS:10273

Chapter 15

Testing and Evaluation of Straw Reaper Combine

Straw reaper combine is also known as "straw reaper", which indeed is a misnomer. Hence a straw reaper combine is a machine designed for harvesting, collecting and conveying the chopped straw left behind, in the field during harvesting operation of a combine harvester. It is being increasingly used for bruising of straw of the wheat crop. The machine is driven by the PTO shaft of tractor. While in motion, it cuts the stubble partially and lifts the straw left over by the combine harvester and chops it in to small pieces to convert it into bhusa which is blown out into the trolley attached behind the machine. The bhusa is widely used as animal feed. A grain trap is also provided in the machine to collect the threshed grain recovered from the straw. A schematic diagram of the straw reaper combine is shown in Fig. 15.1, 15.2 and 15.3.

15.1 Terminology

a) *Header:* The portion of straw reaper-combine comprising of the mechanism for gathering, cutting and stripping or picking the straw and delivering it to the cylinder.

b) *Header working width:* The distance between the center lines of the outermost divider points, expressed in mm.

c) *Knife frequency:* The number of cycles the knife makes in a unit time, expressed in whole cycles/min.

d) *Knife registration:* The alignment of centre line of knife section with the centre line of guard.

e) *Knife stroke:* The distance that a point of the knife travels from one extreme position to the other extreme position, expressed in mm.

f) *Turning space:* The diameter of the circle described by the outer most point of the straw reaper combine, when the prime mover, straw reaper combine and trolley in combination takes shortest turn in mm.

g) *Turning radius:* The distance from the turning center to the center of ground contact of the wheel describing the largest circle while the prime mover, straw reaper-combine and straw storage trailer combination is taking its shortest turn, expressed, in mm with standard trolley size of 2.5 x 5 m (with and without brake).

h) *Grain recovery:* It is expressed as percentage of grain recovered by straw reaper from the field to the leftover grains after operation of grain combine in field.

i) *Height of stubbles:* a) Before Harvesting — it is defined as height of wheat stubbles left over after operation of grain combine in field, expressed in mm, b) After Harvesting — it is defined as height of wheat stubbles in the field after operation of straw reaper-combine, expressed in mm.

j) *Awn:* any bristle like fiber or fibers (the long slender extension of lemma) in wheat. It plays an important role in protection against animals and as a mechanism of seed dispersal.

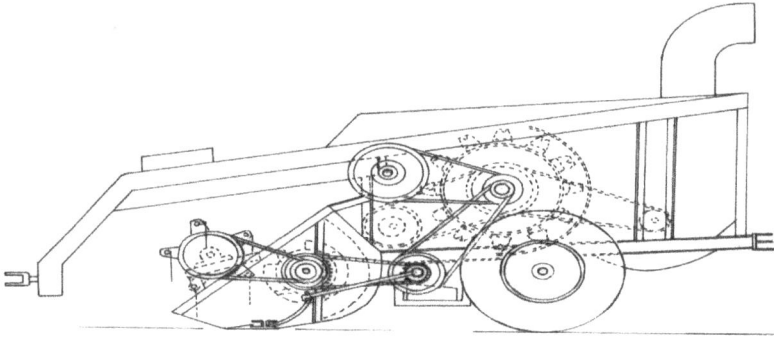

Fig. 15.1: A schematic diagram of straw reaper combine

Courtesy: Swan Agro

Fig. 15.2: Straw reaper combine

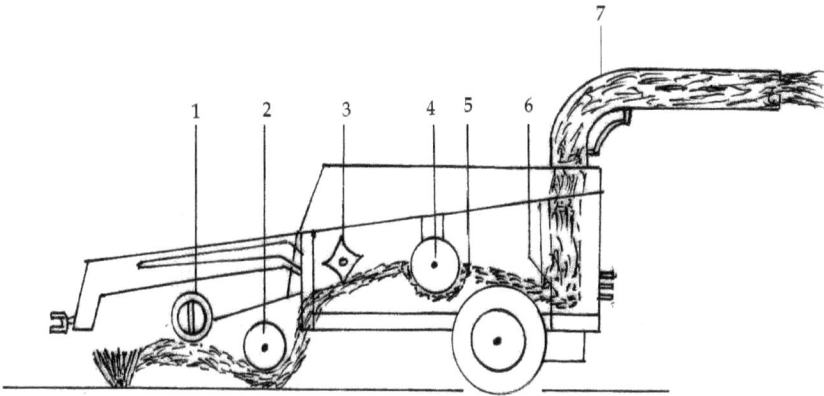

1. Cutter and reel assembly 2. Feeding auger 3.Guid drum 4. Threshing drum 5.Concave 6. Blower 7. Straw outlet duct

Fig. 15.3: A schematic diagram of straw flow in straw reaper combine

k) **Split straw length:** It is defined as the length of split straw after bruising / passing through the machine expressed in mm.

l) **Straw recovery:** It is expressed as percentage of straw recovered by the machine from the field while in operation.

m) **Straw mass density:** It is the mass of stubble/m² recovered manually (kg/m²) at a standard moisture content of 8 percent (dry basis) and at a cut height of 35 mm.

15.2 Scope of Test

The scope of the test is to check and assess the followings:

15.2.1 Lab tests

a) Checking of specifications

b) Chemical analysis and hardness testing of critical components.

c) Visual observations and checking of provision for adjustments.

d) Turning space and turning radius.

e) Checking of wear (dimensional and mass basis)

15.2.2 Field tests

a) Rate of work

b) Quality of work

    i)   Straw recovery percent

    ii) Length of cut straw

    iii) Split straw percentage

    iv) Grain recovery percentage

c) Fuel consumption of prime mover and adequacy of power

d) Soundness of construction

15.2.3 Referred standards /test code

    a)   IS 15805-1: 2008 [65]

    b)   IS 15805-2: 2008. [66]

15.3 Method of Testing

15.3.1 Lab tests

a) Specification checking

The specification of straw reaper combine given by manufacturer shall be checked and compared with the original sheet & reported.

b) Pre-test Checks

The straw reaper combine is to be positioned on a plain surface and examined for pre-test checks like:

    i)   Bearings

    ii)   Drives and other moving parts.

    iii)   Proper adjustments of various assemblies.

    iv)   Correctness of various controls

    v)   Tightness of nuts & bolts.

    vi)   Greasing point should be greased.

c) Wear test

Wear of critical parts/component like, blade of cutter bar, cylinder and concave on weight basis shall be taken before and after test. The percentage wear will be calculated and reported. Hardness of these parts shall also be taken and reported.

d) Chemical analysis

The various critical components to be tested for chemical composition is given in Annexure 15.1

e) Turning space and turning radius

    The test shall be carried out with specified prime mover, straw reaper combine and standard straw storage trolley, combined at a travel speed to avoid any slippage or skidding.

f) Visual observation for ease of operation & adjustments

The straw reaper combine shall be thoroughly checked with particular attention to bearings, drives, and other moving parts for correctness. The observations shall be reported in the test report.

15.3.2 Field test

a) Crop and field conditions

Crop conditions like plant population nos./m^2, straw mass/unit area (kg/m^2), straw mass density (kg/m^3), moisture content of straw, percentage of loose straw, height of stubble before & after harvest shall be recorded & reported.

Field conditions like type of soil, topography, moisture content of soil, number of blade in the direction of travel, height of blades (mm) & frequency of blades/100 meter length (both directions) shall be recorded & reported.

b) Performance parameters

The straw reaper combine shall be tested in the field for a minimum duration of 50 hours for harvesting of wheat straw left over by the combine, to assess field performance of straw reaper combine with regard to quality of work, rate of work, fuel consumption of prime-mover, safety and soundness of construction etc.

i) *Quality of work*

It refers to average length of straw, split straw percentage, dust and awn percentage in the wheat straw produced by straw reaper combine. For measuring length of straw, take sample of 50 - 100 straw pieces coming out from the straw outlet and take the average length. For other quality parameters, take 5 to 10 samples of 10 gm from straw outlet. Separate the un-split bhusa, dust, awn and weigh it. Calculate the split percentage, dust and awn percentage.

Split straw percentage is defined as the percentage of straw splited to the total weight of straw sample collected after passing through the machine. The quantity of straw collected is expressed in terms of straw recovery percentage which is defined as the percentage of difference of straw weight before and after machine operation to the initial weight of straw in the randomly selected sample area of test field. The straw recovery mainly depends upon the stubbles height remaining in the field after harvesting by the combine harvester.

ii) *Grain recovery*

It refers to the recovery of grain from the leftover wheat straw in field by straw reaper-combine. Mark 1 m² area in the field at minimum 3 places. Collect the un-threshed grain (threshed grain lying on the ground not to be included) and ears present in straw, weigh it. After operation of straw reaper combine, again collect the grain and ear at 3 places after marking 1 m² area. Find the percentage of grain recovered by straw reaper-combine as given in Annexure 15.1.

iii) *Rate of Work*

Rate of work of straw reaper-combine consist of (a) Area covered per unit time (b) Straw recovery. The time taken to cover a specified area should be net time, stoppage and breakdown time for any reason should not be added in net time. The stoppage, breakdown time and reason should be noted.

To determine straw recovery mark 3 or 4 plots of 1 m² (this excludes the area containing loose straw) and harvest the straw manually and find the average weight/m². In order to include the loose straw mark 3 or 4 plots of 5 m run of the machine. Collect the loose straw and find the average weight per meter run of the machine. Multiply this by the total run of the machine in the given area to determine the total loose straw. In order to find the loose straw /m^2, divide the total loose straw by the given area. Add straw recovered (excluding loose straw) and the loose straw/m^2 to determine the total straw available before harvesting.

iv) Overlap percentage

Run the machine for at least 3 consecutive runs. Measure the width of 3 consecutive runs. Find the average width of the machine. Find the theoretical width of the machine. The difference between theoretical width and average effective width will give the overlap.

v) Fuel consumption

The fuel consumption of tractor to operate straw reaper-combine with trolley (reaper + tractor + trolley) shall be recorded and expressed in terms of litres per tonne of straw as well as in litres per hour.

c) Ease of handing during operations

Any problem encountered during the operation shall be reported in the test report.

d) Labour requirement

Labour requirement shall be assessed for daily maintenance and during operation & reported.

e) Soundness of construction

All repairs and breakdowns during the entire test shall be recorded and reported in the test report.

Summary of field test of straw reaper combine shall be prepared as shown in Annexure 15.2.

Annexure 15.1

Data Sheet for Straw Reaper Combine Testing

I. Chemical Analysis of Critical Components

S. No.	Critical Component	Chemical Composition				
		Carbon	Manganese	Phosphorus	Sulphur	Silicon
1.	Knife blade					
2.	Knife guard					
3.	Cylinder blade					
4.	Concave blade					
5.	Serrated					
6.	Counter/ serrated blade					
7	Feeder/beater					

II. Grain Recovery Percentage

S. No.	Grain (un-threshed) ear-head Recovered before harvesting the straw	Grain left-over after harvesting of straw	Grain recovered (A-B)	Grain recovery, % (C/2) x 100
	A	B	C	D
1.				
2.				
3.				
4.				
...				
10.				

III. Field Testing of Straw Reaper Combine

Fuel Consumption: **Wheat straw recovered:**

S. No.	Variety	Field area		Time of start, h	Time of finish, h	Run time, h	Remarks		Fuel consumption, l
		Length, m	width, m				Stop time, h	Reason	
1									
2									
3									
4									
5									

IV. Rate of Work

Average speed of the machine:

S. No.	Crop variety	Area, m2	Run time, h	Speed of operation, km/h	Time for covered area (including crop), h	Field capacity, ha/h	Remarks Stoppage time, h	Reason
1	2	3	4	5	6	7	8	9
i)								
ii)								
...								

V. Straw Recovery

S. No.	Crop variety	Wt. of wheat straw recovered manually (stubbles only) g/m2	Wt. of loose straw in l meter length, g/m	Weight of loose straw in 1 m2 (4 x total straw length in the field/ area)	Wt. of wheat straw (stubbles + loose) before harvesting (3+5)	Wt. of straw recovered before harvesting /m2	Straw recovery (6 – 7)	Straw recovery % (8/6) x 100
1	2	3	4	5	6	7	8	9
i)								
ii)								
...								

VI. Percent Overlapping

S. No.	Crop variety	Machine width in simple runs (theoretical), m	Machine width in n runs	Effective machine width in one run (4/n)	Overlap, m (3-5)	Overlap, % (6/3)x100
1	2	3	4	5	6	7
i)						
ii)						
...						

Annexure 15.2

Summary Sheet of Field Performance Test

Crop variety	Rate of work		Fuel consumption			Average length of bhusa, mm	Straw split percent	Straw recovery percent
	ha/h	h/ha	l/h	l/tonne	l/ha			
(i)								
(ii)								
(iii)								
(iv)								
(v)								

Chapter 16

Testing and Evaluation of Root Crop Harvesters

The tropical root and tuber crops are comprised of crops. These are staple foods in many parts of the tropics, being the source of most of the daily carbohydrate intake for large populations. These carbohydrates are mostly starch based, found in storage organs, which may be enlarged roots, corms, rhizomes, or tubers. Many root and tuber crops are grown as traditional foods or are adapted to unique ecosystems and are of little importance to world food production. Root and tuber crops consist of root crops, such as beets and carrots, and tuber crops, such as potatoes, groundnut and sweet potatoes.

Root crops (roots and tubers) are harvested with diggers, digger elevator conveyors, digger-pickers, and combines, which often pull up clods, stones, and vines with the crop. Though some large machines carry workers who manually sort out extraneous material, this task is increasingly being performed mechanically. Modern potato and sugar-beet harvesters lift the whole tuber/root from the ground, clean the earth from it, and deliver it to a bin or wagon. Sometimes the crop tops are removed before harvest of the roots and used for cattle feed. Peanuts (groundnuts) are lifted along with vines and are allowed to dry before removal of the pods.

In this chapter, testing and evaluation procedures of potato harvesters, groundnut diggers and sugarbeet harvesters have been discussed with scope, methods of tests and data sheets. The main criteria in harvesting of root crops is quality of the crop, i.e. without bruising, skinning and damage and maximum recovery. This is more prominent factor in automatic root harvesters as in semi-automatic methods, collection by manual labour increases the recovery percentage and reduces the damage. The pertinent test parameters and procedure for conducting the test are included in the respective equipment section.

16.1 Potato Digger Harvesters

Potato harvesting is one of the most important operations that has to be performed precisely to have a good potato production. Harvesting is the complete recovery of potato tubers out of ridge and separating them from bulk of soil, roots, stones, haulms without causing excessive damage. It has a direct effect on the potato bruising and results in poor quality for potato marketing. Potato is harvested when it attains the physiological maturity. Various harvesting methods, such as manual digging of ridges with a narrow spade /fork/khurpi followed by manual picking by a labor group, use of animal drawn potato digger followed by manual picking, and use of tractor drawn elevator digger followed by manual picking or potato harvester which automatically collects the potatoes in a container and transfer to the transport trailers or wagons, are in use according to the availability of machine or labour. Small to medium farmers follow first two methods which is time consuming and labor intensive [600-700 man-h/ha] and these methods provide maximum recovery and minimum damage. Mechanical harvesting of potatoes using digger or elevator [Fig. 16.1] is partial and selective and is followed by farmers having large land holding (more than 8 ha) [241, 249].

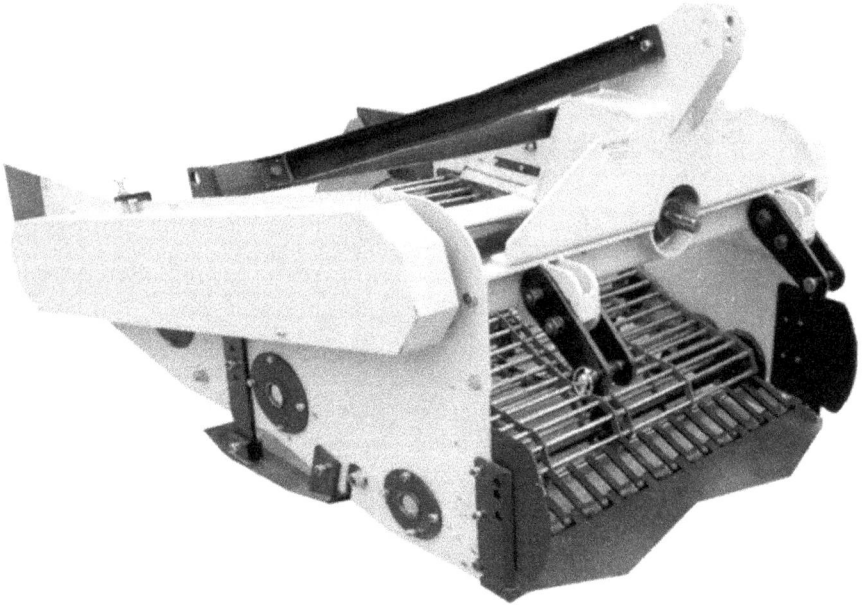

Courtesy: Swn Agro

Fig. 16.1: Potato digger shaker

16.1.1 Terminology

a) **Row Spacing:** Centre to centre distance between adjacent ridges, expressed in millimeter.

b) **Potato Mass:** Mass of potato, expressed in grams. The average mass of potato in a batch shall be determined by weighing at least 50 potatoes.

c) **Potato Index (I):** A constant specifying the relationship between the length, width and thickness of potatoes, and is given by the formula:

$$I = \frac{L^2}{WT} \times 100$$

Where,

I   - potato index

L   - greatest length

W   - greatest width

T   - greatest thickness, averaged from measurements of at least 50 tubers

The tuber shape will be decided based on tuber index as follows:

Tuber shape	:	Round	Oval	Long	Very long
Potato Index	:	100 - 160	161 - 240	241 - 340	above 340

d) **Conveyor Index:** It is the ratio between the speed of elevator conveyor to forward speed of the potato digger/ shaker.

e) **Conveyor Loss:** Percentage of potatoes entering the elevator that get sieved through along with the ones that may fall through either side of the conveyor.

f) **Soil-Potato Ratio:** i) Before Digging - It is the ratio of soil and potato mass taken from the ridge at optimum depth; ii) After Digging - It is the ratio of soil and potato mass collected at the end of the elevator chain conveyor.

g) **Separation Index:** It is the ratio between the difference of soil potato ratio before (f(i)) and after (f(ii)) digging to the soil-potato ratio before digging, given by the formula:

$$I_s = 1 - \frac{r_a}{r_b}$$

Where,

I_s - separation index

r_a - soil-potato ratio after digging

r_b - soil-potato ratio before digging

h) **Optimum Depth of Potato Zone:** The depth where the losses of potatoes are minimum, expressed in millimeter.

i) **Soil-Potato Mix:** Mass per unit volume of the soil potato mix taken from the ridge at optimum depth prior to digging, expressed in kg/m³.

16.1.2 Scope of test

16.1.2.1 Laboratory test

a) Checking of specification

b) Checking of material

c) Visual observations and provision for adjustments

d) Safety provision

e) Test Conditions

    i)   Crop conditions

    ii) Field conditions

f)   Test at No-Load

    i)   Power consumption, and

    ii) Visual observation

16.1.2.2 Field test

Operate the potato digger shaker on the crop row. During the operation, collect the samples randomly from 10 places for the area having width equal to crop spacing and length as 20 m, within the test plot. For field operations, the machine shall be adjusted in accordance with the manufacturer's recommendations. This test shall be repeated a minimum of five times. The following parameters will be assessed

  i)   Rate of work

    1.   Field capacity

    2.   Percentage of exposed potatoes

    3.   Soil-potato ratio

  ii) Quality of work

    1.   Losses

       a)   Percentage of conveyor loss

       b)   Percentage of cut potatoes

          i)   Deskinned

          ii) Damaged, removable by normal pealing

          iii) Badly damaged

       c)   Percentage of exposed potatoes

    2.   Efficiency

       a)   Field efficiency

       b)   Separation index

       c)   Depth of cut and soil potato ratio

  iii) Power consumption

  iv) Visual observations

  v)   Long duration test

Operate the potato digger shaker for at least 25 hours at load. Each test should be of continuous run of atleast 5 hours. Major break-down, defects developed and repairs made during this period shall be reported.

16.1.3 Methods of Tests

a) Measurement of Soil-Potato Ratio

Install a spool with the machine carrying a canvas having width 30 cm more than that of conveyer and length approximately 2 m. Wherever the soil-potato ratio is to be measured, hold the canvas from the two corners and as the spool starts unwinding, the soil potato mass will start falling on it. The soil-potato ratio shall be computed from the quantity received on the canvas.

b) Measurement of Conveyor Loss

While measuring the soil-potato ratio as described above, measure the quantity of potatoes that have been buried below the canvas by picking the potatoes from underneath the canvas. Compute the conveyor loss based on the mass of the potatoes received from the given area.

c) Measurement of Exposed/Unexposed Potatoes

As soon as the digger shaker completes harvesting, remove all the potatoes that are visible to the eye from a row of known length and weigh it to find the mass of exposed potatoes. The tubers which are still contained in the soil but not visible shall be removed by manual digging and weigh it to find the mass of unexposed potatoes.

d) Percentage tuber loss and damaged

i)   Skinned tubers (%)

ii) Cut tubers (%)

iii) Bruised tubers (%)

Data sheet for testing of potato digger shaker is shown at Annexure 16.1.

16.2 Groundnut Digger Shaker

Groundnut growing farmers are facing an arduous task of collecting groundnut pods under the soil which are picked up manually, consuming lot of labor and time. The yield of the crop is affected as up to 20 percent of pods are left underground during harvest [243, 250]. (Fig. 16.2)

16.2.1 Terminology

a)   *Total Loss:* The sum of the following losses in the field during the test.

   i)   *Exposed Pod Loss* - The detached pods lying on the surface with respect to total pods, expressed as percentage by mass.

   ii) *Un-exposed Pod Loss* - The detached pods remain burried in the soil with respect to total pods, expressed as percentage by mass.

   iii) *Undug Pod Loss* - Pods lost due to the plants remaining undug or detached pods remain burried in the soil with respect to total pods, expressed as percentage by mass.

b)   *Damaged Pods*: The hull damaged and shelled pods including hulls and kernels with respect to total pods, expressed as percentage by mass.

c)   *Digging Efficiency*: Percentage by mass of the undetached pods obtained from the plants dug with respect to total pods.

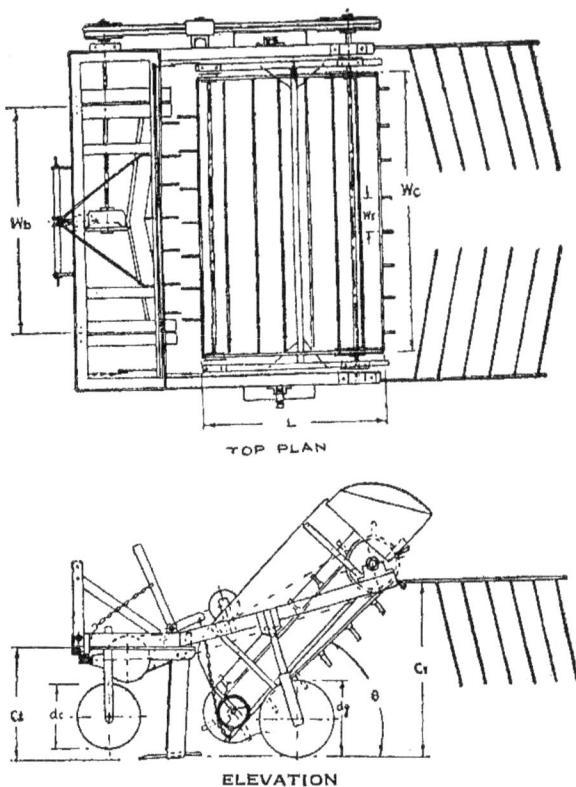

Fig. 16.2: Schematic view of groundnut digger shaker

16.2.2 Scope of test

16.2.2.1 Laboratory test – Refer 16.1.2.1

16.2.2.2 Field Tests

 a) Test parameters

  i) Width and depth of cut

  ii) Percentage of damaged pods

  iii) Percentage of exposed pod loss

  iv) Percentage of unexposed pod loss

  v) Percentage of undug pod loss

  vi) Digging efficiency

  vii) Power requirement

  viii) Field efficiency and labour requirement

  ix) Ease of operation and adjustments

  x) Soundness of construction

 b) Long duration test

16.2.3 Methods of test

a) Pre-test observations

i) *Moisture Content of the Soil* - The soil samples shall be taken from at least five random places in the field and tested for moisture content in accordance with any one of the methods preferably oven-drying method.

ii) *Moisture Content of Pods and Vines* - Take the samples of pods and vines from at least five random places.

iii) *Determination of Pod and Vine Ratio* - Take five samples of the crop each weighing one kilogram. Separate the pods from vines manually for each sample. Take the mass of pod and vines separately for each sample, and calculate their ratio. The average of the five samples shall be taken as pod-vine ratio.

b) Operate the groundnut digger on the crop row. During the operation, collect the samples randomly from 10 places for the area having width equal to crop spacing and length as 2.0 m. At least three series of field tests shall be carried out under different soil conditions.

c) Width and Depth of Cut - Measure the width and depth of cut for minimum at five places and obtain the average of the readings.

d) Analysis of Samples - The different samples collected from the ten randomly selected areas should be analyzed for following:

i) Damaged pods -The plants dug should be collected and all the mature pods should be detached manually and analyzed for the damaged pods.

ii) Exposed pods - After collecting the plants, visualize the crop row and collect the detached pods from 2 m length lying exposed on the soil surface.

iii) Unexposed pods-The area of the soil having length as 2 m should be analyzed for the detached pods lying buried in the soil. For the collection of pods the soil should be dug out manually.

iv) Undug pods - Collect the undug plants, if any, and detach the mature pods from the plants and weigh it.

e) Calculations

A total quantity of pods collected from the plants in the test plot can be calculated as under

$$A = B + C$$

Where,

A - total quantity of pods collected from the plants in the sample area

B - quantity of clean pods collected from the plants dug in the sample area, exposed pods lying on the surface and the buried pods

C - quantity of damaged pods collected from the plants in the sample area

G   - quantity of detached pods lying exposed on the surface

H   - quantity of detached pods remained inside the soil in the sample area

K   - quantity of pods remained undetached from the undug plants in the sample area

Percentage of damaged pods = (C/A) x 100

Percentage of exposed pod loss = (G/A) x 100

Percentage of unexposed pod loss = (H/A) x 100

Percentage of undug pod loss = (K / A) x100

Digging Efficiency = 100 - Total percentage of pod loss

Total percentage of pod loss =   Percentage of exposed pod loss +

Percentage of unexposed pod loss +

Percentage of undug pod loss

f) Power Requirement

g) Field Efficiency and the Labour Requirement

h) Ease of Operation and Adjustments

i) Soundness of Construction

j) Long duration test

The digger shall be operated for at least 25 hours. Each test shall be of continuous run of atleast 5 hours under short-run test. The major breakdown, defects developed and repairs made during this period shall be reported.

The data sheet for testing of groundnut digger shaker is shown at Annexure 16.1.

16.3 Sugarbeet Harvester

The sugarbeet crop is another crop to be mechanized. During the mechanical harvesting of the crop, the plants are topped-up in the field to remove almost all green shoot material, but crown tissue is left to avoid overtopping. Achieving uniform topping-up of sugar beet is difficult and the depends on plant size, initial size of the biological crown, harvesting conditions, and operator skills in setting up and using the harvesting machinery. Test procedures approved by International Institute for Beet Research (IIRB), Germany [www.iirb.org] is described in this section. This test procedure can be used as a reference in testing and evaluation of sugarbeet harvesters and other root crops like onion, radish, and carrot etc.

16.3.1 Scope of test

a) Crop and field conditions

   i)   Soil type
   ii) State of soil (e.g. wet, stony, etc.)
   iii) Cultivation technique (ploughing, no-ploughing, strip-till, ...)
   iv) Cultural practices (crop drilled to stand or thinned)

v) Primary tillage, last crop rotation

vi) Variety

vii) Type of seed (e.g. pelleted, monogerm, etc.)

viii)Seed spacing (cm)

ix) Row width (cm)

x) Plant density (plants/m²)

xi) Diameter of the beets

xii) Potential yield of clean beets (t/ha)

xiii)Other details if any

b) Sugarbeet harvester specifications

i) Name and address of the manufacturer

ii) Harvester characteristics (topping, lifting, conveying, cleaning)

iii) Overall dimensions

iv) Tyre dimensions, make and type, inflation pressure

v) Weight gross and net, axle and wheel load with lifted and unlifted header, fuel tank full and without driver

vi) Other specifications e.g. empty weight, tank capacity

c) Root diameter

The maximum diameter is highly correlated with the beet root mass. Therefore the measurement of the root diameter is optional.

The maximum diameter of each root in a 500 root sample shall be measured. The distribution of frequencies of the largest sugar beet diameter can be shown in the form of a histogram in Fig. 16.3 and the diameters classified into the following categories, 4.5 - 7 cm, > 7 - 9 cm, > 9 - 11 cm, > 11 - 13 cm, > 13 - 15 cm and > 15 cm to provide frequency distributions. These measurements may be done on the same bulk of roots taken for the determination of root breakages.

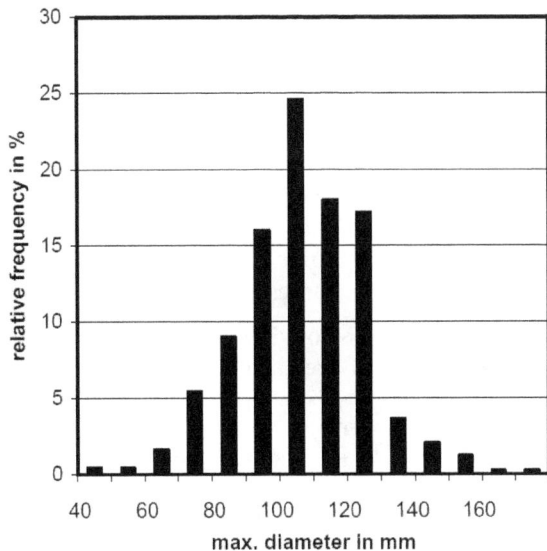

Fig. 16.3: Histogram of max. beet root diameter

d) Performance assessments ('harvesting quality')

The test is performed in a regular crop stand prepared for 3 test passes. After the first test run, manufacturer has to decide whether the first run is accepted or a second or third test run will be necessary due to unforeseen occurrences. The harvesting quality of each machine is determined based on the following:

i) Harvesting losses (lifting, cleaning and conveying losses)

ii) Root breakage

iii) Topping quality

iv) Soil tare

v) Superficial damage

16.3.2 Methods of tests

a) Lifting, cleaning and conveying losses

The lifting losses include all whole roots left above and under the ground, as well as the upper parts of all broken roots with a diameter larger than 4.5 cm that are left on and in the soil. The lower parts of broken roots are not included because these are recorded in the root breakage assessments. The soil will need to be drag or spring-tine harrowed twice at a depth of 15 - 20 cm to recover roots left in the ground. Lifting losses should be measured on four areas that are 6 rows wide and at least 10 m long, to give a total sampled area of at least 100 m². The tines of the cultivator should work both in and between the rows.

The beetroots are collected, topped by hand and weighed to calculate the specific mass loss per hectare in t/ha (manually harvested yield as reference).

b) Root tip breakage

Root breakage assessments are made on 5 samples of 100 root samples taken from the beet lifted from the assessment area (minimum 6 rows, each 200 m long). A representative sample is taken: a) with a catch frame holding bags or buckets of a capacity of at least 25 kg at least twice for each unloading of the tank of the harvester or the conveyor of a loader, b) from the pile after unloading the harvested beets within the test area and increasing the sample number to 10 x 100. The diameter of each root at the point of breakage (if any) is measured and recorded in the following diameter classes as shown in Table 16.1.

Table 16.1 Calculation of yield losses caused by root tip breakage from measurement and classification of breakage diameter

Number of categories	1	2	3	4	5
Categories of dia. of root breakage, cm	0 – 2	>2 - 4	>4 - 6	>6 - 8	>8
Relative frequency (b_i), %					
Factor of losses (c_i^*), g	0	23	60	130	230
Relative yield losses per category ($b_i \times c_i$), g	-				

*centre of classes 1, 3, 5, 7, 9 cm
** for fangy roots, maximum root diameter is recorded

Plant density at harvest (PD_h) = _____/ha

Root tips broken off in the sample (r_b) = _____ g

Relative yield losses due to root tip breakage = $\sum_{i=2}^{5}\left[\dfrac{(b_i \times c_i) - r_b}{b_i \times 10^6} \times PD_h\right]$

= _____ t/ha

Yield of clean beets = _____ t/h

Relative yield losses by root tip breakage = [(yield of clean beets x abs. yield losses)]/100

= _____ %

c) Topping quality

A subjective assessment of topping quality is made at the same time as the root breakage measurements, using the same 500 beet sample (or 10 samples of 100 roots, respectively). They are classified into one of the following six categories (Fig. 16.4). The numbers of roots within each topping category should be recorded and relative frequency of each class may be determined.

d) Soil tare

Soil tare (i.e. the soil adhering to the roots, loose soil and stones) should be measured and reported. For this assessment, an extra sample of 5 x 100 beets, taken as per 16.3.2 (b). The mass of soil is measured by a washing process or the determination is based on tare house washing with regular settings of the existing washing equipment. An average soil tare is calculated for at least 500 kg with a sufficient number of replications.

Soil tare = (gross beet weight - washed beet weight) / gross beet weight

A true soil tare may be calculated by dividing the single mass of dirty samples minus mass of washed (clean) beets by the mass of washed (clean) beets.

Soil tare $_{clean\ beet\ base}$ = (gross beet weight - washed beet weight) / washed beet weight

Under topped	Well topped	Over topped
untopped	Under and correctly topped	Over and angled topped

defoliated with petioles >2cm <	Correctly defoliated without lesions	Completely defoliated with lesions

1. Untopped with petioles longer than 2 cm	3. Under topped with no petioles	5. Over topped with under half of maximum diameter of bundle rings visible
2. Under topped with petioles shorter than or equal to 2 cm	4. Correctly topped	6. Angled topped

Fig. 16.4: Assessment classes for topping and defoliation quality

e) Superficial damage

It is not a mandatory element in performance assessments, but it is important since it provides information on the way in which roots are handled as they pass through the harvester. The damage is classified into the four classes:

i) Non-damaged

ii) Wounded (significant mass loss)

iii) Bruised (scratched surface, visible crack with no mass loss)

iv) Wounded and bruised

The number of roots occurring in each category of damage should be recorded using the procedure, 16.3.2 (c), and presented as relative frequency in each category.

The superficial damage can be measured by storage trials of the beet harvested by the harvesters in test. In assessments of superficial damage, linear measurements could be made of the length and width of the damaged area for calculating the damaged area. The measurements should be made on ten samples of 100 roots, and the results expressed as the total area (cm²) per 100 roots.

Data sheet for testing of sugarbeet harvester is shown in Annexure 16.2.

Annexure 16.1

Data sheet for Field Performance of Potato Diggers

Sl. No.	Description	1	2	3	4
Soil Potato Ridge					
1.	Row to row spacing				
2.	Ridge height				
3.	Base width				
4.	Depth of potato zone				
Soil potato Ratio					
1.	Ridge length				
2.	Depth of cut				
3.	Mass of Soil- Potato Mix Collected at the Canvas				
4.	Mass of Soil				
5.	Soil-Potato Ratio				
Operational Performance					
Conveyor belt length:		No. of revolutions or rear tyre (N):			
Tractor rear tyre diameter:		No. of revolutions of the conveyor (n):			
1.	Depth of cut				
2.	Length of run				
3.	Gear engaged				
4.	Throttle setting				
5.	Fuel consumption (ml)				
6.	Consumption time, min				
7.	Exhaust gas temperature, °C				
8.	Distance travelled for N				
9.	Time taken for N, min				
10.	Wheel slippage, %				
11.	Forward speed, km/h				
12.	Time for n, min				
13.	Belt conveyor speed index				
Separation Index / Conveyor Loss					
1.	Mass of Potato Soil on Canvas				
2.	Mass of potato				

3.	Mass of soil
4.	Separation index
5.	Mass of Potato Soil under Canvas
6.	Conveyor loss, %

Uncovered / Cut Potatoes

1.	Ridge length, m
2.	Exposed, %
3.	Unexposed, %
4.	Cut, %
5.	Bruised, %
6.	Skinned, %

Annexure 16.2
Data sheet for Test of Beet Harvesters

Location: Test team: .... Date: ....

Machine: ....   Plant density :   .... plants/ha

Yield potential (clean beets):   .... t/ha   Working speed: .... km/h

Losses

Root tip breakage category	Rel. Frequency, %	Factor of losses, g/100 beets	Rel. losses, kg/100 beets
0 - 2 cm			
>2 - 4 cm			
>4 - 6 cm			
>6 - 8 cm			
>8 cm			
Total			

Root tip breakage losses: ....% ....t/ha

Lifting, cleaning and conveying losses: ....% ....t/ha

Total losses: .... % ....t/ha

Topping quality		Superficial damages	
Category	Rel. Frequency, %	Category	Rel. Frequency, %
1		non damaged	
2		wounded	
3		bruised	
4		wounded & bruised	
5			
6			

Soil tare: .... %, number of samples: ...., per .... kg of harvested beets

Additional information

Trial identification	:	....
Weather conditions	:	....
Soil type	:	....
Regularity of crop	:	....
Presence of fangy beets	:	....
Miscellaneous	:	....

Chapter 17

Testing and Evaluation of Power Thresher

A machine operated by a prime mover such as electric motor, engine, tractor or power tiller used for threshing the whole crop, ear heads or cobs is called power thresher. Threshing of grain crop is one of the most important crop processing operation to separate the grains from the ear heads of the plants and to prepare it for the market.

Historically, the threshing was done by manual flailing operation, replaced later by treading under animal hooves. These were time consuming methods involving drudgery and exposure of the crop to natural hazards of rain and fire and also loss of grains by animals, birds and insects etc.

The threshing of wheat crop involves detaching the grains from the earhead as well as breaking up of the straw into 'Bhusa' of an acceptable quality. The threshing of wheat by treading under animal hooves accomplished both the detaching of grains as well as making of bhusa. The use of wooden and steel rollers pulled by a pair of animals improved the output. The introduction of 'Olpad thresher' was a breakthrough which enabled a four-fold increase in output from a pair of bullocks. The 'Olpad' thresher had its origin at a place named 'olpad' in south Gujarat. This thresher has three gangs of notched discs mounted on a frame. The farmer can sit on it while driving the animals over the crop.

The use of mechanical power for threshing started with use of chaff cutters in Punjab. It was used to chop up the wheat crop which also partially threshed the heads. The crop was later trampled to complete the threshing in a comparatively shorter period. A modified chaff cutter with the enclosed flywheel having a corrugated periphery and a concave underneath eliminated the use of animal treading and thus paved the way for development of wheat power thresher. This type of thresher was named as "syndicator". In order to obtain clean grain the threshed crop was to be winnowed in a separate operation. M/s Friends Own Foundry, Ludhiana under the guidance of Sh. S.K. Paul, the then Agri. Engineer, (Dept. of Agri., Punjab) produced the first power wheat thresher in the mid-fifties which could thresh, clean and bag the grains in a single operation. In this thresher, the threshing was carried out by a hammer mill type threshing head and separation and cleaning was accomplished through the aspirating action of the air passing through the sieve by a blower which also sucked and flew away the bhusa to some distance. Later on, the thresher with spike tooth type cylinder and a closed concave, aspirators and sieves was developed. Majority of the present day threshers have adopted the main design features of this thresher and have added accessories such as automatic feeding etc.

Presently, various types of threshers are available in the market like drummy thresher, beater type thresher, rasp-bar type, spike-tooth type, chaff-cutter type thresher meant for threshing a specific crop. Recent trend is towards adoption of multi-crop threshers to take care of various crops grown under prevailing crop rotation. This type of thresher can thresh a variety of crops like wheat, barley, sorghum, soyabean, maize and even paddy with suitable adjustments. Axial flow threshers introduced by IRRI have been modified to serve as multi-crop thresher, which can handle paddy also. The details of most commonly used types of threshers are shown in Fig. 17.1, 17.2 and 17.3.

There have been alarming cases of accidents during the use of power threshers. Critical analysis of the causes of accidents has revealed that about 73% are caused due to human factors such as carelessness, intoxication, etc., 13% due to machine factors like improper feeding system and use of inferior material of construction, 9% due to crop factors like abnormal crop moisture content and earhead threshing, and, 5% due to environmental factors like poor light during night operation, crowded surrounding or excessively hot weather conditions. With a view to control such accidents every manufacturer of power thresher should ensure that the machine and every part there of complies with the standard specified clauses of BIS.

17.1 Terminology

a) **Power thresher:** A machine operated by a prime mover to detach and separate the grains from the straw.

b) **Clean grain:** Threshed, mature, unbroken grain and free from foreign matter.

c) **Damaged grain:** Threshed grain, which is partially or wholly broken.

d) **Un-threshed grain:** Whole grain attached to straw after threshing.

e) **Chaffed straw:** Straw being discharged from threshing chamber, which is usually crushed, cut.

f) **Foreign material:** Inorganic and organic material other than grain which includes sand, gravel, clay, metal chip, chaff and straw, weed and other inedible grains.

g) **Grain Straw Ratio:** Ratio of grain to straw by weight.

h) **Maximum Input Capacity:** The maximum feed rate at which no choking occurs in the thresher and no stalling occurs in the prime mover and at the speed specified by the manufacture.

i) **Output capacity:** The mass of the grain mixture received at main grain outlet(s) when collected at optimal input capacity.

j) *Percentage of blown grain:* The clean grain lost along with chaffed straw 'bhusa' with respect to total grain input expressed as percentage by mass.

k) *Percentage of broken grain:* The broken grain from main grain outlet with respect to total grain mixture received at main grain outlet expressed as percentage by mass.

l) *Percentage of spilled grain:* The clean grain dropped through the sieve and overflown from sieve along with tailings with respect to total grain input, expressed as percentage by mass.

m) *Percentage of un-threshed grain:* The un-threshed grain from all outlets with respect to total grain input expressed as percentage by mass.

n) *Threshing efficiency:* The threshed grain received from all outlets with respect to total grain input expressed as percentage by mass.

o) *Cleaning efficiency:* The cleaned grain received from main grain outlet with respect to total grain mixture received at main grain outlet, expressed as percentage by mass.

17.2 Threshing

Threshing is accomplished by rubbing, cutting, bruising and impact of the rotating pegs, spikes or bars (mounted on the cylinder) on the ear heads, which force the grain out from the sheath holding it. In the process of threshing of the wheat crop, the straw may or may not be fully or partially bruised and broken up by the impact of the rotating members of the threshing cylinder.

Converting the straw into bhusa is a design feature of power threshers to utilize the bruised straw as a cattle feed. Different grain crop and varieties of the same crop have varying characteristics; require different threshing speeds of the cylinder and concave clearance for achieving optimum results of threshing. Therefore, accurate adjustment of cylinder speed and concave clearance is essential. Some of the recommended speeds of cylinder and concave clearance for threshing different crops are given in Table 17.1 and 17.2.

17.3 Bhusa Making

Wheat bhusa is an important roughage for cattle feed. Therefore, bhusa making is an essential requirement of a thresher, particularly in the northern and central India. The threshers developed in India for threshing of wheat are based on the principle of repeated beating the crop in an enclosed chamber until the straw is reduced to a size that would pass through the concave grate. In the syndicator thresher, the straw is chopped up on being fed in the threshing chamber by the chaff cutter blades, which have serrated edge to prevent slippage of dry straw while shearing it. Generally, the length of cut is adjusted to about 20 mm. The cut straw is further crushed and bruised between the periphery of the corrugated flywheel and the concave grate. The rubbing action completes the threshing operation and finally grain and bhusa is separated.

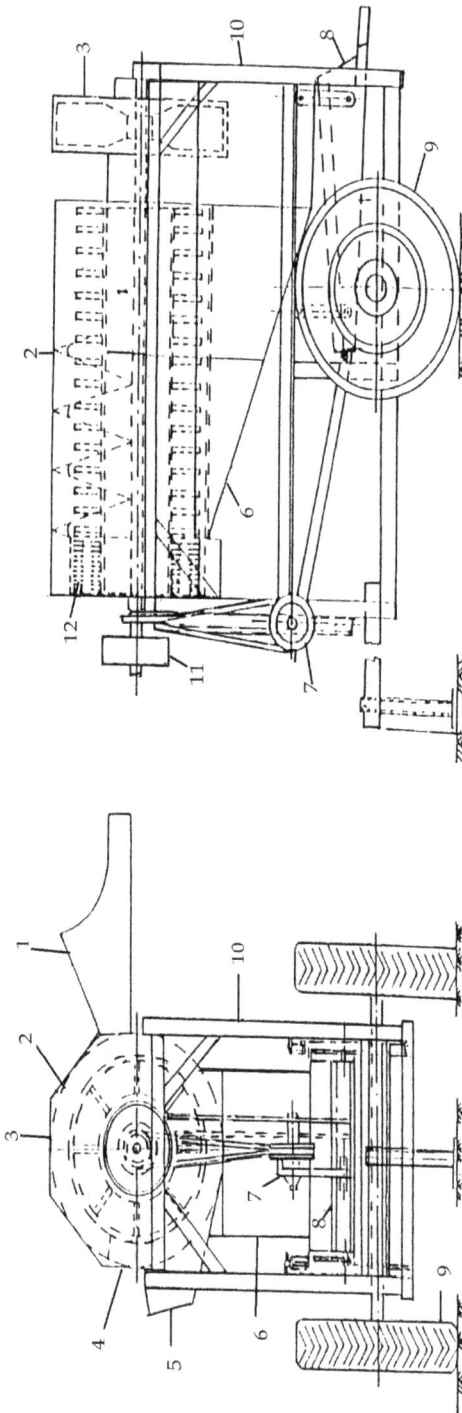

Front view

Side view

Fig. 17.1: A schematic diagram of Axial Flow Thresher

1. Feeding chute, 2. Threshing cylinder, 3. Aspirator blower, 4. Chaff outlet, 5. Straw outlet, 6. Hopper, 7. Cam for oscillating sieves, 8. Oscillating sieves, 9. Transport wheels, 10. Frame, 11. Main Pulley, 12. Louvers

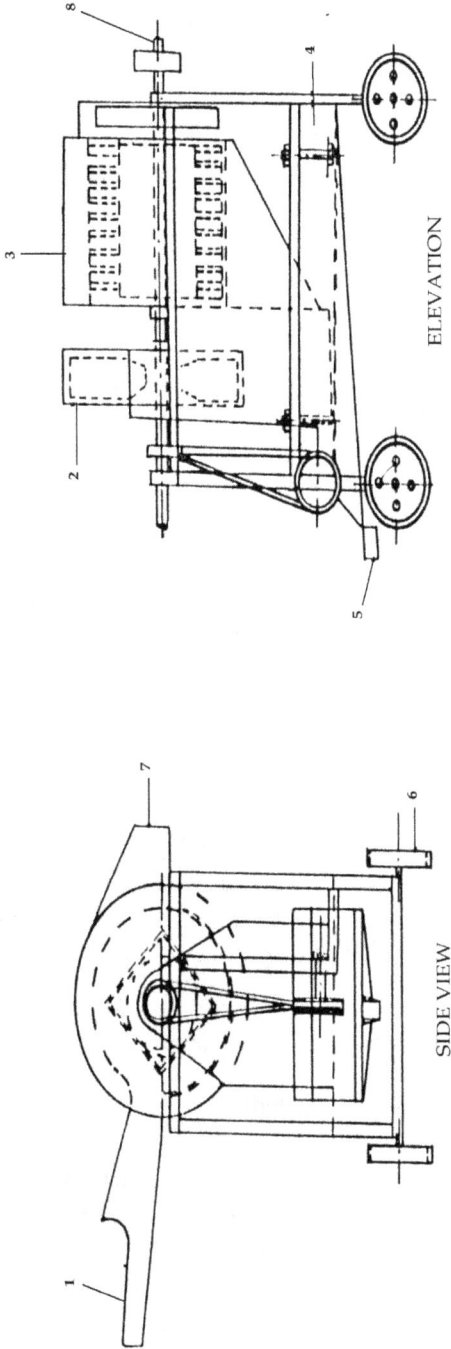

ELEVATION

SIDE VIEW

Fig. 17.2 : A Schematic view of Spike-tooth type Thresher

1. Feeding chute, 2. Blower, 3. Threshing cylinder, 4. Oscillating sieves, 5. Grain outlet, 6. Transport wheel, 7. Bhusa outlet, 8. Shaft

Table 17.1: Recommended peripheral speeds and rpm of threshing cylinder for selected crops [131]

S. No.	Crop	Cylinder speed	
		m/s	rpm
1.	Wheat	20-30	550-1100
2.	Paddy	15-25	675-1000
3.	Jowar	12-20	400-675
4.	Bajra	10-16	400-550
5.	Gram	12-22	400-750
6.	Peas	13-22	430-750
7.	Barley	20-26	740-1080

Table 17.2: Recommended range of concave clearance and grate opening for different crops

S. No.	Crop	Concave clearance, mm	Concave grate opening, mm
1.	Wheat	15-25	6.35-8.5
2.	Paddy	20-30	8.5-10
3.	Jowar	15-25	6.35-8.5
4.	Bajra	15-25	6.35-8.5
5.	Gram	20-30	8.5-12.5
6.	Maize	20-30	25
7.	Barley	15-25	6.35-8.5
8.	Peas	20-30	8.5-10
9.	Soybean	20-30	8.5-10

The clearance between the rotating threshing cylinder and the concave has significant effect on the threshing efficiency, the size of 'bhusa' obtained and the power consumption. However, a suitable combination of cylinder-concave clearance, concave grate/screen opening size and cylinder speed for different crops and their condition has to be arrived. Generally, the concave is made of 6 mm square bars with grate opening of 6.35 mm to 12.5 mm for various crops except maize, which has recommended grate opening of 25mm. In the peg type cylinder, the concave clearance is adjustable by changing the length of peg.

17.4 Cleaning

The threshed grain mixed with bhusa falling through the grate openings of the concave, has to be separated and cleaned. The aspirator system is most suited when the grains are mixed with "bhusa". The first stage of cleaning is done by the material falling through an upward, blowing column of air, which carries away the lighter bhusa with it but allows the heavier grain and straw pieces to fall through. The air is blown by a blower whose inlet is connected to the aspirator column, which carries the bhusa inside the casing of the blower and is blown out through the outlet. Air velocity of 4 m/s to 5 m/s is needed for the main aspirator to remove the straw that is already suspended in air. Those pieces of straw, which are large, and nodes fall down on the sieve below it.

The top (upper) shaker sieve is generally of 4.9 to 10.9 mm perforated sheet according to type of crop through which the grains pass down along with other smaller particles of soil etc. Larger pieces of straw and other materials like nodes slides over the top of the screen and are separated. The lower sieve is also a perforated sheet with hole size of 1.5 mm to 2mm diameter, through which the weed seeds of smaller size, soil and sand particles fall out and only the grains along with some chaff remain on top, which slide out either for bagging or in most cases go under a secondary aspirator which sucks up the chaff and leaves the grains clean. The shaker sieves of the aspirator type cleaning system have a speed of 8m/mts from a crank having a throw of 20 to 25 mm and rotating at a speed of 300 to 400 rpm.

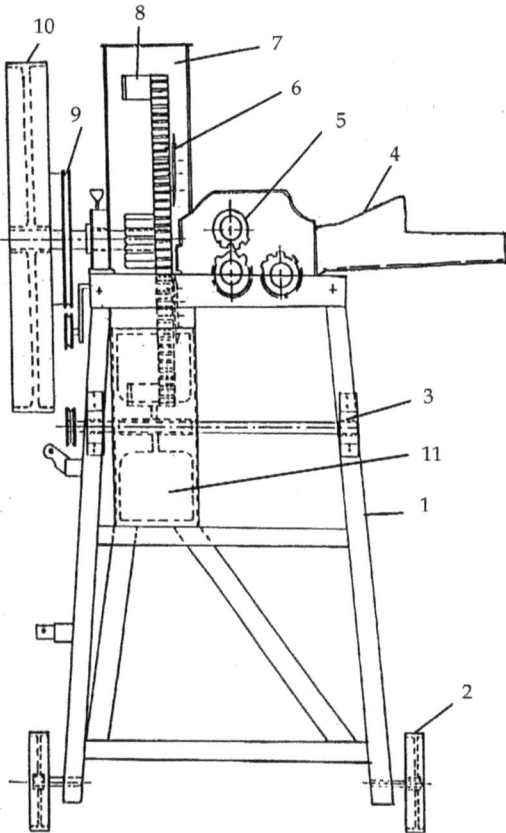

1. *Main Frame, 2. Transportation wheel; 3. Bearing; 4. Feeding chute, 5. Feed regulating rollers, 6. Cutting blade with teeth, 7. Cylinder, 8. Beaters, 9. Pulley, 10. Flywheel, 11. Fan*

Fig. 17.3: A Schematic view of Chaff-cutter type Thresher

17.5 Scope of Test

The power thresher to be tested is normally selected at random from the series production by a representative of the testing institute. The purpose of test is to check and assess the following parameters:

a) Specifications and other data furnished by the manufacturer.

b) Checking of material, visual observation and provision for adjustment

   c) Rate of work

   d) Quality of work

   e) Labour requirement

   f) Adequacy of power of prime mover

   g) Ease of operation and adjustments

   h) Safety provisions

17.5.1 Referred standards and test codes

   a) IS 6320: 1995 [79]   c) IS 9020: 2002 [92]

   b) IS 6284: 1999 [76]   d) IS 9019: 1979 [91]

17.6 Testing of Threshers

17.6.1 General tests

   a) Checking of specification

   b) Checking of material

   c) Visual observations

   d) Safety provision

17.6.2 Test at no load

   a) Power consumption

   b) Visual observation

17.6.3 Test at load

17.6.3.1 Quality of work

i)Losses

   a) Percentage of broken grain

   b) Percentage of blown grain

   c) Percentage of un-threshed grain

   d) Percentage of spilled grain

ii) Efficiency

   a) Threshing efficiency

   b) Cleaning efficiency

17.6.3.2 Power consumption

17.6.3.3 Capacity

   a) Rated input capacity

   b) Output capacity

   c) Corrected output capacity

17.6.4 Long duration test

   a) Effect on the performance at recommended speed over a long period of use

   b) Convenience of handling and transportation

   c) General durability

17.6.5 Safety provisions

17.7 Test Procedure

17.7.1 Pre-test observations

   a) Determination of grain-straw ratio

   Five samples, each weighing one kilogram of crop, will be taken and separated for grains and straw. Weigh the grains and straw separately for each sample and calculate their ratio. The average of five samples shall be taken as grain straw ratio.

   b) Moisture content of grain and straw

   Take suitable sample of grain and straw for moisture content analysis.

17.7.2 Running-in and preliminary adjustments

The power thresher shall be run-in for at least one hour before test. The adjustments for the speeds of different shafts, concave clearance, speed of prime mover, screen slope etc. shall be done as per manufacturer's recommendations.

17.7.3 Crop description

The following characteristics of the crop used for thresher testing shall be reported.

   a) Name and variety of the crop

   b) Moisture content of the grains

   c) Moisture content of the straw

   d) Length of plants (average of 10 plants)

   e) Grain-straw ratio

17.7.4 Verification of specifications

The main objective of this test is to measure preciously the specifications of the machine/critical component. The specifications not confirming to the values declared by the manufacture as given in the relevant Indian Standard(s) need to be highlighted in the test report.

17.7.5 Performance Requirements

Grain moisture variation and other crop parameter and filed condition plays a significant role in determining the performance of thresher. The variation in the crop characteristics may call for precise adjustments. This is specially required in multicrop thresher. The main performance criteria for wheat threshing each shown in Table 17.3.

Table 17.3 Acceptable performance norms for wheat threshers

S. No.	Item		Performance
1.	Threshing efficiency	>	99%
2.	Cleaning efficiency	>	96%
3.	Grain breakage	<	2%
4.	Total Grain Loss	<	5%
5.	Split straw	>	95%
6.	Average length of Bhusa	-	12-15mm

17.7.5.1 Thresher output

The output of thresher depends on the condition of crop mainly in terms of crop moisture content and the grain-straw ratio. Generally, a smaller thresher should give a better output due to the evenness of feeding. Also, an electric motor operated thresher should give better results as there are less fluctuations in the speed of operation and better overload carrying capacity.

17.7.5.2 Operation and collection of data at recommended speed:

The thresher shall be installed on level ground preferably on a hard surface and set the clearance, screen slope, etc. in accordance with manufacturer's recommendations. Attach the thresher with a suitable prime mover preferably an electric motor fitted with an energy meter.

17.7.6 Power consumption

Run the thresher at no-load for at least 10 minutes at the specified recommended speed of threshing unit and record the readings of the energy meter. Calculate the power consumption at no load for one hour. Similarly, calculate power consumption on load (minimum three readings) at maximum input capacity in order to find out energy/power requirements.

In case of dynamometer fitted prime mover, the average of readings taken shall give the average torque required from which power requirement is computed.

$$\text{Power (kW)} = \frac{\text{torque (kgf.m) ṅ speed (rpm)}}{973.363}$$

17.7.7 Collection and analysis of sample

During the test period at the maximum input capacity and at recommended speed, the following samples shall be collected:

Three set of samples at an interval of 20 minutes at the following outlets shall be taken. The actual time for collection of samples shall be recorded.

a) Main grain outlet   60 seconds
b) Sieve over flow   60 seconds
c) Bhusa outlet   10 seconds

Analysis of sample:

Main grain outlet: Analyze for clean, un-threshed, broken grains and foreign materials on the basis of 100 g sample.

Bhusa outlet: Analyze for unthreshed and clean grains

Sieve overflow: Analyze for unthreshed and clean grains

Sieve under flow: Take (1/60)th of the quantity of the material dropped through the sieve and analyze for clean, un-threshed grains.

17.7.8 Performance at varying feed rates

Effect of feed rate on the performance is observed by feeding the machine at different feed rates. Minimum three feed rates should be selected suitably based upon the capacity, tested and analysed for total losses, threshing efficiency and cleaning efficiency for different crops as recommended.

17.7.9 Long duration test

The thresher is run continuously for a total atleast 20 hours. The duration of each run should be about 3 - 5 hours. The thresher is run using an electric motor of recommended horsepower.

Samples from the bhusa outlet, grain outlet and tailings outlet, sieve overflow and sieve underflow are taken after every hour to check the performance. Energy consumption of the motor is also recorded. The convenience in making field adjustments, service, maintenance, change of parts and transportation, etc. are carefully observed and reported.

Any breakdown and major adjustments required during the test (both laboratory and field) are reported. After the test, major components are opened to check for any abnormal wear and tear, which must be shown in the report. The labour requirement in operation of thresher shall also be assessed and reported.

Datasheets for thresher testing and summary sheet for test results of thresher have been shown in Annexure 17.1 & 17.2 respectively.

17.8 Computation of Results

17.8.1 Performance parameters

From the analysis of samples, the performance parameters are calculated as follows:

a) Total grain input per unit time (A)

$$A = B + C + D$$

Where,

A - total grain input per unit time by weight

B - weight of threshed grain (whole and clean grain) per unit time collected from all grain outlet

C - *weight of broken grain per unit time collected from all outlets*

D - *Weight of un-threshed grain from all outlets per unit time*

b) Percentage of broken grain

$$\text{Percentage of broken grain (\%)} = \frac{E}{A} \times 100$$

Where,

E - quantity of broken grain from all outlets per unit time

A - total grain input per unit time

c) Percentage of blown grain

$$\text{Percentage of blown grain (\%)} = \frac{G}{A} \times 100$$

Where,

G - quantity of clean grain collected at bhusa outlet per unit time

A - total grain input per unit time

d) Percentage of spilled grain

$$\text{Percentage of spilled grain (\%)} = \frac{K}{A} \times 100$$

Where,

K - quantity of clean grain collected at sieve overflow and underflow per unit time

A - total grain input per unit time

e) Percentage of unthreshed grain

$$\text{Percentage of unthreshed grain (\%)} = \frac{H}{A} \times 100$$

Where,

H - quantity of unthreshed grain per unit time obtained from all outlets

A - total grain input per unit time

17.8.2 Determination of efficiencies

a) Threshing efficiency (%) = 100 - percentage of unthreshed grain

b) Cleaning efficiency (%) = $\frac{M}{F} \times 100\%$

Where,

M - quantity of clean grain per unit time taken from sample taken at the main grain outlet

F - total quantity of the sample per unit time at main grain outlet

c) Determination of corrected output capacity

To avoid the variation of moisture content of grain, the output capacity obtained may be corrected at the standard moisture content of grain. Corrected output capacity can be specified and calculated by following formula:

$$W_c = W \dot{n} r \left[\frac{\ddot{} -M_1}{-m_1} + \frac{(100-R)(100-M_2)}{R(\ddot{} -m_2)} \right]$$

Where,

W_c	- corrected output capacity
W	- output capacity observed
M_1	- moisture content of grain
M_2	- moisture content of straw
R	- observed grain-crop ratio in percentage

Table 17.4 Standard moisture content of grain, straw and grain crop ratio in percent

Crops	m_1	m_2	r
Wheat	9	7	40
Paddy	20	22	40
Bengal gram	8	7	50
Sorghum (earhead)	8	9	75
Soybean	9	9	40

m_1 - standard moisture content of the grain

m_2 - standard moisture content of straw

r - standard grain-crop ratio in percentage

17.9 Checking for Safety

Feeding chute provided shall be checked from safety point of view. Check the dimensions of chute as per details shown in Fig. 17.4 and compare with the relevant BIS standards (Table 17.5). However, the recommended range of various dimensions for different type of threshers is given below:

A	=	440 to 590 mm
B	=	Minimum 900 mm
C	=	350 to 530 mm
D	=	Minimum 450 mm
E	=	50 to 60 mm
F	=	125 to 210 mm
α	=	10° to 15°
β	=	10° to 30°

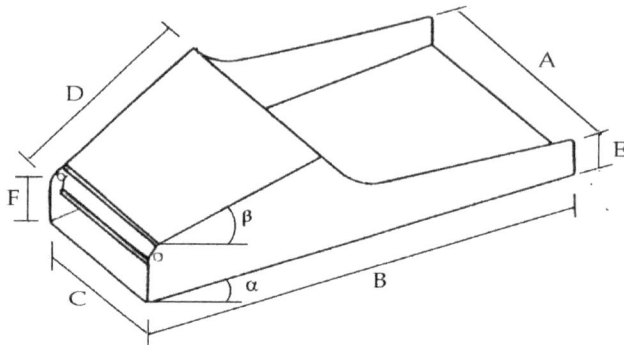

Fig. 17.4: Safe feeding chute

Table 17.5 Recommended dimensions of safety chute for spike-tooth cylinder threshers

S. No.	Size of prime mover for thresher (kW)	Dimension, mm			
		A	C	E	F
1.	3.7	440	350	60	190
2.	5.5	480	400	60	190
3.	7.5	540	480	60	190
4.	11 and above	590	530	60	210

Annexure – 17.1

Data Sheet for Thresher Testing

Name of Thresher:

Date of Test:

Place of Test:

Observations

Recorded by Supervisor:

Name of the Crop:

Variety:

Straw grain ratio:

Moisture content (%):

a) Grain
b) Straw

Speed of drive Pulley

at no load (rpm):

Speed of driven

Pulley at no load (rpm):

Cylinder concave clearance (mm):

Sieve Clearance (mm):

Air output (m³/sec.):

S. No.	Time			Net operation time (h)	Total quantity of crop fed during operation (kg)	Threshing Cylinder Speed, rpm	Power Consumption		
	Start (h)	Finish (h)	Stoppage, if any				Energy meter, reading at start of test (kWh)	Energy meter, reading at end of test (kWh)	Power consumption for test duration (kWh)
i)									
ii)									
iii)									
iv)									

S. No.	Samples of various outlet(s)						Total qty. of grain mixture collected at main grain outlet (kg)	Total qty. of grain mixture collected at sieve under flow (kg)
	Main grain outlet(s)		Sieve overflow		Straw busha outlet(s)			
	Qty. (kg)	Time (sec)	Qty. (kg)	Time (sec)	Qty. (kg)	Time (sec)		
i)								
ii)								
iii)								
iv)								

Annexure – 17.2
Summary Sheet for Test Results of Thresher

S. No.	Item	Test no. 1	2	3	4	5	6
1.	Cylinder Speed (rpm)						
	No Load						
	On Load						
2.	Feed Rate (kg/h)						
3.	Grain Straw Ratio						
4.	Moisture content (%)						
	i) Grain						
	ii) Straw						
5.	Power Consumption (kWh)						
	i) At no load						
	ii) At load						
6.	Broken grain (%)						
7.	Unthreshed grains (%)						
8.	Blown grain(%)						
9.	Spilled grain (%)						
10.	Threshing efficiency (%)						
11.	Cleaning efficiency (%)						
12.	Optimum input capacity (kg/h)						
13.	Total grain output capacity (kg/h)						
14.	Corrected output capacity (kg/h)						

Chapter 18

Advances in Instrumentation and Sensors

There have been significant developments in the field of instrumentation. It is utmost important that testing institution/departments must be well equipped with appropriate measuring instruments/equipments for testing of agricultural machinery. Recently, use of digital instrumentation and computers have revolutionized for accuracy and repeatability of data obtained in testing.

An instrument is a device that transforms a physical variable into a form that is suitable for recording the measurement. An example of a basic instrument is a ruler with which length of object can be meaured. Fig. 18.1 presents a generalized model of a simple instrument. The physical process to be measured is in the left of the figure and the measurand is represented by an observable physical variable X. Note that the observable variable X need not necessarily be the measurand but simply related to the measurand in some known way. For example, the mass of an object is often measured by the process of weighing, where the measurand is the mass but the physical measurement variable is the downward force the mass exerts in the Earth's gravitational field. There are many possible physical measurement variables. A few are shown in Table 18.1.

If the signal output from the sensor is small, it is sometimes necessary to amplify the output. The amplified output can then be transmitted to the display device or recorded, depending on the particular measurement application. In many cases, it is necessary for the instrument to provide a digital signal output so that it can interface with a computer-based data acquisition or communication system. If the sensor does not inherently provide a digital output, then the analog output of the sensor is converted by an analog to digital converter (A/DC).

The digital signal is typically sent to a computer processor that can display, store, or transmit the data as output to some other system, which will use the measurement. Sensors are often transducers that convert input energy of one form into output energy of another form. Sensors can be categorized into two broad classes depending on how they interact with the environment they are measuring.

Passive sensors do not add energy as part of the measurement process but may remove energy in their operation. Active sensors add energy to the measurement environment as part of the measurement process.

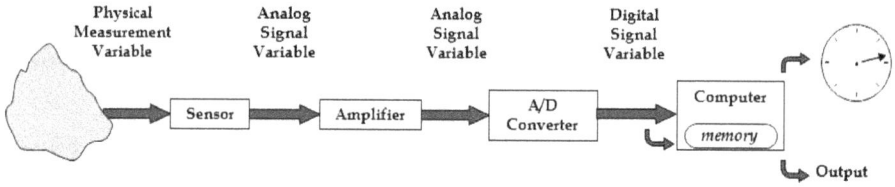

Fig. 18.1: Generalized model of instrument

18.1 Accuracy, Precision, Resolution & Sensitivity

Instrument manufacturers usually supply specifications for their equipment that define its accuracy, precision, resolution and sensitivity. Unfortunately, not all of these specifications are uniform from one to another or expressed in the same terms. Some specifications are given as worst-case values, while others take into consideration the actual measurements.

The ISO 5725-1 uses the terms accuracy (trueness) & precision to describe this same concept, with accuracy being the ability of a measurement system to give a correct result, and precision being the ability of the measurement system to replicate a given result.

Table 18.1 Common physical variables

Common physical variables	Typical signal variables
• Force	• Voltage
• Length	• Displacement
• Temperature	• Current
• Acceleration	• Force
• Velocity	• Pressure
• Pressure	• Light
• Frequency	• Frequency
• Capacity	• ....
• Resistance	
• Time	
•	

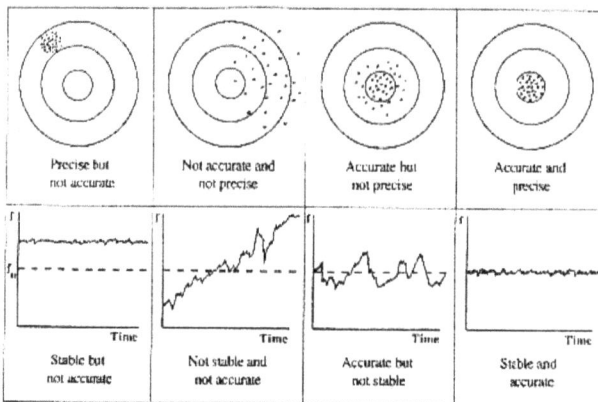

Fig. 18.2: Schematic representation of accuracy and precision

Accuracy can be defined as the amount of uncertainty in a measurement with respect to an absolute standard. Accuracy specifications usually contain the effect of errors due to gain and offset parameters. Offset errors can be given as a unit of measurement such as volts or ohms and are independent of the magnitude of the input signal being measured.

Precision is often described as the ability of a measurement method to replicate a given result. Repeatability and reproducibility are subsets of precision and both help categorize the sources of variability in a measurement system. For example, measure a steady state signal many times. In this case if the values are close together then it has a high degree of precision or repeatability. The values do not have to be the true values just grouped together. Take the average of the measurements and the difference is between it and the true value is accuracy.

Resolution can be expressed in two ways: a) It is the ratio between the maximum signal measured to the smallest part that can be resolved - usually with an analog-to-digital (A/D) converter, and b) It is the degree to which a change can be theoretically detected, usually expressed as a number of bits. This relates the number of bits of resolution to the actual voltage measurements. A technique called averaging can improve the resolution, but it sacrifices speed. Averaging reduces the noise by the square root of the number of samples, however, this technique cannot reduce the effects of non-linearity, and the noise must have a Gaussian distribution.

Sensitivity is an absolute quantity, the smallest absolute amount of change that can be detected by a measurement.

18.2 Measurement Parameters

18.2.1 Revolutions

The measurement of the number of revolutions is simply a counting procedure. The number of revolutions of land engaging wheels on tractors and machines on test rigs and in the field and pedals and handles of machines such as small threshers, grinders and water pumps, can be counted by eye. For shafts turning at higher speeds, some form of counting apparatus is required. Where the end of the shaft is accessible, a mechanical, electrical or electronic counter (tachometer) may be used driven directly from the end of the shaft. However, if the shaft is connected at each end, an electronic counter (non-contact tachometer) can be used which incorporates a light emitter and a sensor working from a reflective strip attached to the rotating part. Some shaft torque measuring devices have integral revolution counters to enable direct power readings to be obtained.

Testo 470 (make Testo India) is an example of optical and mechanical tachometer in one instrument. The specifications are:

	RPM – Optical (non-contact)	RPM – Mechanical (contact)
Measuring range	1 to 99999 rpm	0.1 to 19999 rpm
Accuracy	±0.02 % of mv	±0.2 % of mv
Resolution	0.01 rpm (1 - 99.99 rpm)	
	0.1 rpm (100 - 999.9 rpm)	
	1 rpm (1000 - 99999 rpm)	

18.2.2 Temperature

The measurement of temperature may also require an instrument to cover a wide range from ambient air values to those of the exhaust gasses from internal combustion engines. The SI unit of temperature is the Kelvin (K), degree Celsius (°C) is also recognized for use in conjunction with the SI and has the same datum and value as the Kelvin. Measuring equipment is usually graduated in degrees Celsius. Temperature can be measured via a diverse array of sensors and all of them infer temperature by sensing some change in a physical characteristic. Different types of sensors are: liquid-in-glass, thermocouples, resistive temperature devices (RTDs and thermistors), infrared radiators, bimetallic devices, liquid expansion devices, and change-of-state devices. Mercury-in-glass thermometers may be adequate for some ambient and fuel measurement, however, the thermocouple type of measuring device has a large temperature range and is commercially available and widely used. The electrical device can be built into a hand-held unit or coupled to additional monitoring or recording equipment. Infrared sensors are non-contacting devices and sense temperature by measuring the thermal radiation emitted by a material. *OS530E infrared thermometer* (make OMEGA, India) is an example of non-contact temperature. Specifications are: a) temperature range (-23 to 871°C), b) emissivity adjustable from 0.1 to 1.00 in 0.01 steps, c) 1 mV/degree analog output, d) repeatability: ±(1% reading + 1 digit), e) resolution: 1°C or 0.1°C, f) response Time: 100 ms, and g) data storage and computer connectivity.

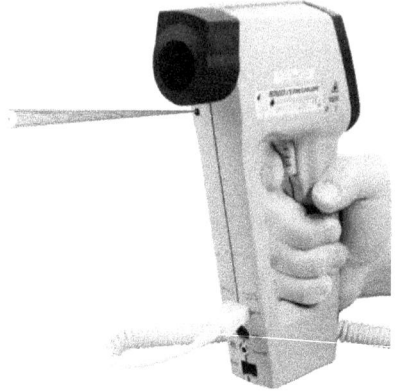

18.2.3 Electrical

During tests of machines incorporating electrical equipment, it is often necessary to measure the units of electric potential (volt), current (ampere) and resistance (ohm). Standard commercial instruments are available with meters adequately covering ranges and units suitable for direct current (DC) and alternating current (AC) supply. It is recommended that measurements on electrical equipment should only be made by experienced personnel.

18.2.4 Force

The newton (N), the SI unit of force, which is defined as the force which, when applied to a body of one kilogram mass, gives it an acceleration of 1 m/s^2). The kilogram force (kgf) is the force that when applied to a body of one kilogram mass gives it the standard acceleration due to gravity.

Spring balances may be used for measurement of force on hand tools and light implements drawn manually or by animals. However, under field conditions because of lack of damping, they may be difficult to read with accuracy.

A load cell is a transducer which converts force into a measurable electrical output. Although there are many varieties of load cells, strain gage based load cells are the most commonly used type. Strain-gage load cells convert the load acting on them into electrical signals. The gauges themselves are bonded onto a beam or structural member that deforms when weight is applied. In most cases, four strain gages are used to obtain maximum sensitivity and temperature compensation. Two of the gauges are usually in tension, and two in compression, and are wired with compensation

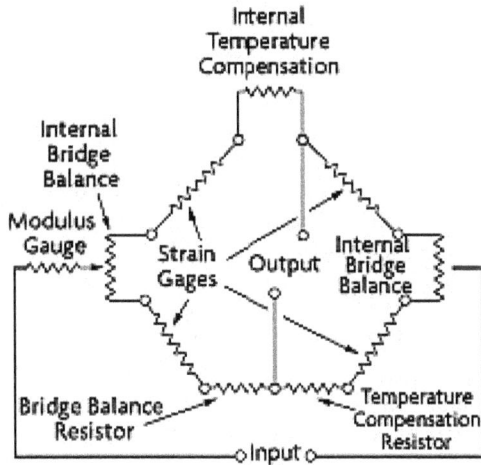

Fig. 18.3: Wheatstone Circuit with Compensation

adjustments as shown in Fig. 18.3. When weight is applied, the strain changes the electrical resistance of the gauges in proportion to the load. Other load cells are fading into obscurity, as strain gage load cells continue to increase their accuracy and lower their unit costs.

18.2.5 Pressure

Pressure measurements are required to be made in tests of power units, tractors, seed drills, planters, sprayers and water pumps. These cover a wide range and can be positive or negative. Lower levels such as depressions in engine inlet manifolds and in water pump suction pipes can be measured using simple water or mercury filled manometers. In this case, the pressure in the system is calculated using the difference in height of the liquid column. Pressure and vacuum gauges are manufactured in various ranges and graduated and scaled in units such as bars, kilogram per square centimeter (kg/cm^2), Pascals (Pa). Pascal is the preferred unit by the SI. Electronic systems using pressure sensors may be obtained together with their associated display and data acquisition equipment.

18.2.6 Speed

The SI unit of speed of rotation is radians per second (rad/s), this unit is used for calculation of power from machines such as engines and tractors. Test measurements are made in revolutions (2π radians) for a given time, per minute (rev/min) or per second (rev/s). Mechanical revolution counters and totalisers will require time to be measured by stopwatch. Electrical and electronic units automatically count the number of revolutions over a period of time and the results are displayed and continually updated.

The unit of linear speed is the meter per second (m/s). However, the unit normally used when presenting the travelling speed of tractors and machines is the kilometer per hour (km/h) which is derived by calculation.

A method of measuring travel speed during field trials of machines is to set up marker poles across the width of the work at say 20m apart to form a rectangle. An observer will then be able to sight across the poles and measure the time taken for the machine to travel the known distance. When drawbar tests and slip measurements are being made in the field, the distance for a tractor or machine to travel for a number of wheel revolutions is measured. If the travel time is also recorded, the travel speed may be calculated. [Refer 7.4.2 (b)]. The most recent developments in measurement of linear speed/forward speed are RADAR and GPS sensor [190].

a) RADAR Sensor

These advanced ground speed sensor delivers the truest velocity measurement, which uses the Doppler principle for measuring the forward speed. It consists of a robust, externally-mounted transmitter-receiver assembly, which transmits on a radio frequency of 24.3 GHz. The frequency difference between the transmitted and received signal is directly proportional to the tractor forward speed (V).

Frequency difference (F_d) is calculated by using the following formulae:

$$F_d = \frac{2V}{\lambda}\cos\theta$$

Where,

λ *- wavelength of transmitted signal*

θ *- angle between ground and sensor* (37°)

The accuracy of measurement is affected by pitch, radar sensor should be mounted as near as possible to tractor instantaneous center of rotation.

Some of the typical specifications of a radar:

- Velocity range: 0.53-96.6 km/h
- Mounts at 35±5° angle and at least 610 mm height from target surface
- Achieves velocity errors of ±1-3% through in-field calibration

b) GPS sensor

GPS sensor uses GPS satellites to convert the signal to velocity/speed. The GPS sensor is a speed sensor targeted to farmers needing a distance sensor and a true ground speed for the tractor. It works with many of the cabin monitors that uses radar. The mushroom-like receiver

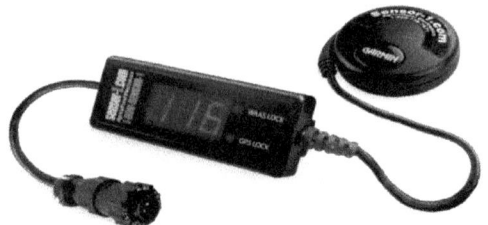

measures only 2.4 inches in diameter, weighs just under two ounces and is sealed to the environment. Quick installation by simply connecting the GPS sensor to the monitor/control equipment, perform the set-up calibration.

18.2.7 Torque

Torque is defined as force x moment of application, the SI unit is the newton metre (Nm). Dynamometers for engine and tractor testing are basically devices which apply a measured and variable torque to a rotating shaft (engine crankshaft, pto shaft, etc). Friction loads applied mechanically, hydraulically or electrically within the dynamometer act upon an arm of known length attached to a load measuring device.

Selected torque values may be to the drive shaft by variation of the internal loading. In most proprietary machines the length of the arm is designed to give constants of whole numbers in the power calculations. Transmission dynamometers that can be fitted into an engine/machine or tractor power take-off/machine drive line (Fig 2.16) are available to cover a large range of torque and speed requirements. These units are portable and comprise an internal strain gauged shaft or tube with suitable monitoring and read-out equipment, they can be mains or battery operated.

18.2.8 Work and Power

The SI unit of energy and work is the joule (J) which is defined as the work done when a force of one Newton acts through a distance of one metre in the direction of the force. The joule (J) is equivalent to one Newton metre (Nm). The unit of power is the watt (W) which is equal to one joule per second (J/s) or one Newton metre per second (Nm/s). The unit most commonly used is the kW, which is equal to one kilo Newton metre per second (kNm/s).

For measurement of linear power, for a tractor drawbar test for example, the following expression is used:

$$\text{Power (kW)} = \frac{\text{torque (kgf.m) } \dot{n} \text{ speed (rpm)}}{973.363}$$

For rotating mechanisms such as engines,

$$\text{Power (W)} = \text{torque (Nm) } \dot{n} \text{ speed of rotation (rad/s)}$$

If the speed of rotation is measured in revolutions per minute (R), then:

$$\text{Power (kW)} = \text{torque}\left(\frac{\text{Nm}}{1000}\right)\dot{n}\,\text{speed of rotatio}\left(\frac{2\pi R}{60}\right)$$

For measurement of power of tractor hydraulic systems, the following formula is used:

$$\text{Power (kW)} = \text{flow}\left(\frac{\overset{..}{}^{3}}{s}\right)\times\text{pressure}\left(\frac{}{m^{2}}\right)\times\frac{}{1000}$$

However, the following formula is in common use:

$$\text{Power (kW)} = \text{flow}\left(\frac{1}{\text{min}}\right) \times \text{Pressure (bar)} \, \acute{\text{n}} \, \frac{1}{600}$$

For water pumps, the equation is again modified to include the suction & pressure head and density of the liquid:

$$\text{Power (kW)} = \text{flow}\left(\frac{\text{m}^3}{\text{s}}\right) \times \text{head (m)} \, \acute{\text{n}} \, \text{density of liquikd}\left(\frac{\text{kg}}{\text{m}^3}\right) \times \frac{1}{102}$$

For machines driven by electric motors, power requirement may be determined by fitting a torque measuring device in the drive and measuring speed of rotation.

If that is not possible, a good approximation can be made by measuring the input voltage (V) and current (A) being consumed by the motor and calculating the average power in watts (W). For this, the relationship power (W) = voltage (V) x current (A) is used. However, it is possible to obtain various types of watt meters which will give direct power readings when connected between the supply and the motor.

It should be understood that measuring the power in the drive line will give the exact requirement of the machine. The watt meter measures the electrical power input to the motor and the efficiency of the motor to transmit that power to the machine should be taken into account. The motor efficiency valves will depend on the level of loading and will range from approximately 70% to 90%.

18.2.9 Noise measurement

There has been a growing concern for hazard of loud noise which affects the hearing of human. So in this direction various type of sound level meters have been designed and are being used with a suitable external microphone for compliance with the relevant standard. The test shall be conducted on a standard test track in all gears corresponding to nominal traveling speed nearest to 7.5 km/h as well as near the operator's ear level. As per BIS code the sound level meter is placed at a distance of 7.5 m from the center line of the tractor and at height of 1.2 m. The sound is measured in db (A) and data so obtained is compared with the relevant standard and reported.

18.2.10 Vibration measurement

The mechanical vibration measurement can be measured with the help of vibration meter. The test shall be conducted on the standard test track at no load and at load corresponding to 85 % of maximum power at various critical points of the machine e.g. steering, gear shift lever, clutch padel, brake padel, sheet, accelerator lever, mudguard, foot rest, head light etc. The vibration shall be measured in microons at horizontal and vertical level in order to specify the horizontal displacement and vertical displacement of vibration. Now a days digital vibration meters are also available for accuracy and repeatability of results.

18.2.11 Measuring wheel slip

Wheel sleep cannot be measured directly as it is the result of comparing two variables: the actual forward travel speed of the vehicle, and the tyre or track surface speed. Wheel slip is expressed as a percentage (i.e. how much travel distance has been reduced due to slippage); it can be calculated using the equation below:

$$\text{Slip (\%)} = \left(\frac{\text{tyre or track surface speed ș true vehicle forward speed}}{\text{tyre or track surface speed}} \right) \times 100$$

Option 1: Use a performance monitor

The preferred method of determining wheel slip is to track it continuously via the tractor's performance monitor. These systems typically require radar to provide accurate readings of slip and field performance. The radar is used in conjunction with the speed input from the tractor's engine/transmission to determine the wheel slip and performance data displayed to the operator [11, 262].

GPS positioning can also be used to reference the tractor's true ground speed. However, GPS systems may require clear weather to work and can be inaccurate at low speeds [12].

Option 2: Measure tyre rotations over a set distance

Alternatively, wheel slip can be calculated from tyre rotations. To do this, one must first measure the number of tyre rotations that occur when the tractor travels over a set distance, at working speed and under no load. Then repeat the measurement while the tractor is under load. The following equation can then be used to calculate slip:

$$\text{Slip (\%)} = \left(\frac{\text{no. of rotations}_{\text{with load}} \text{ ș no. of rotations}_{\text{no load}}}{\text{no. of rotations}_{\text{with load}}} \right) \times 100$$

An implementation of this way of measuring wheel slip is known as the as the '10-turn method' [13, 14].

Be aware that this method of measurement will provide you with the wheel slip only for the specific working conditions tested and therefore, the measurement will have to be repeated if any of the following parameters changed:

- tractor weight or ballasts
- type of implement or working depth
- working speed
- soil conditions (soil type, moisture, hardness).
- tyre pressures

Option 3: Installation of a 'bolt-on' system

Bolt-on' wheel-slip monitors could be of use if you are using very old tractors, which may not have the appropriate components or controls to enable the use of a typical performance monitor.

18.2.12 Fuel Consumption

When considering the fuel consumption of engines, the ability of an engine to convert fuel into useful work will vary with engine type and design and with speed and loading. The measured fuel consumption should be related to the power output and expressed as "specific fuel consumption" in l/kWh. Measurement of specific fuel consumption can be used to highlight areas of higher fuel efficiency in terms of power and speed. Generally, for tractor operation, it is more economical to work in the highest gear ratio possible, adjusting the throttle to maintain the required load and forward speed.

18.2.13 Rate of Flow

Volume of flow is measured to establish the fuel consumption of petrol and diesel engines, and the output of sprayers and pumps. The basic function for flow rate, volume per unit of time is used in each case. Fuel consumption of engines is measured in millilitres or litres per second (ml/s or lps). Proprietary meters giving direct read-out of flow rate are used for tests on hydraulic systems. In cases of tractors with "closed-circuit" systems, the meter must be of a type to withstand the pressure of the boost pump supply into which the return flow is connected. Before fitting test equipment to this type of tractor hydraulic circuit, information on testing methods should be sought from the tractor manufacturer.

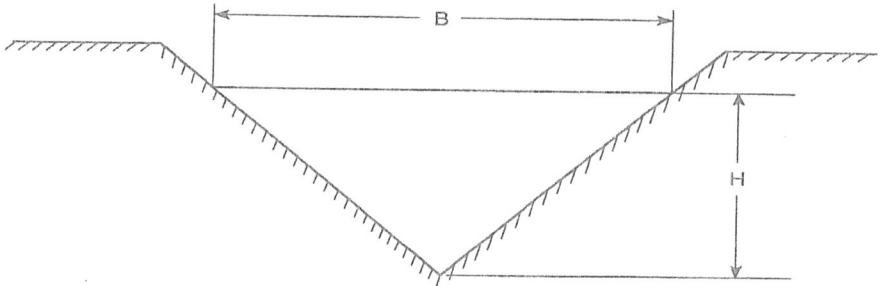

Fig. 18.4 V-notch for rate of flow measurement

Suitable flow meters may also be fitted into water pump outlets in preference to measuring the time to fill tanks of known volume. A further technique for measuring larger amounts of liquid flow is to use a 'V' notch fitted into the flow channel (Fig. 18.4). This consists of a plate with a 90° 'V' notch across the line of flow through which the liquid flows. The expression for a true 90° notch is:

$$\text{Rate of flow, } Q = k \, \dot{n} \, \sqrt{2g} \times \frac{8}{15} \times H^{2.5}$$

Where,

Q	- rate of flow (m³ /s)
G	- acceleration due to gravity (9.8067 m/s²)
H	- height of liquid level (m)
K	- plate coefficient

For each notch, the coefficient k must be determined by experiment to give the correct relationship between rate of flow and liquid height. This information may be supplied by a plate manufacturer.

18.3 Specific Instruments/Test Rigs for Agricultural Machinery

Commercial manufacturers of agricultural machinery and agri-electronics industry developed specific test-rigs or instruments for accurate calibration or testing of the agricultural implements or critical sub-systems. This section describes some of the recent developments.

18.3.1 Seed meter test-rigs for Planters

Some of the major seed sowing machinery manufacturers like John Deere, Kinze and Precision planting Inc. developed seed meter test stands for testing the uniformity of seed metering and quick calibration (Fig. 18.5 and 18.6). These systems are fitted with advanced electronic sensors for measuring shaft rpm, vacuum, pressure and seed sensing. Also, they are equipped with real time data analysis and display along with data storage for further analysis. In addition to testing, demonstration of seed metering systems can be done in these test rig for training purposes.

Courtesy: Kinze, Germany

Fig. 18.5: Kinze Seed meter stand

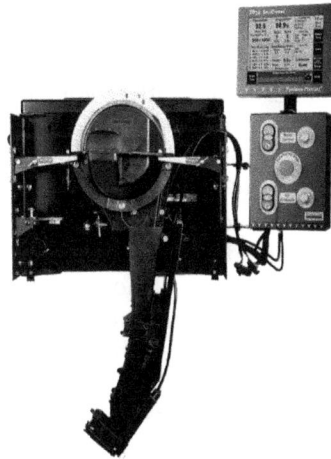

Courtesy: Precision Planting Inc., USA

Fig. 18.6: The MeterMax ultra Test Stand

The test stands are capable of:

- Measuring the seed singulation percentage
- Measuring the misses, doubles
- Seed spacing (average and coefficient of variation)
- Measuring pressure, vacuum, rotation speed of seed disc etc.
- View histograms of results

18.3.2 Sprayer and Nozzle Testing

Measurement parameters for sprayer and nozzle testing are as follows:

- Droplet Size
- Droplet Velocity
- Volume Flux
- Volume Distribution

- Spray Angle
- Droplet Shape
- Spray distribution

Three commonly used techniques to measure the spray parameters are explained below:

a) Phase Doppler Interferometer (PDI)

This instrument measure the drop size, velocity, and volume flux at a point within a spray plume by characterizing each droplet that passes through the probe volume and building up accurate ensemble statistics. The probe volume is formed by two (or four) intersecting laser beams, which provide a non-intrusive, high spatial-resolution measurement point. Spray pattern characterization can be performed by moving either the instrument or the nozzle itself; collecting data at numerous specified locations. Further analysis allows combination of the information from the many individual data points to provide volume flux and area weighted average spray pattern (Fig. 18.7).

Principle of operation: The laser transmitter unit emits two coherent laser beams of the same wavelength (color). At the intersection of these lasers, an interference pattern is formed by the constructive and destructive interference of the two lasers, and a known interference wave frequency is generated. A droplet, acting like a prism, refracts this interference pattern and it is detected by the receiver-unit and analyzed to determine the droplet characteristics. The frequency shift of the pre-determined interference pattern is proportional to the droplet's velocity. This method is commonly known as Laser Doppler Velocimetry (LDV). Also, there are three detectors within the receiver unit that detect the interference burst pattern and the relative phase shift of this signal at each of the detectors is proportional to the droplet diameter. The PDI count and analyze each interference pattern as it is detected, and thus characterize each individual droplet that enters the probe volume formed by the intersecting laser beams.

b) Laser Imaging

The Laser Imaging method provides instantaneous images of a spray that allow for post-processing of droplet and velocity characteristics from the images. The images are spatially calibrated and thus have an associated length scale reference with each image frame in order to measure the droplet/particle size (Fig. 18.8).

Principle of operation: Laser imaging uses a diffused laser light source, and an in-line camera to image a small measurement volume (typically one square centimeter). The laser is pulsed to illuminate the droplets for only a fraction of a second, resulting in an instantaneous spray image. The backlit imaging method allows for clear identification of droplets or particles as they block some percentage of the laser light from reaching the camera. These images are then analyzed to recover droplet size and velocity information (velocity is determined using two images with a known time delay between the pair). Image analysis is used to determine sphericity as well as a number of other characteristics of the investigated spray.

Courtesy: LaVision

Fig. 18.7: Phase Doppler interferometer

Courtesy: Oxford lasers

Fig. 18.8: Laser Imaging

c) Automatic Spray Patternator

For measuring and plotting the liquid distribution under a spray boom or an individual nozzle, patternators are used. The liquid distribution reflects the quality of the sprayer and its distribution under field conditions. The distribution is measured automatically, independently and with high precision. The advanced automatic spray scanner from AAMS-Salvarani makes the complete measurement process automatic and it works autonomously under the spray boom until the entire spray boom is measured. Automatic spray scanner is shown in Fig. 18.9.

Some of the specifications are:

- The spray scanner has a measuring surface of 80 X 150 cm.
- The measuring plane is horizontal (no slope in the upper surface).
- Data is stored in a memory box that can be transferred to a PC. With the software, the distribution can be reported solely or a complete inspection report can be executed and printed.
- Working length of rail – 28m (9 sections of rail of each 3.2 m length).

With the display unit, the scanner can be activated and - if necessary - navigated. A part of spray boom can be measured again without completion of the rest of the spray boom (e.g. when parameters or parts of the spray boom have been changed).

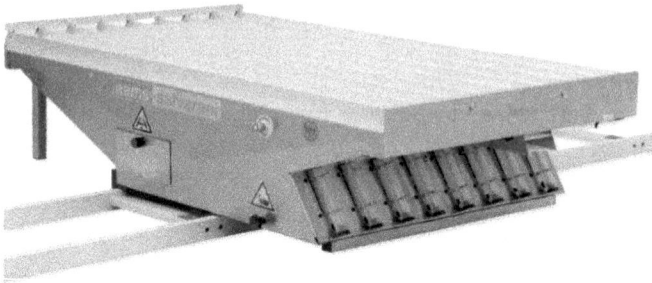

Courtesy: AAMS-Salvarani

Fig. 18.9: Spray Scanner

18.3.3 Engine Fuel Consumption

Measurement of fuel consumption is one of the most important parameter during the testing and evaluation of tractors, power tillers, and other agricultural equipment. Gravimetric method having separate fuel tank or topping up the tank before and after test is used for measuring the fuel consumption during the test. However, with the entry of digital fuel flow meters, the engine fuel consumption can be measured using flow meters with a choice of simple mechanical totalisers, digital flow Indication (local or remote) and data logging with remote telemetry via SMS, GSM and Internet Browser.

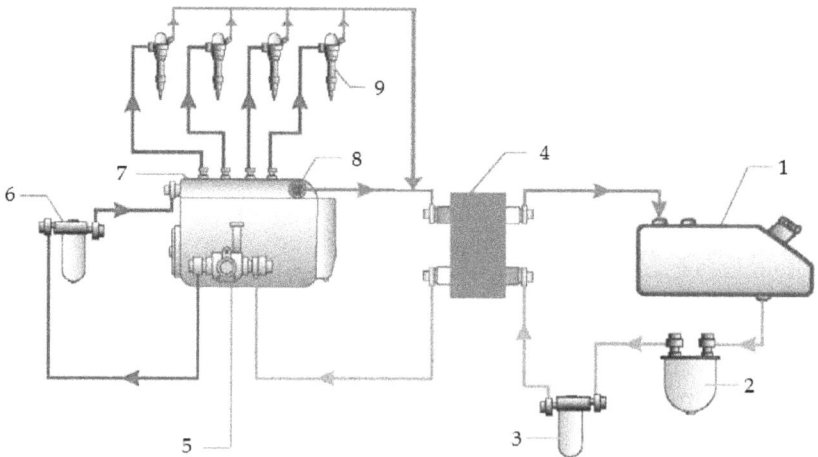

1. Fuel tank, 2. Pre filter, 3. Extra fuel filter, 4. Differential fuel meter, 5. Fuel supply pump, 6. Fine fuel filter, 7. Fuel pump, 8. Relief valve, 9. Injectors

Fig. 18.10: Installation of differential flow meter in fuel supply line with feedback to tank

Most engines have supply and return flow and to obtain the fuel consumption data, it is necessary to measure both fuel flows and calculate the difference.

Different engines will require different solutions based upon their size, age and level of sophistication required as modern engines use the engine fuel to cool the injector nozzles. This means that the unconsumed fuel flows back to the tank and because of the long fuel lines and the time it spends in the fuel tank, the fuel can cool down again. Meter accuracy is critical in this situation as any inaccuracies are compounded by the differential low flows in question and the combined meter calculation. Differential fuel flow meters are the best solution in such situations and a typical layout of installation of the meter is given in Fig. 18.10.

Another method is to use special meters for differential measurements as standard meters have a large measuring range and a maximum permissible error of ±1%. For differential measurements, the piping remains unchanged, with circulation back into the tank. A flowmeter is installed in both supply and return pipes. The consumption is determined as the difference between the amount in the supply section and the amount in the return section.

In the engine variants without feedback to the tank, the fuel also needs to be cooled (via fuel cooler), however only one meter is required here, making this design of system cheaper to monitor and more accurate.

Fuel flow meters are used to monitor the supply/consumption of liquid fuels like, diesel, biodiesel, petrol, and various other bio-fuels, fossil fuels and alcohol based fuels. These fuel consumption flowmeters can be used with third party instrumentation. A choice of pulse and analogue signal outputs are available from the flow meters. Some common fuel flow meters are given in Table 18.3.

The following may be considered for the selection of the fuel meter:

- Operating temperature
- Viscosity of the medium
- Operating pressure
- Flow rate
- Resistance of the material against fuel to be metered and working conditions

Table 18. 3 Single and differential fuel flow meters

I. Model: Fuel flow meter – VZO 4 (*Aquametro, France*)

- Ring piston meter with pulse value of 0.005 litres/pulse.
- Temperature (max.) : 60°C
- Measuring range : 1 - 80 l/h
- Nominal pressure: PN 32 bar
- Repeatability: ± 0.2%

II. Model: Differential flow meter (*Bellflowsystems, UK*)

- Max. pressure 2.5MPa
- Accuracy ±1%
- Temperature range -40°C to +85°C
- Flow range 10 to 100 l/h (in each chamber)
- Scaled pulse output

18.3.4 Tractor PTO torque and speed measurement

The tractor provides necessary power to the equipment through, a) draft power through three-point hitch or drawbar, b) fluid power through hydraulic outlets, c) rotary power /torque through the power take-off (PTO) shaft, and d) electrical power through multiple electrical outlets. In all of these, power transmission (~90% of net engine power) through PTO is the most efficient [7].

Power and speed requirements of implements are calculated depending on the drivetrains and implement load, and then matching tractors are provided with necessary PTO power or depending upon the maximum PTO power transmitted at rated engine speed for each PTO type [143]. For example, almost all tractor manufacturers offers the 540 rpm, 6 spline, 35 mm shaft as the standard PTO type. A torque meter secured to a cart was used to measure PTO power [256]. Limitations of such a cart were the increase in overall machinery length and a possible safety hazard. However, benefits like ability to connect multiple PTO types using different shafts, and avoiding bending or shear stresses on the sensor shaft. Also, the implement PTO shaft was modified to include a built-in slip ring torque sensor for energy mapping [150, 193].

Commercial slip-ring torque sensors with flanged ends are available, however, for their installation, manufacturing of couplers and shaft with close tolerance is difficult. Ready-to-use PTO torque sensors were available from two suppliers (Datum Electronics, United Kingdom and NCTE AG, Germany). These sensors had PTO couplers and shafts mated directly to the measurement shaft instead of having flanged ends [192].

Datum 420 Tractor PTO Drive Shaft Torque and Power Monitoring System (Fig. 18.11) has the ability to monitor and log the torque, shaft speed and transmitted power

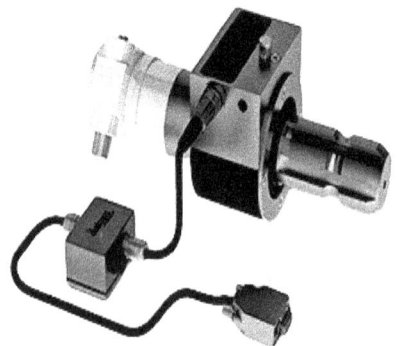

Courtesy: Datum Electronics, UK

Fig. 18.11: Tractor PTO Drive Shaft Torque and Power Monitoring System

from all standard PTO shafts [128]. This has a non-contact transmission system, a complete transducer with bearings to support the stator unit that provides a digital output directly proportional to torque. The system acts as an extension spline adaptor, with the male end replicating the male end of the tractor PTO. The torque and speed signals are transmitted from the shaft to a static cover assembly. Readings of power, torque and speed can be either logged to a handheld indicator (T310TSP) or to a PC or laptop with TorqueLog software.

Agricultural dynamometers for PTO output

Portable PTO power testing is also possible by the use of air-cooled eddy current dynamometer absorbers. This results in lower maintenance costs and fewer reliability issues than those normally asso-

Courtesy: PowerTest, USA

Fig. 18.12: Mobile PTO dynamometer

ciated with agricultural dynamometers that typically rely on a friction brake or hydraulic pump design. These dynamometers comprises Digital Power Meter (DPM) with remote load control and PC interface featuring the DPM Data Logger software, a variety of drive shafts and adapters to fit most applications (Fig. 18.12).

18.4 Calibration of Test Equipment

The reliability of data from measurements made during tests and evaluations will depend on the accuracy of the instruments used. This assumes that the use of such equipment is understood by the testing personnel and operated correctly with accurate data recording. The level of accuracy will depend on the purpose of the tests, greater accuracy is generally required for smaller quantities (example: engine fuel consumption or dimensions of small components). Such accuracy is not required for larger quantities, for example- field size.

Instruments with high levels of accuracy will generally be more expensive. The range of the measuring equipment used should be consistent with the range expected during the test (Ex: expected draft forces of up to 5 kN should be measured with a tensile link/dynamometer of 5 kN rated load). Simple measuring devices such as rules, tapes, and graduated cylinders and thermometers will not require regular calibration as changes due to damage and wear will be obvious and the units are relatively inexpensive to replace.

Most of the mechanical based test equipment has now been superseded by electric and electronic measuring devices which can be either AC or battery operated and are suitable for laboratory or field use.

Testing engineers without suitable knowledge to understand all the basic processes of the new systems will need to have total reliance on the continued accuracy of the equipment. Because of this reliance on accuracy, all test measuring equipment should have periodic calibration checks especially if there are any doubts about the results of measurements. Manufacturers may include standards for calibration within the equipment such as a standard noise source for sound level meters and means for measuring frequency of light for revolution counters.

Fig. 18.13: Pressure gauge calibration tester

Many checks can be made within the testing organization using "standard" equipment for comparison. Graduated cylinders may be used to check the volume or output of fuel measuring devices and the capacity of various containers, pressure gauges, balances and timers can be checked against further units of the same type. However, these should only be used for "spot" checks and instruments should be returned to the manufacturer or to a suitably equipped standard test laboratory.

Fig. 18.13 shows a laboratory set-up for testing a pressure gauge (dead weight pressure gauge tester) where the pressure in the system is provided by known standard weights acting on a very accurately machined piston and cylinder. The reservoir and pump are fitted to ensure that the system is full and that the piston is supported by the liquid column when measurements are made.

Strain gauge devices used on engines, tractors and machines in the field are particularly vulnerable to damage and adverse conditions and will require more frequent checks. A well-equipped laboratory may have a specially designed machine for calibrating compression or tension links. However, standard weights of sufficient quantity may be used for calibrating a tension link used for draft force measurement.

All test equipment and instrumentation should be stored in a clean environment and records kept of periods of use, frequency of calibration and any breakdowns or repairs.

Chapter 19

Testing for Safety and Ergonomic Assessment of Agricultural Machinery

Tractors and agricultural machines need to be designed, manufactured, installed and operated in such a mariner that they do not endanger the safety of the operator and the persons working around them. To achieve this objective, every operation and maintenance of the machine needs to be carried out in accordance with the manufacturer's instructions. The basic safety requirements ought to be met by proper design of the machine components. Additionally the machine needs to be equipped with special safety provisions and protections. These may include safety guards, shields, covers, safe location of the dangerous parts, safety cabins and frames etc. Functional components required to be shielded to the maximum extent permitted for the intended function of the components. Additionally, under certain circumstances warning signals and labels against possible danger/hazard need to be provided on the machine. Therefore, testing and evaluation of safety provisions on tractors and agricultural machinery assume special significance in order to keep down the accidents to the minimum.

19.1 Types and Causes of Agricultural Machinery Accidents

- Overturns: Machine working at steep slopes, high speed and quick turning, quick starting etc.
- Run over: Person inspecting without engine stopping
- Trap: Person wearing loose clothes wrapped in moving belts, Chains, shafts etc.
- Cut: Body parts touching the cutter knives, sharp edges etc.
- Crush: Person seating/standing between machines and attachments; machines and walls under attachments etc.
- Fall: Person slipping from platforms steps
- Burn: Body part touching on the exhaust pipes.

- Fire: Careless use of fire, smoking, exhaust, carbon particle etc.
- Health hazards: Noise, vibration, pesticide spray etc.
- Hits: Thrown objects.

19.2 Technical Requirements for Ensuring the Machinery Safety

a) *Safety guards:* It is a protective device designed and fitted to minimize the possibility with machinery hazards as well as to restrict access to other hazardous areas. Different types of guards are shown in Fig. 19.1.

b) *Safe distance:* A means of providing guard where the possibility of contact with the hazard is minimized by the combination of the guard configuration including openings. The separation dimensions of pinch points in relation to body parts should be as shown in Fig. 19.2 and 19.3.

c) *Safety devices:* A device provided to minimize the incidence of machinery hazards like unexpected movement e.g. safety starting device, emergency stopping switch etc.

d) *Safety signs:* Information affixed on the machine to alert persons to hazards which can cause personal injury. The sign/caution plate should be permanently fitted/attached to the machine so that it can be readily seen and should not be easily removed.

e) *Ease of operation:* The operating controls shall be so designed and installed that they can be operated safely and easily from the operator's seat. Maximum actuating force required to operate various controls are shown in Table. 19.1.

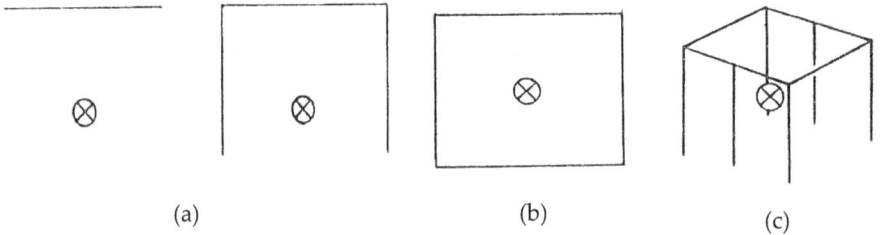

(a) (b) (c)

a. Shield and cover: protection of the side or sides, b. Casing: protection of all sides, c. Enclosure: protection by rails, fences, frames etc.

Fig. 19.1: Different types of guards

1. Rectangular or Slot

Limb	Illustration	Width of Aperture (mm)	Safety Distance to Danger Source (mm)
Finger Tip		$4 < a \leq 8$	$b \geq 15$
Finger		$8 < a \leq 20$	$b \geq 120$

Hand	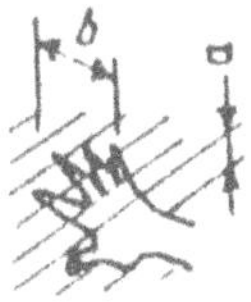	$20 < a \leq 30$	$b \geq 200$
Arm	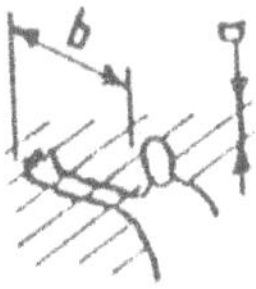	$30 < a \leq 135$	$b \geq 850$

2. Square or Hole

Limb	Illustration	Width of Aperture	Safety Distance to Danger Source
Finger Tip		$4 < a \leq 8$	$b \geq 15$
Finger	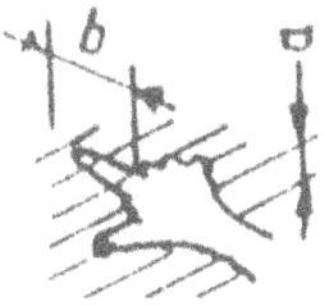	$8 < a \leq 25$	$b \geq 120$
Hand	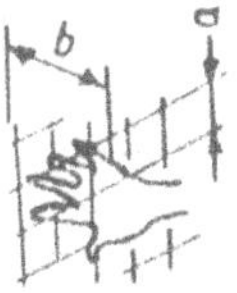	$25 < a \leq 40$	$b \geq 200$
Arm		$40 < a \leq 250$	$b \geq 850$

Fig. 19.2: Safe reach dimensions through openings, mm

Limb	Body	Leg	Foot	Arm	Hand wrist Fist	Finger
Illustration			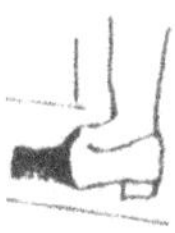			
Minimum Separation Distance Required	500	180	120	120	100	25

Fig. 19.3: Minimum separation distances for pinching points, mm

Table 19.1 Maximum Actuating Force Required to Operate Control as per ISO Recommendations

Control	Type of control	Maximum actuating force to operate control (N)	Type	Remarks
Service brake	Pedal Hand lever	600 400	Pressure Traction	It should be possible to achieve effective braking performance
Parking brake	Pedal Hand lever	600 400	Pressure Traction	when these forces are applied
Clutch Dual clutch	Pedal	350 400	Pressure Pressure	
Power take-off Coupling Manual steering system	Pedal Hand lever Steering wheel	300 200 250	Pressure Traction	
Power-assisted steering system with failure of the power-assisted Steering force	-do-	600		Force required to achieve a turning circle of 12m radius
Hydraulic power lift system	Hand leveler	70	Pressure & Traction	

19.3 Testing of Agricultural Machinery for Safety

The main objective of safety testing is to ensure the safety of the machinery operator and prevent accidents. All technical safety requirements as specified by the standards be checked thoroughly and reported.

Checking tools and devices		
Scales	:	Checking of machine dimensions, safety distance and opening etc.
Push pull tester	:	Checking of actuating forces required to operate controls.
Sound level meters	:	Checking of noise level at operator's ear level.
Vibration level meter	:	Checking of vibration level of operating controls and workplaces.
Surface thermometer	:	Checking of surface temperature of hot parts.

19.4 Methods of Safety Testing - Safety Devices and System

a) Guards for moving parts

i) Parts to be guarded in order to prevent danger to persons

- All shafts (including joints, shaft ends and crank shafts, universal joints. keys, pins and set screws etc.) that protrude from moving parts.

- Pulleys, flywheels, gears, cables, chains, sprockets, belts, clutches and couplings
- Working parts like rotary tynes, digging blades, cutting, binding, cutter knives and conveyors etc.
- Ground wheels, tyres and track adjacent to the operator's position.

ii) Guards should be strong.

iii) For guards formed from a mesh or grill, the permissible opening sizes are as follows:

For rectangular opening	$D \geq 200$ and $x \leq 30$
	$D \geq 120$ and $x \leq 20$
	$D \geq 15$ and $x \leq 8$
	$D \leq 15$ and $x \leq 6$
	$D \geq 850$ and $x \leq 135$
For circular or square opening	$D \geq 850$ and $x \leq 250$
	$D \geq 200$ and $x \leq 40$
	$D \geq 120$ and $x \leq 25$
	$D \leq 15$ and $x \leq 8$
	$D \geq 15$ and $x \leq 6$
For blower of air blast sprayer or liquid chemicals sprayer	$D \geq 200$ and $x \leq 40$
	$D \geq 40$ and $x \leq D/5$
	$D \leq 40$ and $x \leq 8$

Where,
D = distance between the guard and moving part (mm)
x = width or diameter of opening (mm)

b) Guards for PTO shafts

- The PTO shaft should be guarded by casing cap which should be firmly screwed or bolted to the machine body.
- The PTO drive shafts as well as the universal shaft should be guarded by a casing throughout their length. The casing should be secured firmly and held in stationary position.
- Guards should be sufficiently strong.

c) Safety devices

- Every stationary machine should have provision to disengage the power drive shaft. The control of the device should be located within easy reach of the operator.
- A brush cutter /stubble shaver should be provided with adequate means for disengaging the power to the knife blades easily and promptly
- Combine harvester should be equipped with a device to disengage the power of knife bar automatically in the event of clogging or fastening.
- A portable type power unit such as the power knapsack sprayer should be provided within a quick release clutch to enable the operator to disengage the power.

- All machines with lifting members should be provided with a locking device to keep the member in raised position.
- Power driven machines should be equipped with a provision to stop the prime mover instantly.

d) Braking device

- Self-propelled machines should be provided with both the service brake (main brake) and the parking brake.
- The towed machine/equipment especially trolleys should be provided with the parking brake.

e) Operator's workplace

- Any machine on which the presence of a worker or operator is necessary, should be provided with handles or handholds to ensure the safety and convenience of operator's mounting and dismounting.
- Any machine on which the operator is required to sit should be provided with a comfortable seat and adequate footrest. Seat will adequately support the operator in all working and operating modes and prevent the operator from slipping off the seat.
- Any platform on which the operator is required to stand during the operation of the machine should be level and have a non-slip surface. It should also be provided with guardrails around the platform.

f) Operating controls

- The operating controls, such as steering wheel or lever, gear shift lever, brake, clutch and switch should be arranged and fitted in such a way as to allow safe and easy control by an operator while standing or sitting in the normal operating position. The function and the operating method of these controls should be marked clearly.

g) Rollover protective structures (ROPS)

- Safety cribs and safety frames (Rollover protective structures: ROPS) are now a mandatory requirement on all agricultural wheel tractors in all developed countries for the protection of the operator in the event of accidental overturning. The testing procedure has been described at 19.6 & 19.7.

19.5 Terminology

a) *Tractor mass*: The mass of the tractor with full fuel tank as well as recommended coolant and lubricating oil and with all components required for normal operation plus ROPS but excluding optional ballast weights and operator.

b) *Tractor reference mass:* The mass not less than the tractor defined in (a) above as determined by the manufacturer for calculation of the energy inputs.

c) **Tractor reference wheel base:** A wheel base, not less than the maximum wheel base as determined by the manufacturer for calculation of the energy inputs.

d) **Operator seat:** Any part of the seat or its structure including suspension and adjustment systems.

e) **Seat reference points:** It is that point where vertical line tangent to the most forward point at the longitudinal seat centerline of the seat back and a horizontal line tangent to the highest point of the seat cushion intersect in the longitudinal seat centerline section. It is determined with the seat unloaded rind adjusted to the highest and most rearward position, as shown in Fig. 19.4.

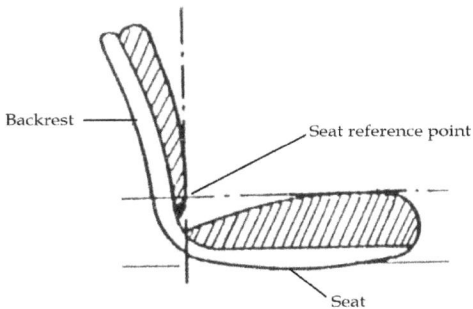

Fig. 19.4: Method of determining seat reference point

f) **Vertical reference plane:** It is the vertical plane, longitudinal to the tractor and passing through the seat reference point and the center of the steering wheel. Normally the vertical reference plane coincides with the median plane of the tractor.

g) **Zone of clearance:** The roll over protective structure (ROPS) which may include overhead protection fenders cab sheet metal, or related ROPS parts outside but near the operator area may be deformed in tests but shall not leave sharp edges exposed to the operator or include on the clearance zone described by the dimensions shown in Fig. 19.5.

The various dimensions are given below:

a = 760 mm at the longitudinal center line

b = not greater than 100 mm to rear edge of crossbar measured from the seat reference points

c = not greater than 305 mm measured from seat reference point to forward edge of crossbar

d = minimum 610 mm

e = 50 mm inside of frame upright to vertical centerline of seat

f = minimum 445 mm

g = 50 mm measured from outer periphery of steering wheel

Imaginary ground plane: The imaginary ground plane (Fig. 19.6) is defined as the surface containing a series of straight lines from the outer edges of the ROPS to any part of the tractor that might come in contact with flat ground and is capable of supporting the tractor in that position if the tractor overturns. For this purpose, the tyres and track width setting shall be assumed to be the smallest standard fitting.

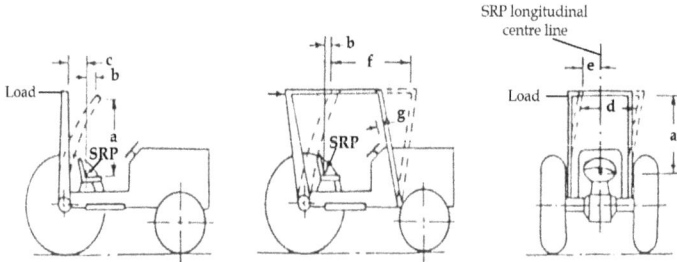

Fig. 19.5: Zone of clearance

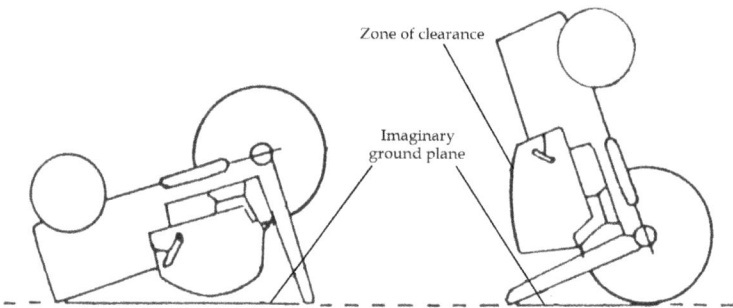

Fig. 19.6: Imaginary ground plane

19.6 Test Procedures for ROPS

19.6.1 Checking of specifications and dimensions

The purpose of this test is to check and record the dimensions, mass and specifications of the ROPS and the tractor to which it is fitted. Following checks and measurements need to be made:

For ROPS:
- Seat reference point, zone of clearance and imaginary ground plane
- Shape, construction, dimensions, mass and assembling method
- Mounting method
- Material used in the construction
- Accessories if any
- Label and caution marks
- Others

For Tractor: ■ Dimensions, mass and wheelbase

■ Static sideways overturning angle of the tractor at minimum track width and with ROPS fitted

■ Others

19.6.2 Strength test

The purpose of this test is to determine the strength characteristics of the ROPS and its mountings by simulating such loads as are imposed on the ROPS when the tractor overturns. It consists of: i) Dynamic strength test, and ii) Static strength test.

19.6.3 General requirements for strength test

Sequence of tests

i.	In the case of dynamic tests	: Impact at the rear
		Crushing at the rear
		Impact at the front
		Impact at the side
		Crushing at the front
ii.	In case of static tests	: Loading from the rear*
		Crushing at the rear
		Loading from the front
		Loading from the side
		Crushing at the front

* *The rear loading needs not be performed on ROPS applied to tractors having four wheels drive.*

iii.	Direction of impacts or loadings	: The side chosen for application of side impact or loading on the ROPS shall be that which in the opinion of Testing Station can result most unfavorable conditions when series of impacts loading occur. The rear impact shall be applied on the corner opposite to the side impact and the front impact on the corner nearest to the side impact or loading.
iv.	Removal of components	: All detachable windows, doors, panels and non-structural fittings that can easily be removed shall be removed.
v.	Setting of track width	: A track width setting chosen for the wheels shall be such that no interference occurs with the ROPS during the tests.

19.6.4 Procedure for dynamic strength test

19.6.4.1 Impact test

The test shall be conducted by applying a dynamic load produced by a pendulum block to the ROPS.

A. Apparatus

i) A pendulum block which shall be suspended by two chains from pivot points about 6 m above the floor shall be used.

ii) The pendulum block shall be fitted in such a way that the position of its center of gravity is constant. The mass of the block shall be 2000+20 kg and its impact face shall have dimensions 680+20 mm square.

iii) The lashings used for restraining the tractor to the ground shall be wire ropes of 13 mm diameter.

B. Test conditions

i) Tyre to be used: Standard pneumatic tyres shall be used.

ii) Tyre pressure and deflections: Inflation pressure and deflections in these tyres shall be, which are used in the various tests as shown in Table 19.2

Table 19.2 Tyre pressure and deflection recommended for different tractors

Type of tractor	Tyre Position	Tyre Pressure, kPa (bar)	Deflection, mm
Four wheel drive but front and rear wheels are of same size	Front	100 (1.0)	25
	Rear	100 (1.0)	25
Four wheel drive but front wheel are smaller than rear wheel	Front	150 (1.5)	20
	Rear	100 (1.0)	25
Two wheel drive	Front	200 (2.0)	15
	Rear	100 (1.0)	25

C. Test procedure

(a) Impact at the rear

i. Positioning of the tractor : The tractor shall be placed in relation to the pendulum block so that the block will strike the ROPS when the supporting chains and the impact face of the block are at an angle of 20 degree to the vertical. If the angle of the member of ROPS at the point of contact and at the moment of maximum deflection is greater than 20 degree to the vertical angle of the block shall be further adjusted by an additional support so that the impact face of the block is approximately parallel to the ROPS.

ii. Suspended height of pendulum block : The suspended height of block shall be so adjusted that the locus of its center of gravity passes through the point of contact.

iii.	Point of impact	:	The point of impact shall be that part of the ROPS which would be likely to hit the ground first in a rearward overturning accident, normally the upper edge.
iv.	Tractor tie-down	:	The tractor shall be lashed down by means of steel wire ropes incorporating tensioning devices to ground rails rigidly attached to the concrete base. The points of attachment of the lashings shall be approximately 2 m behind the rear axle and 1.5 m in front of the front axle. The lashings shall be tightened so that the deflections in the front and rear tyres shall be as indicated in Table 19.2. After the lashings have been tightened a wooden block of 150 mm square shall be clamped in front of the rear wheels and tightened.
v.	Height of lift of the pendulum block	:	The height of lift of pendulum block i.e. the vertical height of its center of gravity above the point of impact shall be calculated using the following formula:

For M less than 2000 kg:

$$H = 25 + 0.07M$$

For M equal to or greater than 2000 kg:

$$H = 2.165 \times 10 - MZ^2 \text{ or } H = 125 + 0.02\ M$$

whichever gives the heavier impact.

where:

H=Height of lift of the pendulum block (mm)

M= Tractor reference mass (kg)

Z= Tractor reference wheelbase (mm)

(b) Impact at the front

i.	Positioning of the tractor	:	Identical to that described in (a)(i)
ii.	Suspended height of pendulum block	:	Identical to that described in (a)(ii)
iii.	Point of impact	:	The point of impact shall be that part of the ROPS which would likely to hit the ground first if the tractor overturns sideways while traveling forward, normally the top of the front corner.
iv.	Tractor tie-down	:	The lashings shall be identical to that specified in (a)(iv) but the wooden block shall be clamped behind the rear wheels.

v. Height of lift of the : The height of lift of pendulum block shall be
    pendulum block   calculated using the following formula
                           For M less than 2000 kg:
                           H = 25 + 0.07 M
                           For M equal to or greater than 2000 kg:
                           H = 125 + 0.02M
                           where:
                           H=Height of lift of the pendulum block (mm)
                           M= Tractor reference mass (kg)

(c) Impact at the side

i. Positioning of the: The tractor shall be placed in relation to the
    tractor   pendulum block so that the block will strike the
                        ROPS when the supporting chains and the impact
                        face of the block are vertical as shown in Fig. 19.7

ii. Suspended height of: Identical to that described in (a)(ii)
     pendulum block

iii. Point of impact   : The point of impact shall be that part of the
                          ROPS, likely to hit the ground first in a sideways
                          overturning accident normally the upper edge.

iv. Tractor tie-down   : The tractor axle on the side to be struck shall be
                          lashed to the ground rails by means of steel wire
                          ropes. After lashing, a wooden block of 150 mm
                          square shall be clamped against the side of the
                          front and rear wheels opposite the blow and then
                          tightened against them. In addition a beam shall
                          be placed against the rear wheel rim opposite the
                          blow and secured to the floor as shown in Fig. 19.7.
                          The length of this beam shall be so chosen that its
                          position against the rim is at an angle of 30±3 degree
                          to the horizontal.

v.   Height of lift of the: The height of lift of the pendulum block (H) as
     pendulum block   shown in Fig. 19.7 shall be calculated using the
                            following formula:

For M less than 2000 kg:

H = 25 + 0.20M

For M equal to or greater than 2000 kg:

H = 125 + 0.15 M

where:

H=Height of lift of the pendulum block (mm)

M= Tractor reference mass (kg)

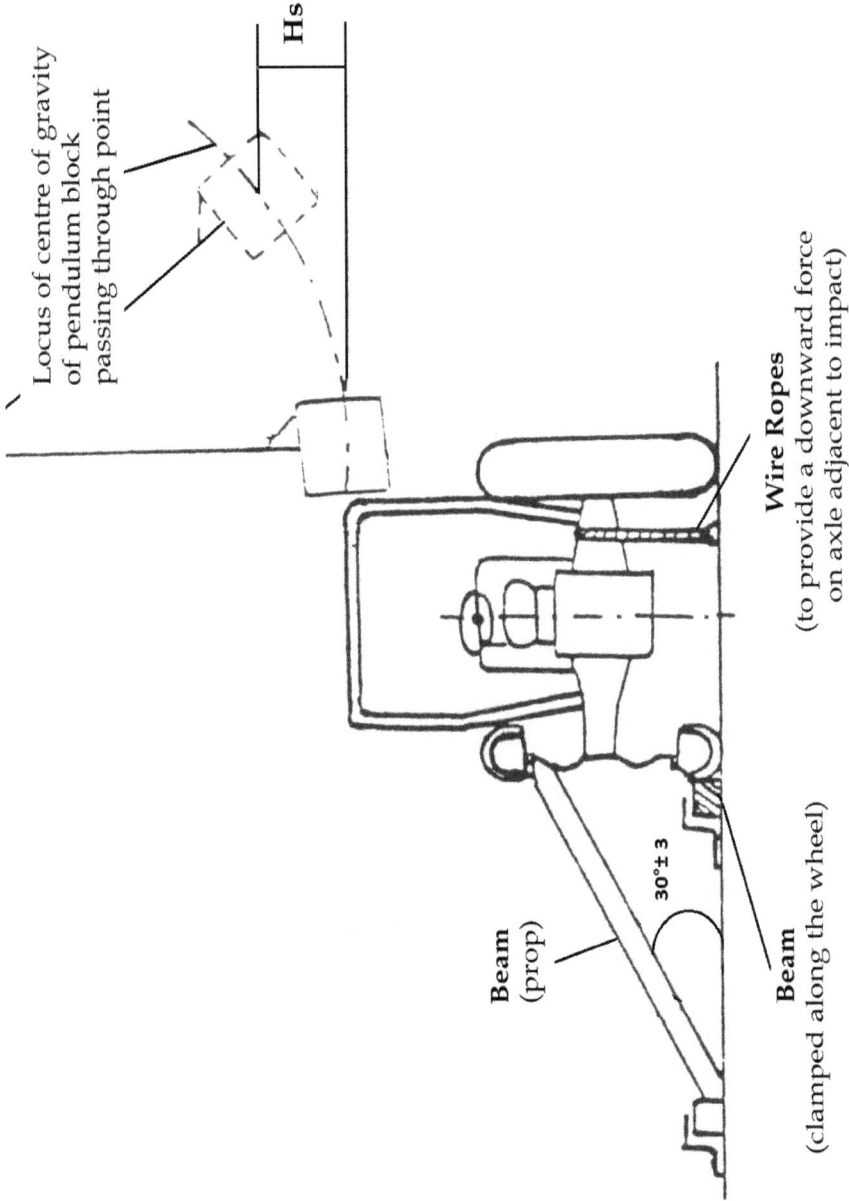

Locus of centre of gravity
of pendulum block
passing through point

Hs

Wire Ropes
(to provide a downward force
on axle adjacent to impact)

Beam
(prop)

30° ± 3

Beam
(clamped along the wheel)

Fig. 19.7: Method of Impact test

D. Measurements to be made

- Defects such as fracture and crack
- Permanent deflections of the ROPS
- Any part entering in the zone of clearance during test.
- Features presenting a serious hazard to the operator
- Any other point

19.6.4.2.Crushing test

The test which is carried out by applying a vertical downwards force on the ROPS through a beam placed laterally across the uppermost members of the ROPS is called crushing test.

A. Apparatus

Test rig consisting of a rigid beam approximately 250 mm wide, connected to the load- applying mechanism by means of universal joints as shown in Fig.19.8.

B. Test procedure

a) Crushing at the rear side

: The tractor shall be positioned such that the rear edge of the beam is over the rear most top part of the ROPS and the median longitudinal plane of the tractor is midway between the points of application of force to the beam.

Blocks shall be placed under the axles of the tractor so that the tyres do not bear the crushing force. The crushing force to be applied shall be calculated using the following formula:

$$F = 20 M$$

Where,

F - Crushing force (N)

M - Tractor reference mass (kg)

The force shall be maintained for approximately 10 seconds after reach of full force as calculated above.

Where the roof of the ROPS is not designed to sustain the full crushing force, the force shall be applied until the roof is deflected to coincide with the plane joining the upper part of the ROPS with that part of the tractor capable of supporting the tractor's mass when overturned.

b) Crushing at the front side

: The procedure is same as followed for crushing at the rear side.

C. Measurements to be made

All items as indicated for impact test.

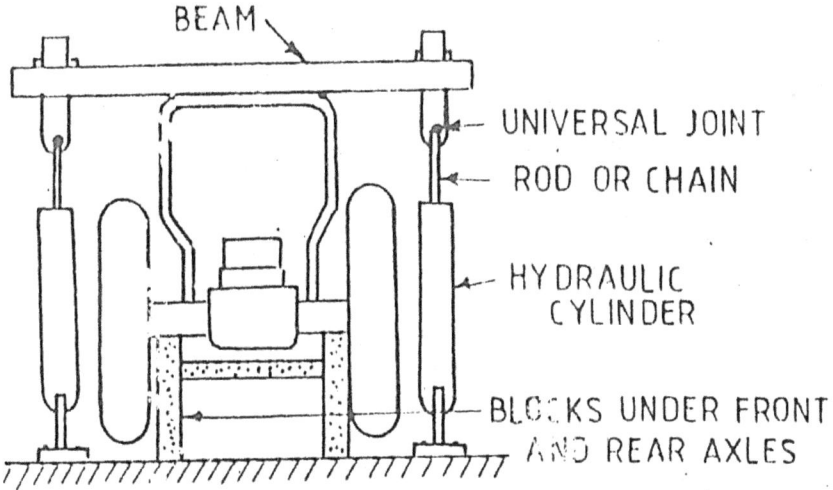

Fig. 19.8: Method of crushing test

19.6.4.3 Static strength test

Horizontal loading test

The test shall be conducted by applying a horizontal static load to the ROPS on test rig as shown in Fig 19.9.

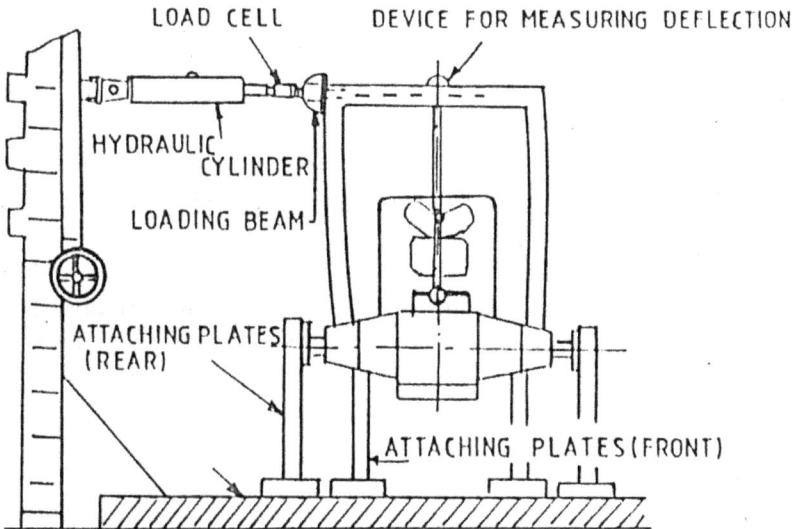

Fig. 19.9: Method of Horizontal loading test

A. Apparatus

- The static testing rig comprises of anchoring rails, supporting plates and hydraulic loading device with load distribution beam.

- The load distribution beam used for applying a horizontal force to the ROPS shall have a vertical face having dimension of 150 mm and a length in the multiples of 50 mm, between 250 mm and 700 mm.

- A read out device for measuring force and deflection in order to compute energy absorbed by the ROPS shall be fitted as shown in Fig. 19.10.

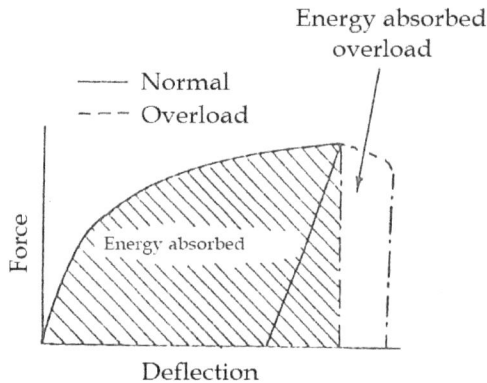

Fig. 19.10: Force deflection curve

B. Test conditions

- The tractor chassis shall be fixed firmly to the ground rails by means of plates, independent of the tyres, to prevent movement during the tests.

- The loads should be applied to the ROPS by means of beam. If necessary a substitute test beam which does not add strength to the ROPS may be utilized.

- The direction of the loading at start of test shall be less than ± 2 degree to the horizontal.

- The rate of load application (deflection rate) shall be less than 5 mm/s.

- The loading shall be stopped when the strain energy absorbed by the ROPS is equal to or greater than the required input energy specified in each test.

C. Test procedure

a) Loading from the rear

i. Point of load application : The point of application of load shall be that part of the ROPS which would likely to hit the ground first in a rearward overturning accident normally the upper edge.

ii. Length of beam : The length of load distribution beam shall not be less than one third of the width of the top of ROPS and not more than 49 mm greater than this.

iii. Required input energy : The energy input to be absorbed by the ROPS shall be calculated using the following formula:

$E = 1.4\,M$ or $Er = 0.143\,M$

Where,

E - Energy input to be absorbed during rear loading (J)

Er - Energy input to be absorbed during rear loading (kgf-m)

M - Tractor reference mass (kg)

b) Loading from the front

i. Point of load application : The point of application of load shall be that part of ROPS likely to hit the ground first if the tractor overturned sideways while traveling forward, normally the upper edge.

ii. Length of beam : Same as in the case of loading from rear side.

iii. Required input energy : The energy input to be absorbed by the ROPS shall be calculated using the following formula:

$E = 500 + 0.5M$ or $Er = 51 + 0.051\,M$

Where,

E - Energy input to be absorbed during rear loading (J)

Er - Energy input to be absorbed during rear loading (kgf-m)

M - Tractor reference mass (kg)

c) Loading from the side

i. Point of load application : The point of application of load shall-be that part of the ROPS likely to hit the ground first in a sideways overturning accidents, normally the upper edge.

ii. Length of beam : The load d1stribution beam shall be as long as practicable subject to maximum of 700 mm.

iii. Required input energy : The energy input to be absorbed by the ROPS shall be calculated using the following formula:

$E = 1.75M$ or $Er = 0178M$

Where,

E - Energy input to be absorbed during rear loading (J)

Er - Energy input to be absorbed during rear loading (kgf-m)

M - Tractor reference mass (kg)

D) Measurements to be made
- Fracture and crack observation
- Permanent deflections of ROPS
- Any part entering the zone of clearance
- Features responsible for serious hazard to operator.
- Any other relevant point

19.6.4.4 Overload test

If any crack is observed during the test which cannot be considered as negligible, then loading equal to 120% of the original required energy shall be applied after the loading test and observation made.

19.6.4.5 Crushing test

This test is similar to dynamic strength test.

19.6.4.6 Final inspection

The ROPS shall be dismantled and inspected for cracks, fractures and defects etc. after completion of all tests.

19.7 Criteria for Acceptance of ROPS

- There shall be no feature representing a serious hazard to the operator during the tests.

- There shall be no serious defects affecting the operation of the tractor fitted with ROPS.

- There shall be no serious fracture or cracks in all major structural members mounting components and tractor parts contributing to the strength of the ROPS during the tests.

- No part of the ROPS shall enter the zone of clearance and come in contact with the seat during the test.

- Machines used for towing or which are towed should be provided with adequate towing device which is fitted and secured properly and are safe.

- Guards for hot parts and fire protection
  - Hot parts which the person may touch and cause burns should be guarded.
  - Hot parts such as exhaust pipe should be designed and fitted in such a manner that these do not cause any hazard to person.
  - Spark arrester may be used with exhaust silencer for arresting glowing carbon particle.

- The tips of divider being sharp and dangerous should be guarded.

- Working parts of machine which may produce thrown objects, such as stones or fragments of crops or cutter knives, should be guarded adequately in such a manner to ensure the safety of the operator in normal operating position.

- Safety Signs: Safety signs should be attached adjacent to the following parts:
  - Dangerous parts which are difficult to protect by safety guard.
  - Other parts which are required to alert the person from danger
- Operational easiness
  - User's manuals should be prepared and supplied with every machine.
  - All machines should be easy to operate.

19.8 Noise measurements

Ears exposed to loud sounds shifts a person's hearing threshold level upward and he or she can hear only the louder' sounds. This frequent exposure will eventually result in a permanent threshold shift and ultimately to hearing loss.

Measuring and evaluation of noise

Sound is measured in decibels. Values range from zero to 140 as shown in Table 19.3. Sound level meter usually has three measuring modes, which are "A", "B" and "F" scales. The "A" scale responds to sound more or less the same as the human ear, and the values are given in units of dB(A). Permissible hours per day that a person can safely be exposed to sound levels are given in Table 19.4.

19.9 Vibration Measurement

Excess mechanical vibration of any part or assembly of the machine or tractor or self propelled machine can be of hazardous to the operator. It is an amplitude of assemblies and components measured in micron, when the engine or prime mover in running at its rated speed, both on loaded and unloaded conditions. It should not exceed 100 microns when measured with vibration meter. The measurement shall be carried out at all important points e.g. steering wheel, foot rest, gear shift lever, brake pedal, clutch pedal, seat, head light, mudguard, PTO lever, hydraulic lever, accelerator lever etc.

Table 19.3 Decibel level of common sounds emitted from different sources

Decibel Level dB(A)	Common sounds from different sources
0	Acute threshold of hearing
15	Average threshold of hearing
20	Whisper
30	Leaves rustling or very soft music
40	Average residence
60	Normal speech. background music
70	Noisy office
80	Heavy traffic or window air-conditioner
85	Inside acoustically insulated protective tractor cab
90	Standard limit; hearing damage when excessive exposure to noise i.e. >90 db
100	Noisy tractor, power mover, motorcycle
120	Thunder clamp, amplified rock music
140	Threshold of pain, shot gun, near jet taking off

Table 19.4 Permissible noise exposure

Hours per day that one can safely be exposed to these sound levels

Duration per day (h)	Sound level dB (A)
8	90
6	92
4	95
3	97
2	100
1.5	102
1	105
0.5	110
0.25 or less	115

19.10 Safety Precautions for Tractors and Agricultural Machinery

19.10.1 General

a) The tractor is designed with adequate safety provisions, there is no real substitute for caution and attention in preventing the mishaps. Once an accidents has occurred, it is too late to think about what one should have done. Therefore, always be careful in operation.

b) Remember that tractor has been designed exclusively for agricultural use. Any other application must first be authorized by the manufacturer.

c) Do not attempt to increase maximum engine speed by tempering injection pump governor.

d) Do not alter relief valve setting of hydraulic system (Power steering, hydraulic lift. remote control valves, etc.)

e) Do not operate tractor, if you feel unwell, suspend work rather than taking a risk.

f) Always use steps and grab handles when getting in or out of the cabin.

g) As far as possible never work without roll-over protection frame or incorrectly fitment on the tractor. Check that fasteners are not loose and that fasteners frame is not damaged in any way. Do not alter cab by welding or drilling.

h) Read manual thoroughly before attempting to start, operate, service or refuel the tractor. A few minutes reading will save trouble later.

19.10.2 Tractor starting

a) Prior to starting the engine check that parking brake is on, gear and PTO levers are in neutral position.

b) Make sure that all implements are fully lowered before starting.

c) Ensure that all guards and protective devices are correctly installed before starting the tractor.

d) Do not attempt to start the tractor unless sitting on the operator's seat.

e) Tractor should be operated only by responsible person suitably trained and having valid driving license.

f) Keep a First-Aid kit handy.

g) Do not work with loose garments that could get caught in moving parts.

h) Check that all rotating parts connected to the PTO shaft are well shielded.

i) Do not run engine in a closed building without adequate ventilation as exhaust gases are dangerous.

j) Ensure that there is no person or obstacle within range before starting the tractor.

19.10.3 Tractor operation

a) Select the track width most suitable to the work in hand, keeping the tractor stability in mind.

b) Engage clutch pedal gradually. Abrupt engagement particularly on uphill or downhill can cause tractor to pitch dangerously.

c) Disengage clutch for a moment, if front wheels start rising.

d) Respect the Highway Code during on-road journey.

e) Do not put foot on brake and clutch pedals continuously.

f) Latch brake pedals during on-road driving, otherwise dangerous skidding may occur when braking.

g) Do not drive downhill in neutral or with clutch disengaged.

h) Operator should always be sitting on the seat while tractor is moving.

i) Do not get on or off a moving tractor.

j) Always depress the clutch pedal gently.

k) Do not take sharp turn at high speed.

l) Safety gauges should be checked time to time.

19.10.4 Towing

a) Adjust the towing attachment correctly to maintain tractor stability.

b) Drive slowly when towing heavy trailers or wheeled implements.

c) Preferably trailers should not be towed unless equipped with an independent' braking system

d) Always use drawbar when towing heavy loads. Do not pull from 3-point hitch links as tractor could pitch.

e) When towing, do not take turn with the different lock in, otherwise you may not be able to steer the tractor.

f) Always operate the tractor at a safe speed. Reduce speed on slopes and curves to prevent roll-over.

g) When working on sloping ground, do not drive too fast, particularly when taking turn.

19.10.5 Using agricultural implements and machinery

a) Always use matching implements or machinery rated with the tractor horse power. Never use machinery designed for more powerful tractors.

b) When hitching impliment, never stand between tractor and implement.

c) Never operate a PTO driven implement without first ensuring that no one is on or too near the machine.

d) Check that all rotating parts connecting to the PTO shaft are well shielded.

19.10.6 Stopping

a) Never leave equipment in the raised position while the tractor is stationary.

b) Ensure that hydraulic system is not under pressure before disconnecting lines.

c) Hydraulic oil escaping under pressure could cause serious personal injury. Thus while tracing oil leaks, one should wear protective shields, glasses and gloves.

d) Return gear lever to neutral position, disengage PTO and apply parking brakes, stop the engine and engage a gear before leaving the tractor seat. Always remove starter switch key before leaving tractor unattended.

e) Park tractor on level ground as far as possible, engage a gear and apply parking brake.

f) Before attempting to inspect, clean, adjust, repair and service the tractor or attach implement, ensure that the engine is stopped, transmission is in neutral position, brakes are applied, PTO is disengaged and all moving parts are stationary.

g) Work on tyres should be carried out only by an experienced personnel using proper equipment. Otherwise tyre fittings could cause a serious accident.

19.10.7 Maintenance

a) If engine just finished the work, then allow engine to cool down before removing radiator cap. Slowly turn cap to release pressure before removing cap completely.

b) Disconnect the battery ground lead before starting any work on the electrical system.

c) Do not fill tank completely when tractor is to be operated in strong sunlight as fuel could expand and escape. Any escaping of fuel should be wiped off immediately.

d) Tractor fuel may be dangerous. Therefore never refuel with tractor in motion, near an open flame or when smoking. Preferably it should be filled in the open area.

e) Always keep a fire extinguisher within reach.

19.11 Ergonomic Assessment

It is essential to ensure that various values and measurements conform to the recommendations given in standards of all prime movers and farm equipment. Deviations, if any, be reported in the test report.

A detailed list of instruments/equipments required in testing and evaluation of agricultural machinery has been shown in Appendix 4.

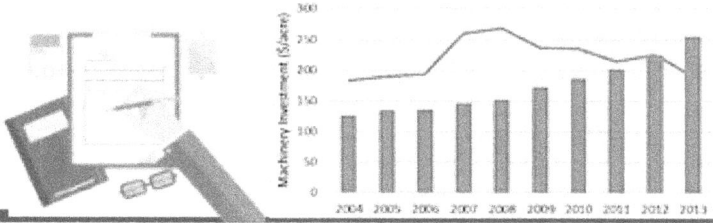

Chapter 20

Estimating Cost of Operation of Farm Machinery

The cost of estimation for operation of farm equipment by using a uniform method assumes significance due to use of farm machinery for self-use as well as for custom hiring services. The later trend is growing rapidly and preferred by small and marginal farmers in developing countries. Direct ownership in such cases is expensive. Larger machines, newer technology, higher initial cost of new machinery, and higher energy cost have led to escalation of operating cost of farm machines and power sources in recent years. The custom hiring services are being organized so that all categories of farmers could reap the benefits of mechanized farming practices, since it is not practical for every farmer to own different farm machines nor it would be economically viable. Every farmer and farm manager must know basic principles of agricultural engineering and economic principles and apply them when deciding to buy, lease, rent or share the equipment. Additionally, capital outlay and running cost can also be reduced considerably by operating farm machinery on more than one farm.

Each equipment must perform under a range of field conditions and production practices with sufficient reliability to provide return on investment. Tillage implements should prepare a satisfactory seedbed while conserving moisture, destroying early weed growth, and minimizing soil erosion. Seed drills and planters should provide consistent seed placement and population as well as proper application of fertilizers and even pesticides. Harvesting machines must harvest clean, undamaged grain while minimizing field losses.

The most accurate method of computing and estimating machine costs is maintaining complete record of the actual expenses incurred. The main goal of estimating cost of farm machinery operation is to serve as a basis for planning and management. Results from cost calculation can be used firstly to compare the income received from a job with the cost of its performance and to suitably adjust the rates; and secondly to pin-point unreasonably high cost and the reasons for them so that timely action can be taken to reduce them. For a farmer, the main purpose of calculating cost may be to compare, for example, the cost of having his own machine with the cost of employing a custom hiring service provider/ contractor, or costs of different types or sizes of machines. It will also help to check the cost of production of a particular crop to ensure its profitability. Fig. 20.1 illustrates the typical effect of machinery size on cost of machine use in their field per acre [118, 119, 196].

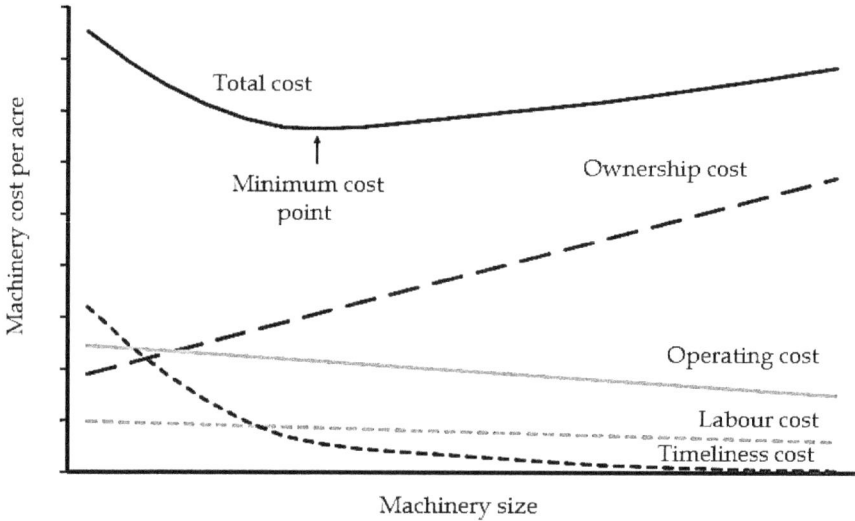

Fig. 20.1: Machinery size vs. machinery cost per acre along with costs [120]

20.1 Scope

The cost of using farm machinery consists of expenses for ownership and operation, and overhead charges. It may also include a margin for profit. Ownership costs are independent of use and are often called as fixed costs. Costs for operation vary directly with use and are referred to as variable costs. A summary of cost items is given below:

20.1.1 Fixed cost

a) Depreciation

b) Interest on investment

c) Insurance and taxes

d) Housing (Shelter)

20.1.2 Variable cost:

a) Fuel

b) Lubricating oil

c) Repair and maintenance

d) Wages and labour charges

20.1.3 Overhead charges

20.2 Calculation for Cost of Operation

20.2.1 Fixed cost

a) *Depreciation* - This cost reflects the reduction in value of a machine with use (wear) and time (obsolescence). While actual depreciation would depend on the sale price of the machine after its use, based on different computational methods, depreciation can be estimated. There are several methods to calculate the depreciation. These include:

   i)   Straight-line method,

   ii)   Estimated value method,

   iii) Sum-of-the year's digits method,

   iv) Declining-balance method,

   v)   Sinking-fund method.

In the straight-line depreciation method, an equal reduction of value is used for each year, the machine is owned. This method can always be used to estimate cost on a specific period of time, provided the proper salvage value is used for the age of the machine. The following formula based on straight-line method is recommended:

$$D = \frac{P-S}{L}$$

Where,

    D   - depreciation cost, Rs. per year

    P   - purchase price of the machine, Rs.

    S   - residual/salvage value of the machine, Rs.

    L   - useful life of the machine, years

The depreciation cost per hour can be calculated by dividing the depreciation cost by the number of hours the machine is expected to be utilized in a year. Residual value of the machines may be taken as 5 - 10 percent of the purchase price. Useful life of some of the commonly used machines under general conditions of usage is given in Annexure 20.1 for guidance.

Estimated value method is the most realistic determination of depreciation as it is calculated based upon difference between the value at the end of each year and the value of machine possessed at the start of year. Machine depreciates much faster in the first few years of ownership than in the later years. Sum of the years' digit method is a much more accurate method estimating the true value of machine at any age because the annual depreciation rate decreases as the machine gets older. In the declining balance method, even though the depreciation rate is same for all years, the depreciation value is different with its age. The straight-line depreciation method is the simplest and most widely used method to calculate the annual depreciation. It lacks in accuracy as compared to other methods due to the fact that it gives constant depreciation value throughout the useful life of the machine, which is not a true depiction of the real situation as the annual depreciation in the initial years are higher compared to the later years.

20.2.1.1 Inflation in depreciation analysis of farm equipment:

Effect of inflation must be included while considering the overall cash flows. Inflation can be defined as an annual percentage rate at which current years prices have increased over the previous year's prices. It has a compounding effect. Because of inflation, the cost of farm equipment or any other product continues to rise. Hence, inflation affects adversely the purchasing power of money [214, 218]. The future price of farm equipment in the nth year at constant inflation rate can be represented by

$$F = P (1+I)^n \qquad\qquad \ldots (1)$$

Where,

F	- future price in the nth year, Rs.
P	- purchase price, Rs.
I	- Inflation rate, fraction

If the inflation changes in each year, the future price of the farm equipment would be

$$F = P (1 + I_1) (1 + I_2) \ldots\ldots\ldots\ldots (1 + I_n) \qquad \ldots (2)$$

The effect of inflation can be included in the depreciation analysis. The depreciation of a machine under constant inflationary condition using straight-line method can be written as

$$CD_n = (n /L) \times [P(1+I)^n - S] \qquad\qquad \ldots (3)$$

Where,

CD_n	- cumulative depreciation charges up to n^{th} year, Rs.
n	- number of years elapsed after the purchase of the machine
L	- life of machine, years
P	- purchase price, Rs.
S	- salvage value of the machine

Annual depreciation charge (D) in the nth year will be

$$D_n = CD_n - CD_{n-1}$$

Future value of machine at the end of nth year (V_n)

$$V_n = \text{Future price at the end of } n^{th} \text{ year} - CD_n$$
$$= P (1+I)^n\, n - (n/L) [P(1+I)^n - S]$$

Taking life of the machine as 15 years and salvage value as 10% of the inflated price of the machine, equation (3) reduces to

$$CDn = (n/15) [P(1+I)^n - 0.1\, P\, (1+I)^n]$$
$$= (n/15)\, 0.9\, P\, (1+I)^n$$
$$= 0.06\, n\, P\, (1+I)^n \qquad\qquad \ldots (4)$$

Similarly, cumulative depreciation charge up to (n-I) year may be written as:

$$CD_{n-1} = 0.06\, (n-1)\, P\, (1+I)^{n-1} \qquad\qquad \ldots (5)$$

Annual depreciation charge (D_n) in the nth year will be

D_n = $CD_n - CD_{n-1}$

= 0.06 n P $(1 + I)^n$ - 0.06 (n - 1) P $(1 + I)^{n-1}$

= 0.06 P $[n (1 + I)^n - (n - 1) (1 + I)^{n-1}]$ ... (6)

The remaining value of the machine after n^{th} year (Vn) may be obtained as:

Vn = Future price at the end of nth year - cumulative depreciation value up to n^{th} year

$\quad$ Vn $\quad$ = P $(1 + I)^n$ - D_n

$\qquad$ = P $(1 + I)^n$ - 0.06n P $(1 + I)^n$

$\qquad$ = P $(1 + I)^n$ [1 - 0.06n] ... (7)

20.2.1.2 Example for calculating depreciation taking inflation as a factor:

A tractor costs Rs.4.8 Lakhs (P) with a total life of 15 years (L). If the rate of inflation is 5% constant throughout its life period, find the future price of machine, total depreciation, remaining value and annual depreciation at the end of 12th year (n). Salvage value = 10% of inflated price.

Solution:

Using equation (1) future price at the end of 12th year will be, F = P $(1 + I)^n$

$\qquad$ = 480000 $(1 + 0.05)^{12}$ = 862011.04

Total depreciation after elapse of 12 years will be, CDn = (n/L) $[(P (1+I)^n) - S]$

$\qquad$ = 12/15 $[(480000 (1+0.05)^{12}) - (0.1 \times 480000 (1+0.05I)^{12})]$

$\qquad$ = 0.8 $\times$ 480000 $\times$ 0.90 $(1+0.05)^{12}$ = 620647.95

Remaining value of the machine at the end of 12th year will be,

$\qquad$ = 862011.04 $-$ 620647.95 = 241363.09

Annual depreciation charge in the 12th year using equation (6) will be

$\qquad$ = 0.06 $\times$ P $\times$ $[n (1 + I)^n - (n-l) (1 + I)^{n-1}]$

$\qquad$ = 0.06 $\times$ 480000 $[12 (1 + 0.05)^{12} - (12-1) (1 + 0.05)^{11}]$

$\qquad$ = 0.06 $\times$ 480000 [21.55 - 18.81]

$\qquad$ = Rs. 78812.44

b) *Interest* – A major contributory item after depreciation in the fixed cost is the interest on capital. Even if cash is paid for purchased machinery, money is tied up that might be available for use elsewhere in the business. Annual charges of interest should be calculated on the basis of the actual rate of interest payable. Interest rates vary considerably but usually are between 12 and 16%. Annual interest is calculated on an average investment by using the following formula:

$$A = \frac{P+S}{2} \times \frac{i}{100}$$

Where,

> A   - annual interest charge, Rs. per year
> P   - purchase price of the machine, Rs.
> S   - residual/salvage value of the machine, Rs.
> i   - interest rate, %

c) *Insurance and Taxes* - Actual amount paid or to be paid annually for insurance and annual taxes, if any should be charged. If the information is not available, it may be calculated on the basis of 2 percent of the average purchase price of the machine, (P+S)/2.

d) *Housing/Shelter* – An average of 1.5 percent of the average purchase price of the machine can be taken as the cost of housing per year.

20.2.2 Variable cost

a) *Fuel* - The cost of fuel and oil must be included in the total machine cost for tractors and other powered farm equipment. Fuel consumption depends on the size of the power unit, load factor and operating conditions. The actual consumption can be observed while the machine is working or may be taken from the results obtained at official testing stations. It is common practice to consider average fuel consumption from the varying load test [IS: 5994 (2)] as approximately equal to fuel consumption on the farm. Average fuel consumption can also be estimated by the following formulae:

For diesel engines: $A = 0.15 \times B$

Where,

> A   - average diesel consumption, l/h,
> B   - rated power, kW

For petrol engines: C $= 0.25 \times B$

Where,

> C   - average petrol consumption, l/h
> B   - rated power, kW

Fuel consumption can also be estimated by the following equation:

$$F = Lc \times Pr \times SFC / 1000$$

Where,

> F   *- fuel consumption, l/h*
> Lc   *- load coefficient factor for the operation*
> Pr   *- rated horsepower of the power source, hp*
> SFC   - specific fuel consumption, ml/kWh)

The values of Lc and SFC for different operations and power sources are given in Table 20.1. The fuel cost is taken as fuel consumption per hour multiplied by cost of fuel per litre.

b) *Oil* - The actual oil consumption should be recorded while the machine is working. In case oil consumption data is not available, oil consumption may be taken as 2.5 to 3 percent of the fuel consumption on volume basis. The cost of filters, replacement of oil and other lubricants is included under repairs and maintenance.

Table 20.1 Values of Lc and SFC for different operations and power sources

Power source	Type of work	Lc	SFC ml/kW h	SFC ml/hp h
Stationary diesel engine	Water lifting Threshing	0.6 0.7	300 300	220 220
Tractor	Light work e.g. transport, water lifting etc. Medium work e.g. secondary tillage, sowing, inter-culture etc. Heavy work e.g. primary tillage, Sheller, cane crusher, combine etc.	0.4 0.5 0.6	285 285 285	210 210 210
Self-propelled combine		0.6	285	210
Small petrol engine	Spraying, dusting etc.	0.8	680	500

c) *Repair and Maintenance* - Repair and maintenance costs are necessary to keep a machine operable due to wear, part failure, renewal of tyres and tubes and accidents. The costs of restoring a machine are highly variable. Good machinery management may keep cost low. Normal wear deterioration is directly related to use, and restoration or repair costs are assumed to be a typical variable costs. Maintenance costs, primarily those related to lubrication, are directly related to use also. Repair costs vary from one geographical region to another because of the differences in machinery use, labour wages and prices of spares. Repair costs increases with the age of a machine but tend to level off as the machine becomes older.

- The accumulated repair and maintenance costs (TAR) at any a machine's life can be estimated from the following formulae:

- Four wheeled and crawler tractors, TAR = $0.100\ X^{1.5}$

- Stationary power unit and two-wheeled tractor, TAR = $0.120X^{1.5}$

- Self-propelled combine, dozer and scraper, TAR = $0.096X^{1.4}$

- Agricultural trailer, TAR = $0.127X^{1.4}$

- PTO driven combine, seed drill, seed-cum-fertilizer drill and sprayer, TAR = $0.159X^{1.4}$

- Plough, planter, harrow, ridger and cultivator, TAR = $0.301X^{1.3}$

Where,

TAR - total accumulated repair and maintenance cost as percentage of initial cost,

X - 100 times the ratio of the accumulated hours of use to the wear out life

The repair and maintenance cost in percentage of purchase price for whole usable life of some of the machines calculated on the basis of formulae given above is given in Annexure 20.2 for guidance.

d) *Labour Charges* - In performing custom hiring work, the cost of at least one operator has to be included and may be one assistant. The yearly cost of the operator is equal to the wages paid plus any allowances to which they may be entitled. Average cost per hour may be computed by dividing the total cost by the number of hours the operator has performed the work. This cost is, of course, higher than the average per hour work on the farm because part of the time will be used for travelling, interruptions and moving machines from one farm to another and this is not paid for directly by the customers. Also, the cost of operator and labour per hour is calculated from the actual cost incurred per day at the prevailing rates in the region.

20.2.3 Overhead charges

This includes charges for supervision and establishment and interest on working capital if applicable. It should be assumed as 20 percent of the sum of fixed and variable costs or say profit margin.

20.2.4 Total cost per hour

The sum of fixed cost, variable cost and overheads per hour shall give the total cost for operation of agricultural machinery.

20.2.5 Total cost per hectare

The total cost per hectare may be obtained on the basis of field capacity of the machine. The data regarding kind of machine, its working width, average speed of travel, size and shape of the fields and travel conditions should be recorded by the contractor since this kind of data would be very useful and sufficiently accurate. However, if, no such data is available, the estimation of field capacity should be made by calculation on the basis of the following formula:

$$C = \frac{S \times W}{10} \times \frac{E}{100}$$

Where,

C - effective field capacity in ha/h,

S - speed of travel in km/h,

W - theoretical width of the machine in m, and

E - field efficiency in percent at theoretical field capacity.

20.2.6 Field efficiency

Field efficiency is a measure of relative productivity of a machine under field conditions. It accounts for constraints to utilize the theoretical operating width of the machine, operator's capability and habits, operating policy and field characteristics. The activities, such as turning and idle travel, materials (seed, fertilizer, chemicals, water, harvested material, etc.) handling, cleaning clogged equipment, machine adjustment, lubrication and refueling and waiting for other machines accounts for a majority of the time lost in the field. Travel to or from a field, major repairs, preventive maintenance and daily service activities are not included in field time or field efficiency [149].

Field efficiency is not a constant for a particular machine, but varies with the type of soil, size and shape of the field, pattern of field operation, crop yield, soil moisture and crop conditions. In absence of any data the field efficiency shall be selected from Annexure 20.3.

20.3 Agricultural Equipment Cost Calculator

This Agricultural Equipment Cost Calculator available at CPCFM website may be used to calculate ownership and operating costs of common farm equipment. Use the drop-down list to choose the power unit or self-propelled machinery that will be used along with suitable implements. The web link of the online calculator is given below:

www.cpcfm.res.in/calculators/agriequipment.asp

20.4 Example for Calculation of Cost for Operation of Thresher

A test was conducted to find out the cost per hour for operation of a 35 hp tractor operated with wheat thresher. Also find the cost of operation of the thresher per tonne.

Given:

Price of tractor	:	5,20,000	Rs.
Life of tractor	:	15	years
Salvage value	:	10	%
Interest rate	:	12	%
Shelter and Insurance	:	2	%
Price of diesel	:	68	Rs./l
Number of labour employed	:	4	no.
Labour charge	:	350	Rs./day
Driver's charge	:	500	Rs./day
Annual use of tractor	:	1000	h
Price of thresher	:	60000	Rs.
Life of thresher	:	10	yrs
Annual use of thresher	:	200	h
Output of thresher	:	800	kg/h

Depreciation= Straight line method

Solution:

a) Tractor

Depreciation, Rs./h	=	(P-S)/LH = (520000-52000) / (15x1000)
	=	31.20
Interest, Rs./h	=	[(P+S)/2 X i]/H
	=	[520000+52000)/2 X 0.12] / 1000
	=	34.32
Shelter and Insurance, Rs./h	=	(2% of P)/H=(0.02 X 520000)/1000
	=	10.40
Total fixed cost, Rs./h	=	31.20 + 34.32 + 10.40 = 75.92
Repair & Maintenance cost		
	=	(Yearly use X 100)/(Total life)
	=	(1000 X 100)/(15 X 1000)
	=	6.67
TAR, %	=	$0.12 (6.67)^{1.5} = 2.07$
Repair & maintenance cost, Rs/h	=	[2.06/(100 X 1000)] X [(520000+20200)/2]
	=	5.91
Fuel Consumption, l/h	=	LCF x RHP x SFC/1000 = 0.6 x 35 x 210/1000
	=	4.41
Fuel cost, Rs/h	=	4.41 x 68 = 299.88
Oil cost, Rs/h	=	0.2 X 299.88 =7.50
Driver's charge, Rs/h	=	500/8 = 62.5
Total variable cost, Rs/h	=	5.91 + 4.41 + 299.88 + 7.50 + 62.50 = 375.78
Tractor operating Cost, Rs/h	=	**75.92 + 375.78 = 451.70**

b) Thresher

Depreciation, Rs./h	=	(P-S)/LH = (60000-6000) / (10x200)
	=	27.00
Interest, Rs./h	=	[(P+S)/2 X i]/H
	=	[(60000+6000)/2 X 0.12]/200
	=	19.80
Shelter and Insurance, Rs./h	=	(2% of P)/H=(0.02 X 60000)/ 200
	=	6.00
Repair & Maintenance cost		
	=	(Yearly use X 100)/(Total life)
	=	(200 X 100)/(10 X 200)
	=	10
TAR, %	=	0.159 (X)1.4
		0.159 (10)1.4 = 3.99
Labour charge, Rs/h	=	(4 x 350)/8 = 175.0
Total Tractor & Thresher Cost, Rs/h	=	**451.70+27.00+19.80+6.00+3.99+175.00**
	=	**683.49**
	Or	**Rs 683/h**

20.5 Example of Calculation for Cost of Operation for Plough

A farmer purchased a 45 hp tractor at a price of Rs.6,10,000.00 and a 3-bottom mould board plough with 30 cm bottom width at Rs.45,000.00. Calculate:

i) Area covered per day for 8 hours

ii) Cost of ploughing per hectare and per hour

Make necessary assumptions.

Given:

Price of tractor, Rs	:	6,10,000
Life of tractor, years	:	15
Speed of tractor, km/h	:	5.0
Salvage value , %	:	10
Interest rate , %	:	12
Annual use of tractor, h	:	1000
Price of diesel , Rs./l	:	68
Price of plough, Rs.	:	45,000
Width of plough, m	:	0.9
Life of plough, years	:	15
Annual use of plough, h	:	200
Field efficiency of plough, %	:	60

Solution:

a) Tractor

Depreciation, Rs./h	=	(P-S)/LH = (610000-61000) / (15x1000)
	=	36.60
	=	[(P+S)/2 X i]/H
Interest, Rs./h	=	[610000+61000)/2 X 0.12] / 1000
	=	40.26
Shelter and Insurance, Rs./h	=	(2% of P)/H=(0.02 X 610000)/1000
	=	12.20
Total fixed cost, Rs./h	=	36.60+40.26+12.20 = 89.06
Repair & Maintenance cost		
	=	(Yearly use X 100)/(Total life)
	=	(1000 X 100)/(15 X 1000)
	=	6.67
TAR, %	=	0.12 (6.67)$^{1.5}$ = 2.067
	=	[2.067/(100 X 1000)] X [(610000+61000)/2]
Repair & maintenance cost, Rs/h	=	6.93

Fuel consumption, l/h	=	LCF x RHP x SFC/1000
	=	0.6 x 45 x 210/1000
		5.67
Fuel cost, Rs/h	=	5.67 x 68 = 385.56
Oil cost, Rs/h	=	0.2 X 385.56 = 77.11
Labour cost, Rs/h	=	(1 x 350)/8 = 43.75
Total variable cost, Rs/h	=	6.93+385.56+77.11+43.75
	=	513.35
Tractor operating Cost, Rs/h	=	**89.06+513.35 = 602.41**

b) Mould board plough

Depreciation, Rs./h	=	(P-S)/LH = (45000-4500) / (15x200)
	=	13.50
Interest, Rs./h	=	[(P+S)/2 X i]/H
	=	[(45000+4500)/2 X 0.12]/200
	=	14.85
Shelter and Insurance, Rs./h	=	(2% of P)/H=(0.02 X 45000)/ 200
	=	4.50
Total fixed cost	=	13.50+14.85+4.50
		32.85
Repair & Maintenance cost		
	=	(Yearly use X 100)/(Total life)
	=	(200 X 100)/(15 X 200)
	=	6.67
TAR, %	=	0.301 (X)1.3
		0.301 (6.67)1.3 = 3.55
Repair & maintenance cost, Rs/h	=	(3.55/100)X(45000+4500)/2
		877.45
		877.45/200 = 4.39
Total cost, Rs./h	=	32.85+4.39 = 37.24
Ploughing operational cost, Rs./h	=	602.41+37.24 = 639.65
Field capacity of plough	=	(0.9 x 5.0)/10 = 0.45 ha/h
Total tractor and plough cost of operation Rs/h	=	639.65 / 0.45 = 1421.44

Annexure 20.1
Useful Life of Commonly Used Farm Machinery [94]

S. No.	Name of machine	Useful life	
		Hours	Years
1.	Stationary Engine	10000	10
2.	Electric Motor	15000	15
3.	Power Tiller	8000 (9600)	10 (12)
4.	Tractor (wheeled and crawler)	10000 (15000)	10 (15)
5.	Combine harvester (self-propelled)	3000	6
6.	Combine harvester (mounted or drawn)	2000	7
7.	Seed drill	2500	10
8.	Seed-cum-fertilizer drill	2000	8
9.	Planter	2000	10
10.	Plough	3000	10
11.	Disc harrow	3000	10
12.	Cultivator	4000	10
13.	Front-mounted dozer attachment for wheeled tractor	3000	10
14.	Towed scraper for wheeled tractor	2000	10
15.	Power sprayer (knapsack and tractor mounted)	2000	8
16.	Seed cleaner	2500	5
17.	Agricultural trailer	3600 (5000)	12 (15)
18.	Power thresher	2500	8
19.	Centrifugal pump	10000	10
20.	Power chaff cutter	5000	8
21.	Rotavator	2400	8
22.	Ridger	1500	12
23.	Blade terracer	2000	10
24.	Puddler	2500	10
25.	Cane crusher	10000	10

NB: Useful life in parenthesis may be used on quality of machinery improved in recent years.

Annexure 20.2

Accumulated Repair Cost in Percentage of Purchase Price

S. No.	Name of machine	Cost in percentage of purchase price for useful life, years									
		1	2	3	4	5	6	7	8	9	10
1.	Stationary engine	3.8	10.8	19.8	30.3	42	55.5	70.3	86.4	102.6	120
2.	Electric motor	2	5.8	10.7	16.5	23	30.1	38.4	46.7	56.1	65.3
3.	Power tiller	3.8	10.8	19.8	30.3	42	55.5	70.3	86.4	102.6	120
4.	Tractor (wheeled & crawler)	3.2	9.0	16.5	25.3	35	46.2	38.6	72	85.5	100
5.	Combine harvester (Self-propelled)	4.9	13	23	34.3	46.8	60.6	--	--	--	--
6.	Combine harvester (mounted & drawn)	6.6	17.3	30.4	44.8	61.9	80	100.3	--	--	--
7.	Seed drill	4	10.5	18.6	27	38.9	49	60.9	73.4	86.5	100.5
8.	Seed-cum-fertilizer drill	5.5	14.4	25.4	38	52.5	67	83.2	100.3	--	--
9.	Planter	6	14.8	29	36.4	48.7	61.6	75.4	88.7	104.4	119.8
10.	Plough	6	14.8	29	36.4	48.7	61.6	75.4	88.7	104.4	119.8
11.	Disc harrow	6	14.8	29	36.4	48.7	61.6	75.4	88.7	104.4	119.8
12.	Cultivator	6	14.8	29	36.4	48.7	61.6	75.4	88.7	104.4	119.8
13.	Dozer	2.4	6.4	11.2	16.9	22.9	29.6	36.8	44.3	52.3	60.6
14.	Scraper	2.4	6.4	11.2	16.9	22.9	29.6	36.8	44.3	52.3	60.6
15.	Power sprayer	5.5	14.4	25.4	38	52.5	67	83.2	100.3	--	--
16.	Seed cleaner	12.6	33.2	58.6	87.3	120.5	--	--	--	--	--
17.	Agricultural trailer	2.5	6.5	11.5	17.2	23.8	30.4	37.7	45.4	53.6	63.4

Annexure 20.3

Recommended Speed and Field Efficiencies for various Machines

S. No.	Machines	Recommended avg. speed of travel (km/h)	Recommended avg. field efficiency (%)
1.	Plough	4.5	80
2.	Disc harrow	6	80
3.	Cultivator	6	80
4.	Seed drill	5	70
5.	Seed-cum-fertilizer drill	5	70
6.	Planter	5	70
7.	Ridger	4.5	90
8.	Puddler	5	75
9.	Rotavator	2.5	80
10.	Combine harvester (Self-propelled) For Paddy	2	75
	For wheat	3.5	75
11.	Combine harvester (Mounted & drawn) For paddy	2	70
	For wheat	3	70

Bibliography

1. Agricultural Machinery Manufacturers Association (AMMA) – India. 2014. Country Presentation Paper, October 2014.

2. Agricultural Statistics at a Glance. Government of India 2013, 2012, 2010, 2007 and 2006.

3. Agriculture Research Data Book. 2016. Chapter 6 - Agricultural Engineering and Produce Management. IASRI. New Delhi. Retrieved on 08.05.2018 from http://www.iasri.res.in/ebook/TEFCPI_sampling/AGRICULTURAL%20 STATISTICS% 20SYSTEM%20IN%20INDA.pdf

4. Alberta Farm Machinery Research Centre; PAMI, 1996. Research Update 725: Ballasting Your Tractor for Performance [Online].

5. All India Report on Agriculture Census 2010-11. 2015. Agriculture Census Division. Department of agriculture, cooperation & farmers welfare, Ministry of agriculture and farmers welfare. Govt. of India.

6. Allam, R.K., Wiens, H. 1982. An investigation of air seeder component characteristics, Alberta: Winter meeting-American society of agricultural engineers, ASAE, No. 82-1505.

7. American Society of Agricultural & Biological Engineers (ASABE). 2015. ASAE 497.7. Agricultural Machinery Management Data.

8. American Society of Agricultural & Biological Engineers. ASAE S281.3 DEC96, Capacity Designation for Fertilizer Pesticide Hoppers and Containers.

9. American Society of Agricultural & Biological Engineers. ASAE S327.2 DEC95, Terminology and Definitions for Agricultural Chemical Application.

10. American Society of Agricultural & Biological Engineers. ASAE S341.3 FEB04. Procedure for Measuring Distribution Uniformity and Calibrating Granular Broadcast Spreaders.

11. American Society of Agricultural and Biological Engineers Standards. ASABE S506 OCT2010 (R2014). Terminology and Definitions for Planters, Drills and Seeder. ASABE. St. Joseph, MI 49085, USA.

12. American Society of Agricultural and Biological Engineers. ASABE: D497.7. 2011. Agricultural Machinery Management Data.

13. Anonymous. 1978. Proceedings of the International Agricultural Machinery Workshop held at International Rice Research Institute, Manila, Philippines.

14. Anonymous. 1986. Perspective for Agricultural Tractor Industry in India. A report prepared by Ministry of Industry, Govt. of India, New Delhi.

15. Anonymous. 1990. Japanese Industrial Standard on Testing Methods for Centrifugal Pumps, Mixed Flow Pumps and Axial Flow Pumps. JIS-B-8301.

16. Anonymous. 1992 Proceedings of the All-India Seminar on Pumping Systems, Selection, Maintenance and Management held at IIT, Roorkee.

17. Anonymous. 1970. OECD. Standard code for official Testing of Agricultural Tractor. Organization for Economic Co-operation and Development, Paris. France.

18. Anonymous. 1976. Mechanical transplanting of paddy. Test Bulletin/ Series-2/76 prepared by Tractor Training & Testing Station (Govt. of India), Budni.

19. Anonymous. 2015. Asian and Pacific Network for Testing of Agricultural Machinery. ANTAM: 2015. Standard codes for testing of power tillers.

20. Automotive Industry Standard. AIS-107. 2009. Requirements of Driver's Field of Vision for Agricultural Tractors. The Automotive Research Association of India. Pune, India.

21. Bayat, A., Yusuf, Z., Ulusoy, M. R. 1994. Spray deposition with conventional and electrostatically - charged spraying in citrus trees. Agriculture Mechanization in Asia Africa and Latin America. 25 (1): 35-39.

22. Bernaki, H., Haman, J., Kanafajski, Cz. 1972. Agricultural Machines - Theory and Construction, Vol. I., Department of Agriculture and the National Science Foundation, Washington D.C, USA.

23. Böttinger, S., Doluschitz, R., Klaus, J., Jenane, C., Samarakoon, N. 2013. Agricultural Development and Mechanization in 2013 – A Comparative Survey at a Global Level. CECE-CEMA Summit 2013, 16-17 October, 2013 at European Parliament, Brussels, Belgium.

24. Bracy, R.P., Parish, R.L., Mc Coy, J.E. 1999. Precision seeder uniformity varies with theoretical spacing. Hort Technology 9: 47-50.

25. Bureau of Indian Standards. 1986. Certification Scheme - Procedure for grant of license.

26. Bureau of Indian Standards. 1988. BIS (Certification) Regulations. Retrieved on 09.05.2018 from http://www.bis.org.in/bs/certif.htm.

27. Bureau of Indian Standards. IS: 10000-1. 1980. Methods of tests for internal combustion engines - Glossary of terms relating to test methods.

28. Bureau of Indian Standards. IS: 10000-12. 1980. Methods of tests for internal combustion engines - Test certificates.

29. Bureau of Indian Standards. IS: 10000-13. 1980. Methods of tests for internal combustion engines - Recommendation on nature of tests required for functional changes in critical components.

30. Bureau of Indian Standards. IS: 10000-2. 1980. Methods of tests for internal combustion engines - Standard reference conditions.

31. Bureau of Indian Standards. IS: 10000-3. 1980. Methods of tests for internal combustion engines - Measurement for testing – units and limits of accuracy.

32. Bureau of Indian Standards. IS: 10000-4. 1980. Methods of tests for internal combustion engines - Declarations of power, efficiency, fuel consumption and lubricating oil consumption.

33. Bureau of Indian Standards. IS: 10000-5. 1980. Methods of tests for internal combustion engines - Preparation for tests and measurements for wear.

34. Bureau of Indian Standards. IS: 10000-6. 1980. Methods of tests for internal combustion engines - Recording of test results.

35. Bureau of Indian Standards. IS: 10000-8. 1980. Methods of tests for internal combustion engines - Performance tests.

36. Bureau of Indian Standards. IS: 10000-9. 1980. Methods of tests for internal combustion engines - Endurance tests.

37. Bureau of Indian Standards. IS: 10134. 1994. Methods of tests for manually operated sprayers.

38. Bureau of Indian Standards. IS: 10233. 2001. Tractor-operated disc ploughs.

39. Bureau of Indian Standards. IS: 10273. 1987. Guidelines for declaration of power and specific fuel consumption and labelling of agricultural tractors.

40. Bureau of Indian Standards. IS: 10274. 1993. Agricultural wheeled tractors - Maximum travel speed - method of determination.

41. Bureau of Indian Standards. IS: 10691. 2001. Share for tractor-operated mouldboard ploughs.

42. Bureau of Indian Standards. IS: 10743. 1983. Method for determination of centre of gravity of agricultural tractors.

43. Bureau of Indian Standards. IS: 11170. 1985. Specification for performance requirements for constant speed compression ignition (diesel) engines for agricultural purposes (up to 20 kW).

44. Bureau of Indian Standards. IS: 11313. 2012. Hydraulic power sprayers.

45. Bureau of Indian Standards. IS: 11442. 1996. Agricultural tractors - Operator's field of vision - test procedures.

46. Bureau of Indian Standards. IS: 11531. 1981. Test code for puddler.

47. Bureau of Indian Standards. IS: 11691. 1986. Specifications for Power thresher.

48. Bureau of Indian Standards. IS: 11859. 2004. Agricultural tractors - Turning and clearance diameters - methods of test.

49. Bureau of Indian Standards. IS: 12036. 1995. Agricultural tractors - Test procedures - power tests for power take off.

50. Bureau of Indian Standards. IS: 12061. 1994. Agricultural Tractors - Braking performance - method of test.

51. Bureau of Indian Standards. IS: 12180. 1987. Method for Noise Measurement of Agricultural Tractors.

52. Bureau of Indian Standards. IS: 12207. 2014. Agricultural tractors - Recommendations on selected performance characteristics.

53. Bureau of Indian Standards. IS: 12224. 1987. Method of test for hydraulic power and lifting capacity of agricultural tractors.

54. Bureau of Indian Standards. IS: 12226. 1995. Agricultural tractors - Power tests for drawbar - Test procedure.

55. Bureau of Indian Standards. IS: 12337. 2009. Manually Operated Fertilizer Broadcaster.

56. Bureau of Indian Standards. IS: 12362. 1994. Agricultural Vehicles-Mechanical connection on Towing Vehicles.

57. Bureau of Indian Standards. IS: 12482. 2009. Methods of test for manually operated dusters.

58. Bureau of Indian Standards. IS: 13064. 1991. Power tillers - Installations and preventive maintenance.

59. Bureau of Indian Standards. IS: 13539. 2008. Power Tillers - Recommendations on Selected Performance Characteristics.

60. Bureau of Indian Standards. IS: 13548. 1992. Agricultural wheeled tractors and field machinery - measurement of whole-body vibration of the operator.

61. Bureau of Indian Standards. IS: 13581. 1993. Agricultural wheeled tractors - Operator's seat - laboratory measurement of transmitted vibration.

62. Bureau of Indian Standards. IS: 14414. 1996. Agricultural tractors - Axle power determination - Test procedures.

63. Bureau of Indian Standards. IS: 14536. 1998 (2009). Selection, installation, operation and maintenance of submersible pump set - Code of practice.

64. Bureau of Indian Standards. IS: 14582. 1998 (2008). Single-phase small AC electric motors for centrifugal pumps for agricultural applications.

65. Bureau of Indian Standards. IS: 15805-1. 2008. Straw Reaper Combine - Test Code: Terminology.

66. Bureau of Indian Standards. IS: 15805-2. 2008. Straw Reaper Combine - Test Code: Performance Test.

67. Bureau of Indian Standards. IS: 15806. 2008. Combine-harvester-thresher ® Selected Performance and other characteristics – Recommendations.

68. Bureau of Indian Standards. IS: 1970. 2009. Crop protection equipment - hand-operated compression knapsack sprayer.

69. Bureau of Indian Standards. IS: 3062. 2006. Crop protection equipment - rocker sprayer.

70. Bureau of Indian Standards. IS: 3652. 2006. Crop protection equipment - foot sprayer.

71. Bureau of Indian Standards. IS: 3906. 2006. Crop protection equipment – hand operated knapsack sprayer, piston type.

72. Bureau of Indian Standards. IS: 4366-1. 2001. Agricultural tillage discs: Part 1 concave type.

73. Bureau of Indian Standards. IS: 4366-2. 2001. Agricultural tillage discs: Part-2: Flat type.

74. Bureau of Indian Standards. IS: 5135-2. 2009. Hand rotary duster - shoulder mounted type.

75. Bureau of Indian Standards. IS: 5994. 1998. Agricultural Tractor - Test code.

76. Bureau of Indian Standards. IS: 6284. 1999. Test Code for Power Thresher for Cereals.

77. Bureau of Indian Standards. IS: 6288. 1999. Test code for mouldboard ploughs.

78. Bureau of Indian Standards. IS: 6316. 1993. Seed-cum-fertilizer drill - Test code.

79. Bureau of Indian Standards. IS: 6320. 1995. Specification for wheat power thresher (hammer mill type).

80. Bureau of Indian Standards. IS: 6595. 2002 (2017). Horizontal Centrifugal Pumps for Clear, Cold Water: Agricultural and Rural Water Supply Purposes.

81. Bureau of Indian Standards. IS: 6635. 2001. Tractor operated disc harrows.

82. Bureau of Indian Standards. IS: 6638. 2001. Tractor-mounted spring loaded cultivators.

83. Bureau of Indian Standards. IS: 6690. 1981. Specification for blades for Rotavator for power tillers.

84. Bureau of Indian Standards. IS: 6813. 2000. Sowing Equipment - Seed-cum-fertilizer Drill.

85. Bureau of Indian Standards. IS: 7353. 1974. Specification for Blade for Tractor Operated Terracer.

86. Bureau of Indian Standards. IS: 7593–1. 2012. Power-operated pneumatic sprayer-cum-duster - knapsack type.

87. Bureau of Indian Standards. IS: 7640. 1999. Test code for disc harrows.

88. Bureau of Indian Standards. IS: 8034. 2002 (2017). Submersible Pumpsets – Specification.

89. Bureau of Indian Standards. IS: 8122-1. 1994. Test Code for Combine Harvester-thresher: Terminology.

90. Bureau of Indian Standards. IS: 8122-2. 2000. Combine-Harvester-Thresher - Test Code: Performance Test.

91. Bureau of Indian Standards. IS: 9019. 1979. Code of practice for installation, operation and preventive maintenance of power threshers.

92. Bureau of Indian Standards. IS: 9020. 2002. Power threshers — safety requirements.

93. Bureau of Indian Standards. IS: 9137. 1978. Indian Standard for acceptance test for centrifugal, mixed flow and axial flow pump.

94. Bureau of Indian Standards. IS: 9164. 1979. Guide for estimating cost of farm machinery operation.

95. Bureau of Indian Standards. IS: 9217. 2001. Test code for agricultural discs.

96. Bureau of Indian Standards. IS: 9253. 2013. Agricultural wheeled tractors - Field performance and haulage tests – guidelines.

97. Bureau of Indian Standards. IS: 9283. 2013. Motors for Submersible Pumpsets.

98. Bureau of Indian Standards. IS: 9694–1. 1987 (2017). Code of Practice for the Selection, Installation, Operation and Maintenance of Horizontal Centrifugal Pumps for Agricultural Applications: Selection.

99. Bureau of Indian Standards. IS: 9694–2. 1980 (2017). Code of practice for the selection, installation, operation and maintenance of horizontal centrifugal pumps for agricultural application: Installation.

100. Bureau of Indian Standards. IS: 9694–3. 1980 (2016). Code of practice for the selection, installation, operation and maintenance of horizontal centrifugal pumps for agricultural applications: Operation.

101. Bureau of Indian Standards. IS: 9694-4. 1980 (2016). Code of Practice for the Selection, Installation, Operation and Maintenance of Horizontal Centrifugal Pumps for Agricultural Applications: Maintenance.

102. Bureau of Indian Standards. IS: 9813. 2002. Specification for Tractor Mounted Blade Terracer.

103. Bureau of Indian Standards. IS: 9856. 1981. Test code for potato planters.

104. Bureau of Indian Standards. IS: 9877-2. 1981. Code of practice for installation, operation and preventive maintenance of grain combine.

105. Bureau of Indian Standards. IS: 9935. 2012. Test code for Power Tiller.

106. Bureau of Indian Standards. IS: 9939. 1981. Glossary of terms relating to agricultural tractors and power tillers.

107. Bureau of Indian Standards. IS: 9980. 2004. Guidelines for Field performance and haulage Tests of power tillers.

108. CEMA - European Agricultural Machinery Industry Association. 2017. Smart Agriculture for All Farms. 23pp.

109. CEMA - European Agricultural Machinery Industry Association. 2017. Connected Agricultural Machines in Digital Farming. CEMA, Brussels, Belgium.

110. Chauhan, A.M. and Ramesh Kumar. 1992. Review of research and development in seed cum fertilizer drills in India and areas for further work. J. Agril. Engg. 9(3):31-43.

111. Cugnasca, C. E., Saraiva, A. M., Hirakawa, A. R., Strauss, C. 2003. Communication Protocols for Application in Agricultural Vehicles. In: Mahalik N.P. (eds) Fieldbus Technology. Springer, Berlin, Heidelberg. https://doi.org/10.1007/978-3-662-07219-6_18.

112. Dafa'alla, A.M., Hummeida, M.A. 1991. Performance evaluation of a sugarcane planter. J. King Saud. Univ. Vol.3. Agric. Sci. (1).5-14.

113. Dass, R.S. 1986. Testing of Irrigation Pumps – Test Codes and procedures and Interpretation of Test Results. Presented at USAID development and Management Training Project held at CFMT&TI, Budni.

114. De, D., Singh, R. S., Chandra, H. 2000. Power Availability in Indian Agriculture. Technical Bulletin No. CIAE/2000/83.

115. Dhir, D. K., Rajan, P., Verma, S.R. 2017. Seed drill discharge rate variation due to varietal differences using an automated calibration test rig. Agricultural Mechanization in Asia, Africa & Latin America (AMA). Accepted for Publication.

116. Doharey, R.S. and Patil, R.N. Status of Agricultural Machinery Manufacturing in India. Souvenir released during 22nd Annual Convention of ISAE held at CIAE, Bhopal.

117. Everett, C. Calibration of fertilizer application equipment on-farm. http://www2.gnb.ca/content/dam/gnb/Departments/10/pdf/Agriculture/CalibrationFertilizer.pdf

118. Extension Bulletin. 2015. Estimating Farm Machinery Costs. PM 710 (A3-29). Iowa State University Extension & Outreach.

119. Extension Bulletin. 2015. Machinery Cost Estimates: Field Operations. Farm Business Management. University of Illinois Extension.

120. Extension Bulletin. 2017. Farm Machinery Selection, PM 952 (A3-28). Iowa State University Extension & Outreach.

121. Food and Agriculture Organization of the United Nations. 2017. Report: Consultative Meeting on a Mechanization Strategy (New Models for Sustainable Agricultural Mechanization in sub-Saharan Africa). BR840E/1/4.17.

122. Garg, I.K. and Sharma, V.K. 1984. Development and Evaluation of a Manually Operated Paddy Transplanter. J. of Ag. Engg. 21(1&2): 17-24.

123. Garg. I.K. and Sharma, V.K. 1987. Riding Type Engine Operated paddy Transplanter. Invention Intelligence, 22 (9-10) : 368-374

124. Gill, G.S. and Mangat, I.S. 1991. Package of Practices in Agricultural Engineering. Directorate of Extension Education, PAU, Ludhiana.

125. Gill, M.S. 1993. Farm Mechanization. Paper presented at National Conference on Farm Mechanization held at Bhopal, March 2, 1991.

126. Government of India (2011 and 2013), Agricultural Statistics at a Glance, Directorate of Economics and Statistics, Ministry of Agriculture, New Delhi.

127. Hebblethwaite, P. 1955. A detailed procedure of testing of combine harvesters. Silsoe, U.K. National Institute of Agricultural Engineering. Technical memorandum 121. 11p.

128. http://www.kvalitest.com/userData/noqom/kdocs/Datum-Power-Take-Off-PTO-Power-Monitoring-System-Handbook-1.pdf Accessed on 05.05.2018

129.https://www.bellflowsystems.co.uk/files/attachments/4480/Direct_Delta%20 Installation%20manual.pdf, Accessed on 05.05.2018

130.IASRI 2006. Study relating to formulating long-term mechanization strategy for each agro-climatic zone/state in India. Indian Agricultural Statistics Research Institute, New Delhi, India.

131.Ingle, C. 2011. Agricultural tractor test standards in America. CMGT 564 - Strategic Standards, The Catholic University of America.

132.International Organization for Standardization Standard. ISO: 3046-3. 2006. Reciprocating internal combustion engines -- Performance: Test measurements.

133.International Organization for Standardization Standard. ISO: 5691. 1981 (2018). Equipment for planting, Potato planters - Method of testing.

134.International Organization for Standardization Standard. ISO: 7256-1. 1984. Sowing equipment, Test methods - Single seed drills (precision drills).

135.International Organization for Standardization Standard. ISO 4254-1:2013. Agricultural machinery -- Safety: General requirements.

136.International Organization for Standardization Standard. ISO: 789 Parts 1 – 13. [1982/1983/1986/1991/1996/2000/2006/2018]. Agricultural tractors - Test procedures.

137.International Organization for Standardization Standard. ISO: 11783 (Parts 1 – 14). Tractors and machinery for agriculture and forestry - Serial control and communications data network.

138.International Organization for Standardization. ISO 7256-2:1984. Sowing equipment -- Test methods: Seed drills for sowing in lines.

139.International Organization for Standardization. ISO: 5702. 1983. Equipment for harvesting — Combine harvester component parts — Equivalent terms.

140.International Organization for Standardization. ISO: 6720. 1989. Equipment for sowing, planting, distributing fertilizers and spraying -- Recommended working widths.

141.International Organization for Standardization. ISO: 7256-1. 1984. Sowing equipment -- Test methods: Single seed drills (precision drills).

142.International Organization for Standardization. ISO: 8210. 1989. Equipment for harvesting — Combine harvesters — Test procedure.

143.ISO 500-1:2014. Agricultural tractors -- Rear-mounted power take-off types 1, 2, 3 and 4 -- Part 1: General specifications, safety requirements, dimensions for master shield and clearance zone.

144.Jat, M.L., Chandna, P., Gupta, R., Sharma, S.K., Gill, M.A. 2006. Laser Land Leveling: A Precursor Technology for Resource Conservation. Rice-Wheat Consortium Technical Bulletin Series 7. New Delhi, India: Rice-Wheat Consortium for the Indo-Gangetic Plains. pp 48.

145.Kachman, D.S., Smith, J.A. 1995. Alternative measures of accuracy in plant spacing for planters using single seed metering. Transactions of the ASAE 38: 379-387.

146. Kale V.N. 2007, Presentation on Authorization, NRFMT&TI, Hissar for CMVR certificate to the combine harvester manufacturers.

147. Kalkat, H.S., Sharma, V.K. and Saini, K.S. 1975. Care and Maintenance of Fertilizer Seed Drill. Progressive Farming.

148. Kaul, R.N. and Ramesh Kumar. 1972. Farm Machinery Testing Centre Test Procedures on Thresher, PAU Publication, Ludhiana

149. Kepner, R.A., Bainer, R., Barger, E.L. 1987. Principles of Farm Machinery, Third Edition. CBS Publishers, India. Pages: 25-32.

150. Kheirella, A.F., Yahya, A. 2001. A tractor instrumentation and data acquisition system for power and energy demand mapping. Pertanika J. Sci. Technology, 9(2), 1-14.

151. Khepar, S.D., Chaturvedi, M.C., Sinha, B.K. 1982. Effect of precise leveling on the increase of crop yield and related economic decision. Journal of Agricultural Engineering. 19(4): 23-30.

152. Kienzle, J., Ashburner, B., Sims, G. 2013. Mechanization for Rural Development: A review of patterns and progress from around the world. Integrated Crop Management Vol. 20-2013. FAO. www.fao.org/docrep/018/i3259e/i3259e.pdf

153. Kumar, A., Singh, J.K., Mohan, D., Varghese, M. 2008. Farm hand tools injuries: A case study from northern India. Safety Science, 46: 54–65.

154. Kyle, J. T. 1960. Agricultural machinery testing. Canadian Society for Bioengineering Journal. 2: 37-38.

155. Laser Leveling Training Manual. 2013. Postharvest Unit, International Rice Research Institute (IRRI). Los Baños, Philippines.

156. Lijedhal. J.B., Jurnquist, P.K., Smith, D.W. and Mokoto Hoki. 1989. Tractor and Their Power Units, Springer, USA.

157. Mahal, J.S., Ahuja, S.S., Shukla, L.N., Singh, K. 2009. Strip-till-drill in Punjab- a study. Indian J. Agric. Res., 43 (4): 243–250.

158. Matsuo Yosuke. 1992. Safety Testing of Agricultural Machinery. Lecture delivered at –BRAIN-OMIYA-JAPAN.

159. Mehta, C.R., Varshney, A.C., Nandy, S.M. 2003. Instrumentation and testing of agricultural machinery (teaching material). Central Institute of Agricultural Engineering (CIAE). Bhopal.

160. Mehta, M.L., Tiwari, R., Patil, V.A. 1988. Farm machinery testing in India. Agri. Engg. Today. Vol. 22, No. 5 & 6.

161. Mehta, M.L. 1984. Standardization in the field of Farm Power. Paper presented in RNAM Training Programme in Standardization of Agricultural Machinery, BIS, New Delhi.

162. Mehta, M.L., Patil, V.A. 1989. Role of Testing and Evaluation of tillage Implements for Minimum Energy Consumption. Paper presented during seminar of minimum Tillage Technology held at NRFMT & TI, Hissar.

163. Mehta, M.L., Patil, V.A. 1991. Agricultural Machinery Testing in India. Agricultural Mechanization in Asia, Africa and Latin America, Japan Vol. 22, No. 3.

164. Mehta, M.L., Sharma, V.K. 1985. Feasibility of delinking cutting and threshing systems of chaff cutter type thresher. J of Ag. Engg. 22(2) : 10-18.

165. Mehta, M.L., Sharma, V.K. 1985. Studies on Threshing System of Chaff Cutter Type thrasher. J. of Research, PAU, 22(4): 735-741.

166. Mehta, M.L., Tiwari, R., Patil. V.A. 1989. Quality Control in Rural Industries Manufacturing Agricultural Implements. "Udyog Yug", Haryana Govt. Publication.

167. Mehta, M.L., Tiwari, R., Patil, V.A. 1989. Studies on Grain losses by some selected Self Propelled. Combines in wheat crop. Paper presented during XXVI Annual Convention of Indian Society of Agricultural Engineers.

168. Mehta, M.L., Suneja, Y., Tiwari, R. 1988. Studies on Testing and Evaluation of a few selected centrifugal pumps. Paper presented during XXV Annual Convention of ISAE, held at Udaipur.

169. Mehta, M.L., Tiwari, R., Patil V.A. 1989. Role of Safety Standards in Farm Machinery. Paper presented at Farm Machinery Safety Day, HAU, Hisar.

170. Mehta, M.L., Tiwari, R, Patil V.A. 1990. Testing of Agricultural Machinery. Agricultural Engineering today. Vol. 14, No. 1 & 2.

171. Mehta, M.L., Tiwari, R., Omkar, S., Patil, V.A. 1988. Studies on Standardization of Seed-cum-Fertilizer Drills. Paper presented during XXVth Annual Convention of ISAE held at Udaipur (Rajasthan).

172. Mehta, M.M. 1986. Status of Farm Mechanization in India, USAID Development & Management Training Project Programme held at Central Farm Machinery Training & Testing Institute, Budni.

173. Misra. S.K. 1991. Formulation and Implementation of Agricultural Mechanization Strategies in India. Agricultural Mechanization Policies and Strategies in Africa. Commonwealth Secretariat, London.

174. Mrema, G., Soni, P., Rolle, R. 2014. A regional strategy for sustainable agricultural mechanization: Sustainable mechanization across agri-food chains in Asia and the Pacific region. FAO Regional Office for Asia and the Pacific Publication. 2014/24. 74 pp.

175. Nakra, B.C. and Chaudhary, K.K. 1989. Instrumentation Measurement and Analysis. Tata McGraw-Hill Publishing Company Ltd, New Delhi.

176. National Standardization Agency of Indonesia. SNI: 0738. 2014. Quality standard and testing method of two-wheel tractors.

177. Ochiai, Y., Sugiura, Y., Shigeta, K. 1992. Testing and Evaluation of Agricultural Tractor (Riding type). Lecture delivered at BRAIN-Japan.

178. OECD Standard Codes for the Official Testing of Agricultural and Forestry Tractors. Organisation for Economic Co-operation and Development (OECD), Paris, France. February 2018.

179. Pandey, M.M. 2009. Country Report India -Indian Agriculture an Introduction. Central Institute of Agricultural Engineering Bhopal, India. Presented in Fourth Session of the Technical Committee of APCAEM 10-12 February 2009, Chiang Rai, Thailand.

180. Pangotra, P.N. 1981. Report on Survey on Farm Mechanization in India. Presented for the Asian Productivity Organization at New Delhi.

181. Pangotra. P.N. 1975. Tractor Testing Today. Paper presented at 13th Annual Convention of Indian Society of Agricultural Engineers held at Allahabad Agril. Instt. Allahabad.

182. Patel, M.K., Bushra, P., Sahoo, H.K, Patel, B., Kumar, A., Singh, M., Kumar, M., Rajan, P. 2017. An advance air-induced air-assisted electrostatic nozzle with enhanced performance. Computers and Electronics in Agriculture. Vol. 135: 280-288. https://doi.org/10.1016/j.compag.2017.02.010

183. Patil, A., A.K. Dave and R.N.S. Yadav. 2004. Evaluation of sugarcane cutter planter. Sugar Tech. Vo1.6 (3):121-125.

184. Peter, E.C. 1986, Development of Trainer's skill in selection of power threshers, operational techniques and preventive maintenance. Paper presented during USAID programme held at Central Farm Machinery Training & Testing Instt. Budni (MP)

185. Philippine Agricultural Engineering Standards. PNS/PAES: 160. 2011. Agricultural machinery, Sugarcane Planter - Methods of Test. ICS 65.060.01

186. Prasad, J and Srivastava, N.S.L. 1982. Impact of Agricultural mechanization on production, productivity, income and employment generation. Agricultural Engineering Today. Vol. 15 & 16 (1-6).

187. Prasad, J. and Misra, S.K. 1980. Guidelines for testing animal drawn implements. Paper presented at IXth Annual Workshop of Coordinated Scheme for Research and Development of Farm Machinery held at IGFRI, Jhansi.

188. Ram. R.B., Singh. S. and Verma. S.R. 1980. Comparative Performance of Some new and Conventional Tillage Equipment. J. of Ag. Engg. 17(1): 7-13.

189. Reed, J., Turner, P. 1993. Slip Measurement Using Dual Radar Guns, Alberta: Alberta Farm Machinery Research Centre. [801]

190. Richardson, N., Lanning, R., Kopp, K., Carnegie, E. 1982. True Ground Speed Measurement Techniques. SAE Technical Paper 821058, 1982, https://doi.org/10.4271/821058

191. RNAM Test Codes & Procedures for Farm Machinery. 1983. Economic and Social Commission for Asia and the Pacific Regional Network for Agricultural Machinery. Technical Series No.12.

192. Roeber, J.B.W., Pitla, S.K., Hoy, R. M., Luck, J. D., Kocher, M. F. 2017. Tractor Power Take-Off Torque Measurement and Data Acquisition System. Applied Engineering in Agriculture. ASABE. Vol. 33(5): 679-686.

193. Rotz, C. A., & Muhtar, H. A. 1992. Rotary power requirements for harvesting and handling equipment. Appl. Eng. Agric., 8(6), 751-757.

194. Sahay, J. 1986. Elements of Pumps and tubewells. Agro Book Agency. Patna.

195. Schmitz, A., Moss, C.B. 2015. Mechanized Agriculture: Machine Adoption, Farm Size, and Labor Displacement. AgBioForum, 18(3): 278-296.

196. Schuler, R.T., Frank, G.G. Estimating agricultural field machinery costs. AG3510. Agricultural and Applied Economics. University of Wisconsin – Madison.

197. Sharma, V.K. Singh, C.P., Gupta, P.K. 1984. Research on Thrashers at PAUA Review. Paper presented at All- India Seminar on Role of Agricultural Engineers in Rural Development at New Delhi.

198. Sharma, V.K., Gupta, P.K. Singh, S., Singh, C.P. 1986. Power Requirements of Different Systems of Spike Tooth Wheat Thrasher. J. of Institution of Engineers, AG Vol. 65, PP-17-20.

199. Sharma. R.N. 1982. Indian Standards on Testing of Agricultural machinery. Paper presented at UNIDO/ICAR Mini Workshop on Testing of Agril. Machinery, P.A.U. Ludhiana.

200. Sharma. V.K., Garg. I.K. 1981. Paddy Transplanters – History of Development and Present Status. A background paper for Inter-Design Workshop Proposed by National Institute of Design, Ahmedabad.

201. Shigeta, Kazuto. 1992 Data processing and Analyzing Lecture delivered at BRAIN-JAPAN.

202. Shreshta, S. 2012. Testing of hand tools and non-motorized machines used in agriculture in the Asia Pacific region. Agricultural Engineering Division, Nepal Agricultural Research Council, available on UNESCAP CSAM website.

203. Shukla, L.N., Chauhan, A.M., Verma, S.R. 1996. Development of minimum till planting machinery. AMA. 27(4).

204. Shukla, L.N., Verma, S.R. 1981. Plant Sugarcane Mechanically with Tractor Mounted Sugarcane Planter, Progressive Farming.

205. Shukla, L.N., Verma, S.R., Singh, Amar. 1982. Two-row Tractor drawn Semi-automatic Sugarcane Planter. Farm Digest.

206. Shukla, L.N., Sharma, M.P., Verma, S.R. 1984. Development of a sugarcane planter for developing countries. Agric. Mech. In Asia, Africa and Latin America, Vol.15 (1): pp.33-42.

207. Sidhu, H.S., Singh, M, Humphreys, E., Singh, Y., Sidhu, S.S. 2007. The Happy Seeder enables direct drilling of wheat into rice stubble. Australian J. Exp. Agric., 47 (7): 844–854,

208. Sims, B., Hilmi, M., Kienzle, J. 2016. Agricultural mechanization: A key input for sub-Saharan African smallholders. Integrated Crop Management Vol. 23-2016. Food and Agriculture Organization of the United Nations, Rome.

209. Singh, B., Singh, T.P. 1995. Development and performance evaluation of zero till fertilizer cum seed drill. J. Agric. Engg., 32(1):13-23.

210.Singh, C.P. Garg, I.K., Sharma, V.K. and Panesar, B.S. 1982. Design, Development and Evaluation of 10-Raw Tractor Mounted Paddy Transplanter. J. of Ag. Engg., 19(3) : 81-89.

211.Singh, G, 2001. Relationship between mechanization and agricultural productivity in various parts of India. Agricultural Mechanization in Asia, Africa, and Latin America. 32(2): 68-76.

212.Singh, G., Doharey R.S. 1999. Tractor Industry in India. Agricultural Mechanization in Asia, Africa, and Latin America. 30(2): 9-14.

213.Singh, K.K., Mehta, C.R 2116. Status on testing of agricultural machinery in India. Bulletin prepared by ICAR.

214.Singh, S. 2007. Farm Machinery- Principles and Applications. Indian Council of Agricultural Research (ICAR). New Delhi.

215.Singh, S. 2014. Agricultural mechanization in India, country paper. Regional Roundtable of National Agricultural Machinery Associations in Asia and the Pacific - Connection for Cooperation and Development. UN ESCAP. 28-30 October 2014. Wuhan, China.

216.Singh, S. 2016. Agricultural Machinery Industry in India. Agricultural Mechanization in Asia, Africa, and Latin America. 47(2): 26-35.

217.Singh, S., Garg, I. K. 2002. Farm mechanization viz-a-viz agricultural production in India. Paper presented at XXV Indian Social Science Congress held at Kerala University, Trivandrum, from Jan. 28 to Feb. 1.

218.Singh, S., Verma, S.R. 2009. Farm Machinery Maintenance and Management. Indian Council of Agricultural Research (ICAR). New Delhi.

219.Singh, S., Verma, S.R. 1977. Guidelines for Safe Operation of Tractors, Progressive Farming.

220.Singh, Santokh, Verma, S.R. 1977. Safe Operation of Sprayers and Dusters. Progressive Farming.

221.Singh, S., Verma, S.R., Singh, M, 1983. Performance studies in manually operated Fertilizer Broadcaster, J. Res. Punjab Agric. Univ. 20(1):81-88

222.Smith, D.W., Sims, B.G., Neill, D.H. 1994. Testing and evaluation of agriculture machinery and equipment. FAO Agriculture Services Bulletin

223.Srivastava, A.C. 1990. Elements of Farm Machinery. Oxford & IBH Publishing, New Delhi.

224.Standardization Administration of China. GB/T: 6229. 2007. Test methods for walking tractor.

225.State of Indian Agriculture, Department of Agriculture Report, 2012-13

226.State of Indian Agriculture. 2009. National Academy of Agricultural Sciences (NAAS). New Delhi, India.

227. Strauss, C., Cugnasca, C. E., Saraiva, A. M., Paz, S. M. 1999. The ISO 11783 Standard and Its Use in Precision Agriculture Equipment. In: P.C. Robert, R.H. Rust, W.E. Larson, editors, Precision Agriculture, ASA, CSSA, SSSA, Madison, WI. p. 1253-1261. *doi:10.2134/1999.precisionagproc4.c28b.*

228. Suneja, Y., Mehta, M.L., Tiwari, R., Patil, V.A. 1991. Standardization in Centrifugal Pumps –A case study. Journal of the Institution of Engineers. Vol. 72, Part AG-1.

229. Suneja. Y., Mehta, M.L., Tiwari, R., Patil, V.A. 1991. Standardization and quality control of centrifugal pumps. AMA. Japan, Vol. 22, No. 2.

230. Takahashi, M. 1992. Testing and Evaluation of Centrifugal Pump. Lecture delivered at BRAIN-OMIYA-JAPAN.

231. Takizawa, N, 1992. Testing and Evaluation of Rice Transplanter. Lecture delivered in BRAIN-JAPAN

232. Tan, L., Cao, S., Wang, Y., Zhu, B. 2012. Direct and inverse iterative design method for centrifugal pump impellers. Proc. of the Institution of Mechanical Engineers, Part A: Journal of Power and Energy. 226. 764-775. doi.10.1177/0957650912451411 [760]

233. Taneja, D.S., Kaushal, M.L., Sondhi. S.K., Murty, V.V.N. 1986. A manual on centrifugal pumps for irrigation. Department of Soil and Water Engineering, PAU, Ludhiana.

234. Test Board Actions. Nebraska Tractor Test Laboratory. Institute of Agriculture and Natural Resources. University of Nebraska–Lincoln. Retrieved on 09.05.2018 from https://tractortestlab.unl.edu/tractortestboard.

235. Tiwari, R., Mehta, M.L., Singh, A.K., Patil, V.A. 1989. Studies on standardization of Hand Operated Knapsack Sprayers. Paper presented at XXVI Annual Convention of ISAE.

236. Tiwari, T.C. 1986. Testing of tillage implements and their evaluation. Paper presented during USAID programme held at CFMT&TI-Budni (MP)

237. Transforming Agriculture Through Mechanisation. 2015. A Knowledge Paper on Indian farm equipment sector. Grant Thornton and FICCI. [505]

238. Tsuga, Kohnosuke. 1992. Rice Transplanter. Lecture delivered in a BRAIN-JAPAN.

239. UGA Cooperative Extension Circular 798. 2012. Calibration of Bulk Dry Fertilizer Applicators. The University of Georgia.

240. Verma, S.R., Singh, J. 1986. Useful Gadgets for Manufacture of Low-cost Agricultural Machinery. AMA, Japan. pp. 58-62

241. Verma, S.R, Kalkat, H.S. 1978. Efficient use of Potato Harvesting Equipment. The Agril. Engg., Vol. No.21.

242. Verma, S.R, Mittal, V.K. 1970. Harvesting Machinery – A Must for the Wheat Farmer, Progressive Farming.

243. Verma, S.R. 1970. Machines for Groundnut Harvesting, Progressive Farming

244. Verma, S.R. 1971. Mechanization in Fertilizer Placement, Farm Extension Digest, Vol.4: 171-179

245. Verma, S.R. 1984. Standardization in the field of harvesting and Threshing Equipment. Proceeding on RNAM Training Programme of Standardization of Agri Machinery. BIS, New Delhi.

246. Verma, S.R. 1985. Problems and Progress in the evaluation and extension of Seed-cum-Fertilizer drills in India. Proceedings of the Silver jubilee International Conference organized by IRRI, Los banos. Philippines.

247. Verma, S.R., Sharma, V.K. 1993. Mechanization in Indian Agriculture, Achievements and Challenges. Changing Scenario of India Agriculture. Commonwealth Publishers. New Delhi.

248. Verma, S.R., Chauhan, A.M., Kalkat, H.S. 1977. Multi-crop Seed Drill-cum Planter. Agri. Engineering today, 1(ii)

249. Verma, S.R., Garg, R.L. 1971. Design and field performance of tractor-mounted single-row elevator-digger for potato., Indian Journal Agric. Sci., 41(8): 666-671

250. Verma, S.R., Garg, R.L., Development and performance of tractor mounted groundnut digger shaker, Indian Journal Agric. Sci., Vol. VIII, No.2

251. Verma, S.R., Kalkat, H.S. 1976. Proper Maintenance of Farm Machinery, Progressive Farming.

252. Verma, S.R., Kalkat, H.S. 1977. Development of Potato Planters in India.

253. Verma, S.R., Rawal, G.S., Bhatia, B.S. 1978. A study on human accidents in wheat threshers. J. on Ag. Engg., Vol. 15(1): 19-23

254. Verma, S.R., Singh, C.P. 1980. Farm Equipment for Small Farms. Progressive Farming.

255. Verma. S.R. 1989 Standardization of Agricultural machinery in Nigeria - Some Suggestions. Proceeding of first national Colloquium on Standardization of Agri. Machinery held at national Centre of Agricultural Mechanization, IIorin, Nigeria

256. Vigneault, C., St. Amour, G., Buckley, D. J., Masse, D. I., Savoie, P., Tremblay, D. 1989. A trailer-mounted PTO torque meter system. Can. Agric. Eng., 3111, 89-91.

257. Walker, T.W., Kingery, W.L., Street, Joe E., Lox M.S., Oldham, J.L., Gerard, P.D., Han, F.X. 2003. Rice yield and soil chemical properties as affected by precision land leveling in alluvial soils. Agron. J. 95:1483-1488.

258. World Bank Indicators, CIA Fact book, Mechanisation and Farm Technology, Division of Department of Agriculture and Cooperation, Trading Economics, FAO Yearbook 2013.

259. World Bank statistics. https://data.worldbank.org/indicator.

260. Xinshen, D., Jed, S., Hiroyuki, T. 2016. Agricultural mechanization and agricultural transformation. IFPRI Discussion Paper 1527. Washington, D.C.: International Food Policy Research Institute (IFPRI). http://ebrary.ifpri.org/cdm/ ref/collection/p15738coll2/id/130311.

261. Zachariah. P.J. 1976. Farm Equipment Standardization for Agril. Development Agri. Engg. Today Vol. I, No. 5 & 6.

262. Zhixiong, L. Zhou, H., Xuefeng, B. 2013. Wheel Slip Measurement in 4WD Tractor Based on LABVIEW. International Journal of Automation and Control Engineering, August, 2(3), p. 113.

Appendice - I

International System of units (SI)

S.No.	Physical Quantity	Unit	Symbol
1.	Length	Meter	m
		Millimeter	mm
		Centimeter	cm
		Kilometer	km
2.	Mass	Kilogram	kg
		Gram	g
3.	Time	Second	s
		Minute	min
		Hour	h
4.	Electric Current	Ampere	A
5.	Area	Square Centimeter	cm^2
		Hectare	ha
		Square Meter	m^2
6.	Force	Kilogram Force	kgf
		Newton	N
7.	Power	Metric Horsepower	Ps
		Kilowatt	kW
8.	Pressure	Kilogram force per square centimeter	kgf/cm^2
		Pascal	Pa
		Millibar	mbar
9.	Speed/Velocity	Meter per second	m/s
		Kilometer per hour	km/h
10.	Volume	Cubic centimeter	cm^3 or cc
		Milliliter	ml
		liter	l
11.	Sound Level	decible (A)	dB(A)
12.	Fuel Consumption	Litres per hour	l/h
13.	Specific Fuel Consumption	Grams per kilowatt hour	g/kw.h

Appendice - II

Conversion Factor to SI Units

S. No.	Quantity	Unit	Conversion Factor
1.	Length	1 in	0.0254 m
		1 ft	0.3048 m
		1 yd	0.9144 m
		1 mile	1609.344 m
2.	Area	1 in^2	6.4516 x 10^{-4} m^2
		1 ft^2	0.092 903 m^2
		1 yd^2	0.836 127 m^2
		1 acre	4046.86 m^2 = 0.404 686 ha
		1 mile^2	2.589 99 x 10^6 m^2 = 258.999 ha
3.	Volume	1 in^3	1.638 71 x 10^{-5} m^3
		1 ft^3	0.028 316 8 m^3
		1 UK gal	0.004 546 092 m^3 = 4.546 092 1
		1 US gal	0.003 785 41 m^3 3.785 41 1
4.	Mass	1 lb	0.453 592 37 kg
		1 UK ton	1016.05 kg = 1.016 05 tonne
		1 short tone	907.185 kg = 0.907 tonne
5.	Velocity	1 ft/s	0.3048 m/s = 1.097 28 km/h
		1 mile/h	0.447 04 m/s = 1.609 34 km/h
6.	Force	1 lbf	4.448 22 N
		1 kgf	9.806 65 N
7.	Torque	1 lbf ft	1.355 82 Nm
8.	Power	1 hp	745.700 W
		1 metric hp	735.499 W
9.	Pressure	1 lbf/in^2	6894.76 N/m^2
		1 std atmos	101.325 kN/m^2
		A bar	105 N/m^2
		A in Hg	3386.39 N/m^2
		A mm Hg	133.322 N/m^2
		1 in H$_2$O	249.089 N/m^2
		1 mm H$_2$O	9.806 65 N/m^2
10.	Temperature	1°F	K or C

Appendice - III

List of Abbreviations

%	:	**Percent**
AICRP	:	All India Coordinated Research Project
ANTAM	:	Asian and Pacific Network for Testing of Agricultural Machinery
ASABE	:	American Society of Agricultural and Biological Engineers
BIS	:	Bureau of Indian Standards
CGIAR	:	Consultative Group on International Agricultural Research
CIAE	:	Central Institute of Agricultural Engineering
CIPHET	:	Central Institute of Post-Harvest Engineering and Technology
cm	:	Centimeter
CMERI	:	Central Mechanical Engineering Research Institute
CMVR	:	Central Motor Vehicles Rules
CPCFM	:	Centre for Precision & Conservation Farming Machinery
CSIR	:	Council of Scientific and Industrial Research
DARE	:	Department of Agricultural Research and Education
ENTAM	:	European Network for Testing of Agricultural Machines
FAO	:	Food and Agriculture Organization of the United Nations
FMT&TI	:	Farm Machinery Training & Testing Institute
GDP	:	Gross Domestic Product
GPS	:	Global Positioning System
ha	:	Hectares
IARI	:	Indian Agriculture Research Institute
ICAR	:	Indian Council of Agricultural Research
IIRB	:	International Institute for Beet Research
IPM	:	Integrated Pest Management
IRRI	:	International Rice Research Institute
ISRO	:	Indian Space Research Organization
kg:	:	Kilogram
Kgf	:	Kilogram Force
KVK	:	Krishi Vigyan Kendra

m	:	Meter
m³	:	Cubic meter
MPa	:	Mega Pascal
NNPSHR	:	Net Positive Suction Head Requirement
OECD	:	Organisation for Economic Co-operation and Development
PTO	:	Power take off
R&D	:	Research and development
ROPS	:	Rollover protective structures
SAUs	:	State Agricultural Universities
SERB	:	Science & Engineering Research Board
T	:	Ton/tonnes
UNDP	:	United Nation Development Programme
USA	:	United States of America
USDA	:	United States Department of Agriculture
VMD	:	Volume Median Diameter

Appendice - IV

List of Recommended Instruments/Equipment and Range for Testing and Evaluation of Agricultural Machinery

S. No.	Name of Equipment	Range
A	*Power measurements*	
1.	Eddy current dynamometer with digital load and speed indicator	10 hp at 1000 rpm 70 hp at 6000 rpm
2.	A. C. Absorption SCR Electrical dynamometer	1.5 hp at 2000 rpm 5.4 hp at 4000 rpm 8 hp at 6000 rpm
3.	Schenck Hydraulic dynamometer (0700-IE)	75 kW at 1000 rpm 700 kW at 7500 rpm
4.	Hydraulic water brake dynamometer (UI-30)	20 hp at 500 rpm to 250 hp at 1000 rpm
5.	Hydraulic schenck dynamometer (D-1100)	10 kW at 500 rpm at 740 kW at 1000 rpm
6.	Loading car for drawbar performance test	
B	*Pull Force Measurement*	
7.	Hydraulic dynamometer (pull type)	0-2000 kgf
8.	Spring type tension dynamometer	0-2000 kgf
9.	Tensiometer for instant indication of load and tension	0-250 kgf
10.	Universal type load cell	0-100 kg and above
11.	Strain gauges	Strain gauges
C	*Material Testing*	
12.	Universal testing machine (UTM)	0-40 tonne
13.	Brinnel & Rockwell hardness testing machine	0-100 HRC
14.	Vickers hardness testing machine Mode1HD-10	5-10 kgf
15.	Charpy impact testing machine	0-240 kgm
16.	Spring testing machine	0 – 200 kg

S. No.	Name of Equipment	Range
D	*Meteorological Instruments*	
17.	Steel tape	3M, 15M, 30M
18.	Outside Micrometer set	0-25 mm,25-50 mm, 50-75 mm,
		75-100 mm, 100-125 mm
		125-150 mm, 150-175 mm
		175-200 mm
19.	Inside micrometer set	50 mm to 300 mm
20.	Cylinder bore gauge	35-60 mm
21.	Height gauge dial	0-600 mm
22.	Steel micrometer	0-25 mm
23.	U-type throat micrometer	0-200 mm
24.	Point micrometer	0-25 mm
25.	Disc micrometer	0-25 mm
26.	Spline micrometer	0-25 mm
27.	Ball micrometer	0-25 mm
28.	Vernier caliper dial type (Digital / Manual)	150 mm, 200 mm, 300 mm, 600 mm, 1000 mm
29.	Vernier depth gauge	
E	*Soil Testing*	
30.	Digital PH meter	0 - 14 pH
31.	Soil shear test instrument	
32.	Cone Penetrometer (Digital)	
33.	Sieve shaker with set of sieve	
34.	Bulk density apparatus	
35.	Sampling auger	
36.	Hydrometer	
F	*Electric Power, Voltage & Current Measurement*	
37.	Insulation tester	
38.	Phase sequence indicator	
39.	Digital frequency indicator	0-1000 Hz
40.	Three Phase balance wattmeter	0-5 kW, 0-10 kW, 0-30 kW, 0-120 kW
41.	Ampere meter	5, 10, 15, 60, 100 A

S. No.	Name of Equipment	Range
42.	Voltmeter	0-500 V
43.	Energy meter	0-10000 kWh
44.	Dimmer state	0-500 V
45.	Potential transformer	0-200 V
46.	Oscilloscope (dual trace)	Freq. range 20 MHz
47.	Digital frequency meter	1 Hz to 10 MHz
48.	Functional generator	upto 01 MHz
49.	Digital multi meter	
50.	Wheat stone bridge	
51.	Kelvin bridge	
52.	Automate voltage stabilizer	0.5 KVA
53.	Power factor meter	
54.	Temperature recorder	20-600 °C
55.	Two, four & six channel recorders	
G	***Vacuum & Pressure Measurements***	
56.	Hydraulic In-line tester	0-75 gpm, 0-300 l/m
57.	Deadweight pressure gauge tester	0-70 kg/cm^2
58.	Manometers for air intake and exhaust gas pressure	
59.	Vacuum gauge tester	0 - 1 kg/cm^2, 0 - 10 kg/cm^2
60.	Bourdon tube pressure gauge	
61.	Oil Pressure gauge	
62.	Tyre pressure gauge	
63.	Atmospheric Barometer	
64.	Compression pressure gauge	
H	***Temperature Measurement***	
65.	Six channel Digital temp. indicator	0 - 200°C
66.	Six channel temperature recorder	0 - 200°C
67.	Thermocouple indicator	0 - 1000°C
68.	Industrial thermometer	0 - 120°C, 0 - 20°C, 0 - 300°C
I	***Speed, Time and Distance Measurement***	
69.	Hand tachometer	0 – 10000 rpm
70.	Strobometer	250-18000 rpm

S. No	Name of Equipment	Range
71.	Digital Stop watch	0 - 15 min
72.	Digital hand tachometer	0 - 9900 rpm
73.	Non-contact type digital tachometer	0 - 9900 rpm
74.	Panel-mounted digital tachometer	1 - 9999 rpm
J	*Mass Measurement*	
75.	Physical balance and weight box	0 - 5 kg
76.	Single pan semi-micro projection balance	0 – 100 g, 0 – 1000 g
77.	Electric pan balance	0 - 1 kg
78.	Semi indicating balance	0 - 1 kg & 0-10 kg
79.	Electronic precision balance	0 - 1 kg
80.	Platform balance	0 - 2000 kg
81.	Top pan balance	0 - 10 kg
82.	Avery platform balance	300 kg (sensitivity- 50 g)
83.	Spring balance	50 kg, 100 kg, 200 kg
K	*Moisture Measurement*	
84.	Vacuum Oven/Electric oven	
85.	Grain moisture tester (Digital)	
86.	Infrared Moisture balance	0 - 100%
87.	Speedy moisture tester for grain/straw and soil	0 - 20%
L	*Fuel Consumption Measurement*	
88.	Digital fuel meter	0 - 20 l/h
89.	Volumetric fuel consumption meter	
M	*Air Velocity*	
90.	Digital Anemometer	0 - 15 m/s
91.	Digital air flow meter	
N	*Relative Humidity*	
92.	Psychrometer	
93.	Hygrometer	
94.	Hygrothermograph	
O	*Flow, Crack & Thickness Detection*	
95.	Ultrasonic flow detector	1 MHz to 10MHz
		50 mm to 1000 mm

S. No.	Name of Equipment	Range
96.	Magnetic crack detector	
97.	Ultrasonic thickness gauge	
P	**Chemical Analysis**	
98.	Spectrometer	
99.	Carbon & Sulphur apparatus	
100.	Muffle furnace	
Q	**Noise Level And Vibration Test**	
101.	Precision integrating sound level meter	2"microphose, 24 to 130 dB 20 kHz to 20 kHz
102.	Octave filter set	10 Hz to 20 kHz
103.	Microphone preamplifier	20 Hz to 20 kHz
104.	Vibration meter	0.3 Hz to 15 kHz
105.	High Sensitivity Accelerometer	Charge sens. 2-1+2% pc/ms voltage sens. 2 Bmu/MJ 5%, 0.2-9100 Hz, 10%, 1-12600 Hz
106.	Tunable hand pass filter	2 Hz to 20 kHz
107.	Level recorder	1 Hz to 20 kHz
108.	Triaxial accelerometer	0.2 to 8700 Hz
109.	Portable tape recorder	Tape speed 33.1 m/s
110.	High resolution signal analyzer	Freq. range 10 Hz to 20 kHz
111.	X.Y. recorder	X-axis 0.02 to 500 mv/mm X-axis 0.02 to 1000mv/mm
112.	High C. Exciter head	Rated force 112N, 5- 10 kHz
113.	Power supply	Load output 7.5 D.C. Open circuit output 12V,DC
R	**Smoke Density**	
114.	Bosch smoke meter	
S	**Ergonomics Measurement**	
115.	Bicycle Ergometer and Ergograph	
116.	Expirograph	
117.	Stethoscope	
118.	Biomedical Telemetry	

S. No	Name of Equipment	Range
119.	Anthropometric instrument	
120.	Thermophygrograph	
121.	Respiration gas meter	
122.	Treadmill	
123.	Hand Grip Strength tester	
124.	Audiometer	
125.	Gas Analyzer	
126.	Dust Particle Analyzers	
T	*Miscellaneous Instruments*	
127.	Water level indicator with sensor	
128.	Suction & delivery head indicator with sensor	
129.	Revolution indicator with sensor	
130.	Torque indicator with sensor	
131.	Personal computer / Laptop and printer	

Index